Eighth Edition

FUNDAMENTALS OF ENGINEERING SUPPLIED-REFERENCE HANDBOOK

*National Council of Examiners
for Engineering and Surveying*

Published by the
National Council of Examiners for Engineering and Surveying®
280 Seneca Creek Road, Clemson, SC 29631 800-250-3196 www.ncees.org

©2008 by the National Council of Examiners for Engineering and Surveying®. All rights reserved.

ISBN 978-1-932613-30-8

Printed in the United States of America
November 2007

PREFACE

The Fundamentals of Engineering (FE) examination was developed by the National Council of Examiners for Engineering and Surveying (NCEES) as the first step toward professional engineering licensure. It is designed for students completing a bachelor degree program in engineering. The FE exam consists of two 4-hour sessions—one administered in the morning and the other in the afternoon. The morning session tests the subject matter covered by the first 90 semester credit hours of engineering coursework, while the afternoon session tests upper-division subject knowledge covering the remainder of required degree coursework.

The *FE Supplied-Reference Handbook* is the only reference material allowed during the examination. A copy of this handbook will be made available to each examinee during the exam and returned to the proctor before leaving the exam room at the conclusion of the administration. Examinees are prohibited from bringing any reference materials with them to the exam. As well, examinees are prohibited from writing in the *Reference Handbook* during the exam administration.

There are no sample questions or solutions included in the *Reference Handbook*—examinees can self-test using one of the NCEES *FE Sample Questions and Solutions* books, CD-ROMs, or online practice exams, all of which may be purchased by calling (800) 250-3196 or visiting our Web site at www.ncees.org. The material included in the *FE Supplied-Reference Handbook* is not all-encompassing or exhaustive. NCEES in no event shall be liable for not providing reference material to support all the questions in the FE exam. Some of the basic theories, conversions, formulas, and definitions examinees are expected to know have not been included. Furthermore, the *FE Supplied-Reference Handbook* may not contain some special material required for the solution of a particular exam question—in such a situation, this material will be included in the question itself.

In the interest of constant improvement, NCEES reserves the right to revise and update the *FE Supplied-Reference Handbook* as it deems appropriate. Each FE exam will be administered using the latest version of the *FE Supplied-Reference Handbook*. To report suspected errata in this book, please e-mail your correction using our online feedback form. Examinees are not penalized for any errors in the *Reference Handbook* that affect an exam question.

For current exam specifications, a list of approved calculators, study materials, errata, guidelines for special accommodations requests, and other information about the exams and the licensure process, visit www.ncees.org or call (800) 250-3196.

Do not write in this book or remove any pages.

Do all scratch work in your exam booklet.

CONTENTS

Do not write in this book or remove any pages.

Do all scratch work in your exam booklet.

EXAM SPECIFICATIONS

Fundamentals of Engineering (FE) Examination

Effective October 2005

- The FE examination is an 8-hour supplied-reference examination: 120 questions in the 4-hour morning session and 60 questions in the 4-hour afternoon session.

- The afternoon session is administered in the following seven modules—Chemical, Civil, Electrical, Environmental, Industrial, Mechanical, and Other/General engineering.

- Examinees work all questions in the morning session and all questions in the afternoon module they have chosen.

MORNING SESSION (120 questions in 12 topic areas)

Topic Area	Approximate Percentage of Test Content
I. Mathematics	15%
A. Analytic geometry	
B. Integral calculus	
C. Matrix operations	
D. Roots of equations	
E. Vector analysis	
F. Differential equations	
G. Differential calculus	
II. Engineering Probability and Statistics	7%
A. Measures of central tendencies and dispersions (e.g., mean, mode, standard deviation)	
B. Probability distributions (e.g., discrete, continuous, normal, binomial)	
C. Conditional probabilities	
D. Estimation (e.g., point, confidence intervals) for a single mean	
E. Regression and curve fitting	
F. Expected value (weighted average) in decision-making	
G. Hypothesis testing	
III. Chemistry	9%
A. Nomenclature	
B. Oxidation and reduction	
C. Periodic table	
D. States of matter	
E. Acids and bases	
F. Equations (e.g., stoichiometry)	
G. Equilibrium	
H. Metals and nonmetals	
IV. Computers	7%
A. Terminology (e.g., memory types, CPU, baud rates, Internet)	
B. Spreadsheets (e.g., addresses, interpretation, "what if," copying formulas)	
C. Structured programming (e.g., assignment statements, loops and branches, function calls)	
V. Ethics and Business Practices	7%
A. Code of ethics (professional and technical societies)	
B. Agreements and contracts	
C. Ethical versus legal	
D. Professional liability	
E. Public protection issues (e.g., licensing boards)	

VI.   **Engineering Economics**   8%

         A.   Discounted cash flow (e.g., equivalence, PW, equivalent annual FW, rate of return)
         B.   Cost (e.g., incremental, average, sunk, estimating)
         C.   Analyses (e.g., breakeven, benefit-cost)
         D.   Uncertainty (e.g., expected value and risk)

VII.   **Engineering Mechanics (Statics and Dynamics)**   10%

         A.   Resultants of force systems
         B.   Centroid of area
         C.   Concurrent force systems
         D.   Equilibrium of rigid bodies
         E.   Frames and trusses
         F.   Area moments of inertia
         G.   Linear motion (e.g., force, mass, acceleration, momentum)
         H.   Angular motion (e.g., torque, inertia, acceleration, momentum)
         I.   Friction
         J.   Mass moments of inertia
         K.   Impulse and momentum applied to:
                 1. particles
                 2. rigid bodies
         L.   Work, energy, and power as applied to:
                 1. particles
                 2. rigid bodies

VIII.   **Strength of Materials**   7%

         A.   Shear and moment diagrams
         B.   Stress types (e.g., normal, shear, bending, torsion)
         C.   Stress strain caused by:
                 1. axial loads
                 2. bending loads
                 3. torsion
                 4. shear
         D.   Deformations (e.g., axial, bending, torsion)
         E.   Combined stresses
         F.   Columns
         G.   Indeterminant analysis
         H.   Plastic versus elastic deformation

IX.   **Material Properties**   7%

         A.   Properties
                 1. chemical
                 2. electrical
                 3. mechanical
                 4. physical
         B.   Corrosion mechanisms and control
         C.   Materials
                 1. engineered materials
                 2. ferrous metals
                 3. nonferrous metals

X.   **Fluid Mechanics**   7%

         A.   Flow measurement
         B.   Fluid properties
         C.   Fluid statics
         D.   Energy, impulse, and momentum equations
         E.   Pipe and other internal flow

XI. **Electricity and Magnetism**   **9%**
  A.   Charge, energy, current, voltage, power
  B.   Work done in moving a charge in an electric field (relationship between voltage and work)
  C.   Force between charges
  D.   Current and voltage laws (Kirchhoff, Ohm)
  E.   Equivalent circuits (series, parallel)
  F.   Capacitance and inductance
  G.   Reactance and impedance, susceptance and admittance
  H.   AC circuits
  I.   Basic complex algebra

XII. **Thermodynamics**   **7%**
  A.   Thermodynamic laws (e.g., 1st Law, 2nd Law)
  B.   Energy, heat, and work
  C.   Availability and reversibility
  D.   Cycles
  E.   Ideal gases
  F.   Mixture of gases
  G.   Phase changes
  H.   Heat transfer
  I.   Properties of:
       1. enthalpy
       2. entropy

AFTERNOON SESSION IN CHEMICAL ENGINEERING
(60 questions in 11 topic areas)

	Topic Area	Approximate Percentage of Test Content
I.	**Chemistry**	**10%**
	A. Inorganic chemistry (e.g., molarity, normality, molality, acids, bases, redox, valence, solubility product, pH, pK, electrochemistry)	
	B. Organic chemistry (e.g., nomenclature, structure, qualitative and quantitative analyses, balanced equations, reactions, synthesis)	
II.	**Material/Energy Balances**	**15%**
	A. Mass balance	
	B. Energy balance	
	C. Control boundary concept (e.g., black box concept)	
	D. Steady-state process	
	E Unsteady-state process	
	F. Recycle process	
	G. Bypass process	
	H. Combustion	
III.	**Chemical Engineering Thermodynamics**	**10%**
	A. Thermodynamic laws (e.g., 1st Law, 2nd Law)	
	B. Thermodynamic properties (e.g., internal thermal energy, enthalpy, entropy, free energy)	
	C. Thermodynamic processes (e.g., isothermal, adiabatic, isentropic)	
	D. Property and phase diagrams (e.g., T-s, h-P, x-y, T-x-y)	
	E. Equations of state (e.g., van der Waals, Soave-Redlich-Kwong)	
	F. Steam tables	
	G. Phase equilibrium and phase change	
	H. Chemical equilibrium	
	I. Heats of reaction	
	J. Cyclic processes and efficiency (e.g., power, refrigeration, heat pump)	
	K. Heats of mixing	
IV.	**Fluid Dynamics**	**10%**
	A. Bernoulli equation and mechanical energy balance	
	B. Hydrostatic pressure	
	C. Dimensionless numbers (e.g., Reynolds number)	
	D. Laminar and turbulent flow	
	E. Velocity head	
	F. Friction losses (e.g., pipe, valves, fittings)	
	G. Pipe networks	
	H. Compressible and incompressible flow	
	I. Flow measurement (e.g., orifices, Venturi meters)	
	J. Pumps, turbines, and compressors	
	K. Non-Newtonian flow	
	L. Flow through packed beds	

V. Heat Transfer 10%

    A.   Conductive heat transfer
    B.   Convective heat transfer
    C.   Radiation heat transfer
    D.   Heat transfer coefficients
    E.   Heat exchanger types (e.g., plate and frame, spiral)
    F.   Flow configuration (e.g., cocurrent/countercurrent)
    G.   Log mean temperature difference (LMTD) and NTU
    H.   Fouling
    I.   Shell-and-tube heat exchanger design (e.g., area, number of passes)

VI. Mass Transfer 10%

    A.   Diffusion (e.g., Fick's 1st and 2nd laws)
    B.   Mass transfer coefficient
    C.   Equilibrium stage method (efficiency)
    D.   Graphical methods (e.g., McCabe-Thiele)
    E.   Differential method (e.g., NTU, HETP, HTU, NTP)
    F.   Separation systems (e.g., distillation, absorption, extraction, membrane processes)
    G.   Humidification and drying

VII. Chemical Reaction Engineering 10%

    A.   Reaction rates and order
    B.   Rate constant (e.g., Arrhenius function)
    C.   Conversion, yield, and selectivity
    D.   Series and parallel reactions
    E.   Forward and reverse reactions
    F.   Energy/material balance around a reactor
    G.   Reactions with volume change
    H.   Reactor types (e.g., plug flow, batch, semi-batch, CSTR)
    I.   Homogeneous and heterogeneous reactions
    J.   Catalysis

VIII. Process Design and Economic Optimization 10%

    A.   Process flow diagrams (PFD)
    B.   Piping and instrumentation diagrams (P&ID)
    C.   Scale-up
    D.   Comparison of economic alternatives (e.g., net present value, discounted cash flow, rate of return)
    E.   Cost estimation

IX. Computer Usage in Chemical Engineering 5%

    A.   Numerical methods and concepts (e.g., convergence, tolerance)
    B.   Spreadsheets for chemical engineering calculations
    C.   Statistical data analysis

X. Process Control 5%

    A.   Sensors and control valves (e.g., temperature, pressure)
    B.   Dynamics (e.g., time constants, 2nd order, underdamped)
    C.   Feedback and feedforward control
    D.   Proportional, integral, and derivative (PID) controller concepts
    E.   Cascade control
    F.   Control loop design (e.g., matching measured and manipulated variables)
    G.   Tuning PID controllers and stability (e.g., Method of Ziegler-Nichols, Routh Test)
    H.   Open-loop and closed-loop transfer functions

XI. **Safety, Health, and Environmental**   5%
    A.   Hazardous properties of materials (e.g., corrosive, flammable, toxic), including MSDS
    B.   Industrial hygiene (e.g., noise, PPE, ergonomics)
    C.   Process hazard analysis (e.g., using fault-tree analysis or event tree)
    D.   Overpressure and underpressure protection (e.g., relief, redundant control, intrinsically safe)
    E.   Storage and handling (e.g., inerting, spill containment)
    F.   Waste minimization
    G.   Waste treatment (e.g., air, water, solids)

AFTERNOON SESSION IN CIVIL ENGINEERING
(60 questions in 9 topic areas)

Topic Area	Approximate Percentage of Test Content

I. Surveying — 11%
- A. Angles, distances, and trigonometry
- B. Area computations
- C. Closure
- D. Coordinate systems (e.g., GPS, state plane)
- E. Curves (vertical and horizontal)
- F. Earthwork and volume computations
- G. Leveling (e.g., differential, elevations, percent grades)

II. Hydraulics and Hydrologic Systems — 12%
- A. Basic hydrology (e.g., infiltration, rainfall, runoff, detention, flood flows, watersheds)
- B. Basic hydraulics (e.g., Manning equation, Bernoulli theorem, open-channel flow, pipe flow)
- C. Pumping systems (water and wastewater)
- D. Municipal water distribution systems
- E. Reservoirs (e.g., dams, routing, spillways)
- F. Groundwater (e.g., flow, wells, drawdown)
- G. Sewer collection systems (storm and sanitary)

III. Soil Mechanics and Foundations — 15%
- A. Index properties and soil classifications
- B. Phase relations (air-water-solid)
- C. Laboratory and field tests
- D. Effective stress (buoyancy)
- E. Retaining walls (e.g., active pressure/passive pressure)
- F. Shear strength
- G. Bearing capacity (cohesive and noncohesive)
- H. Foundation types (e.g., spread footings, piles, wall footings, mats)
- I. Consolidation and differential settlement
- J. Seepage
- K. Slope stability (e.g., fills, embankments, cuts, dams)
- L. Soil stabilization (e.g., chemical additives, geosynthetics)

IV. Environmental Engineering — 12%
- A. Water quality (ground and surface)
- B. Air quality
- C. Solid/hazardous waste
- D. Sanitary sewer system loads
- E. Basic tests (e.g., water, wastewater, air)
- F. Environmental regulations
- G. Water treatment and wastewater treatment (e.g., primary, secondary, tertiary)

V. Transportation   **12%**
- A. Streets and highways
    1. geometric design
    2. pavement design
    3. intersection design
- B. Traffic analysis and control
    1. safety
    2. capacity
    3. traffic flow
    4. traffic control devices

VI. Structural Analysis   **10%**
- A. Force analysis of statically determinant beams, trusses and frames
- B. Deflection analysis of statically determinant beams, trusses and frames
- C. Stability analysis of beams, trusses and frames
- D. Column analysis (e.g., buckling, boundary conditions)
- E. Loads and load paths (e.g., dead, live, moving)
- F. Elementary statically indeterminate structures

VII. Structural Design   **10%**
- A. Codes (e.g., AISC, ACI, NDS, AISI)
- B. Design procedures for steel components (e.g., beams, columns, beam-columns, tension members, connections)
- C. Design procedures for concrete components (e.g., beams, slabs, columns, walls, footings)

VIII. Construction Management   **10%**
- A. Procurement methods (e.g., design-build, design-bid-build, qualifications based)
- B. Allocation of resources (e.g., labor, equipment, materials, money, time)
- C. Contracts/contract law
- D. Project scheduling (e.g., CPM, PERT)
- E. Engineering economics
- F. Project management (e.g., owner/contractor/client relations, safety)
- G. Construction estimating

IX. Materials   **8%**
- A. Concrete mix design
- B. Asphalt mix design
- C. Test methods (e.g., steel, concrete, aggregates, asphalt)
- D. Properties of aggregates
- E. Engineering properties of metals

AFTERNOON SESSION IN ELECTRICAL ENGINEERING
(60 questions in 9 topic areas)

	Topic Area	Approximate Percentage of Test Content
I.	**Circuits**	16%
	A. KCL, KVL	
	B. Series/parallel equivalent circuits	
	C. Node and loop analysis	
	D. Thevenin/Norton theorems	
	E. Impedance	
	F. Transfer functions	
	G. Frequency/transient response	
	H. Resonance	
	I. Laplace transforms	
	J. 2-port theory	
	K. Filters (simple passive)	
II.	**Power**	13%
	A. 3-phase	
	B. Transmission lines	
	C. Voltage regulation	
	D. Delta and wye	
	E. Phasors	
	F. Motors	
	G. Power electronics	
	H. Power factor (pf)	
	I. Transformers	
III.	**Electromagnetics**	7%
	A. Electrostatics/magnetostatics (e.g., measurement of spatial relationships, vector analysis)	
	B. Wave propagation	
	C. Transmission lines (high frequency)	
IV.	**Control Systems**	10%
	A. Block diagrams (feed forward, feedback)	
	B. Bode plots	
	C. Controller performance (gain, PID), steady-state errors	
	D. Root locus	
	E. Stability	
V.	**Communications**	9%
	A. Basic modulation/demodulation concepts (e.g., AM, FM, PCM)	
	B. Fourier transforms/Fourier series	
	C. Sampling theorem	
	D. Computer networks, including OSI model	
	E. Multiplexing	

VI.   Signal Processing   8%
   A.   Analog/digital conversion
   B.   Convolution (continuous and discrete)
   C.   Difference equations
   D.   Z-transforms

VII.   Electronics   15%
   A.   Solid-state fundamentals (tunneling, diffusion/drift current, energy bands,
           doping bands, p-n theory)
   B.   Bias circuits
   C.   Differential amplifiers
   D.   Discrete devices (diodes, transistors, BJT, CMOS) and models and their performance
   E.   Operational amplifiers
   F.   Filters (active)
   G.   Instrumentation (measurements, data acquisition, transducers)

VIII.   Digital Systems   12%
   A.   Numbering systems
   B.   Data path/control system design
   C.   Boolean logic
   D.   Counters
   E.   Flip-flops
   F.   Programmable logic devices and gate arrays
   G.   Logic gates and circuits
   H.   Logic minimization (SOP, POS, Karnaugh maps)
   I.   State tables/diagrams
   J.   Timing diagrams

IX.   Computer Systems   10%
   A.   Architecture (e.g., pipelining, cache memory)
   B.   Interfacing
   C.   Microprocessors
   D.   Memory technology and systems
   E.   Software design methods (structured, top-down bottom-up, object-oriented design)
   F.   Software implementation (structured programming, algorithms, data structures)

AFTERNOON SESSION IN ENVIRONMENTAL ENGINEERING
(60 questions in 5 topic areas)

Topic Area	Approximate Percentage of Test Content
I. Water Resources	**25%**
A. Water distribution and wastewater collection	
B. Water resources planning	
C. Hydrology and watershed processes	
D. Fluid mechanics and hydraulics	
II. Water and Wastewater Engineering	**30%**
A. Water and wastewater	
B. Environmental microbiology/ecology	
C. Environmental chemistry	
III. Air Quality Engineering	**15%**
A. Air quality standards and control technologies	
B. Atmospheric sciences	
IV. Solid and Hazardous Waste Engineering	**15%**
A. Solid waste engineering	
B. Hazardous waste engineering	
C. Site remediation	
D. Geohydrology	
E. Geotechnology	
V. Environmental Science and Management	**15%**
A. Industrial and occupational health and safety	
B. Radiological health and safety	
C. Radioactive waste management	
D. Environmental monitoring and sampling	
E. Pollutant fate and transport (air/water/soil)	
F. Pollution prevention and waste minimization	
G. Environmental management systems	

AFTERNOON SESSION IN INDUSTRIAL ENGINEERING
(60 questions in 8 topic areas)

Topic Area

I. Engineering Economics   15%
   A. Discounted cash flows (equivalence, PW, EAC, FW, IRR, loan amortization)
   B. Types and breakdown of costs (e.g., fixed, variable, direct and indirect labor, material, capitalized)
   C. Analyses (e.g., benefit-cost, breakeven, minimum cost, overhead, risk, incremental, life cycle)
   D. Accounting (financial statements and overhead cost allocation)
   E. Cost estimating
   F. Depreciation and taxes
   G. Capital budgeting

II. Probability and Statistics   15%
   A. Combinatorics (e.g., combinations, permutations)
   B. Probability distributions (e.g., normal, binomial, empirical)
   C. Conditional probabilities
   D. Sampling distributions, sample sizes, and statistics (e.g., central tendency, dispersion)
   E. Estimation (point estimates, confidence intervals)
   F. Hypothesis testing
   G. Regression (linear, multiple)
   H. System reliability (single components, parallel and series systems)
   I. Design of experiments (e.g., ANOVA, factorial designs)

III. Modeling and Computation   12%
   A. Algorithm and logic development (e.g., flow charts, pseudo-code)
   B. Spreadsheets
   C. Databases (e.g., types, information content, relational)
   D. Decision theory (e.g., uncertainty, risk, utility, decision trees)
   E. Optimization modeling (decision variables, objective functions, and constraints)
   F. Linear programming (e.g., formulation, primal, dual, graphical solution)
   G. Math programming (network, integer, dynamic, transportation, assignment)
   H. Stochastic models (e.g., queuing, Markov, reliability)
   I. Simulation (e.g., event, process, Monte Carlo sampling, random number generation, steady-state vs. transient)

IV. Industrial Management   10%
   A. Principles (e.g., planning, organizing) and tools of management (e.g., MBO, re-engineering)
   B. Organizational structure (e.g., functional, matrix, line/staff)
   C. Motivation theories (e.g., Maslow, Theory X, Theory Y)
   D. Job evaluation and compensation
   E. Project management (scheduling, PERT, CPM)

V.   Manufacturing and Production Systems   13%
    A.   Manufacturing systems (e.g., cellular, group technology, flexible, lean)
    B.   Process design (e.g., number of machines/people, equipment selection, and line balancing)
    C.   Inventory analysis (e.g., EOQ, safety stock)
    D.   Forecasting
    E.   Scheduling (e.g., sequencing, cycle time, material control)
    F.   Aggregate planning (e.g., JIT, MRP, MRPII, ERP)
    G.   Concurrent engineering and design for manufacturing
    H.   Automation concepts (e.g., robotics, CIM)
    I.   Economics (e.g., profits and costs under various demand rates, machine selection)

VI.   Facilities and Logistics   12%
    A.   Flow measurements and analysis (e.g., from/to charts, flow planning)
    B.   Layouts (e.g., types, distance metrics, planning, evaluation)
    C.   Location analysis (e.g., single facility location, multiple facility location, storage location within a facility)
    D.   Process capacity analysis (e.g., number of machines/people, trade-offs)
    E.   Material handling capacity analysis (storage & transport)
    F.   Supply chain design (e.g., warehousing, transportation, inventories)

VII.   Human Factors, Productivity, Ergonomics, and Work Design   12%
    A.   Methods analysis (e.g., improvement, charting) and task analysis (e.g., MTM, MOST)
    B.   Time study (e.g., time standards, allowances)
    C.   Workstation design
    D.   Work sampling
    E.   Learning curves
    F.   Productivity measures
    G.   Risk factor identification, safety, toxicology, material safety data sheets (MSDS)
    H.   Environmental stress assessment (e.g., noise, vibrations, heat, computer-related)
    I.   Design of tasks, tools, displays, controls, user interfaces, etc.
    J.   Anthropometry, biomechanics, and lifting

VIII.   Quality   11%
    A.   Total quality management theory (e.g., Deming, Juran) and application
    B.   Management and planning tools (e.g., fishbone, Pareto, quality function deployment, scatter diagrams)
    C.   Control charts
    D.   Process capability and specifications
    E.   Sampling plans
    F.   Design of experiments for quality improvement
    G.   Auditing, ISO certification, and the Baldridge award

AFTERNOON SESSION IN MECHANICAL ENGINEERING
(60 questions in 8 topic areas)

		Approximate Percentage of Test Content
Topic Area		

I. Mechanical Design and Analysis   **15%**
   A.   Stress analysis (e.g., combined stresses, torsion, normal, shear)
   B.   Failure theories (e.g., static, dynamic, buckling)
   C.   Failure analysis (e.g., creep, fatigue, fracture, buckling)
   D.   Deformation and stiffness
   E.   Components (e.g., springs, pressure vessels, beams, piping, bearings, columns, power screws)
   F.   Power transmission (e.g., belts, chains, clutches, gears, shafts, brakes, axles)
   G.   Joining (e.g., threaded fasteners, rivets, welds, adhesives)
   H.   Manufacturability (e.g., fits, tolerances, process capability)
   I.   Quality and reliability
   J.   Mechanical systems (e.g., hydraulic, pneumatic, electro-hybrid)

II. Kinematics, Dynamics, and Vibrations   **15%**
   A.   Kinematics of mechanisms
   B.   Dynamics of mechanisms
   C.   Rigid body dynamics
   D.   Natural frequency and resonance
   E.   Balancing of rotating and reciprocating equipment
   F.   Forced vibrations (e.g., isolation, force transmission, support motion)

III. Materials and Processing   **10%**
   A.   Mechanical and thermal properties (e.g., stress/strain relationships, ductility, endurance, conductivity, thermal expansion)
   B.   Manufacturing processes (e.g., forming, machining, bending, casting, joining, heat treating)
   C.   Thermal processing (e.g., phase transformations, equilibria)
   D.   Materials selection (e.g., metals, composites, ceramics, plastics, bio-materials)
   E.   Surface conditions (e.g., corrosion, degradation, coatings, finishes)
   F.   Testing (e.g., tensile, compression, hardness)

IV. Measurements, Instrumentation, and Controls   **10%**
   A.   Mathematical fundamentals (e.g., Laplace transforms, differential equations)
   B.   System descriptions (e.g., block diagrams, ladder logic, transfer functions)
   C.   Sensors and signal conditioning (e.g., strain, pressure, flow, force, velocity, displacement, temperature)
   D.   Data collection and processing (e.g., sampling theory, uncertainty, digital/analog, data transmission rates)
   E.   Dynamic responses (e.g., overshoot/time constant, poles and zeros, stability)

V. Thermodynamics and Energy Conversion Processes   **15%**
   A.   Ideal and real gases
   B.   Reversibility/irreversibility
   C.   Thermodynamic equilibrium
   D.   Psychrometrics
   E.   Performance of components
   F.   Cycles and processes (e.g., Otto, Diesel, Brayton, Rankine)
   G.   Combustion and combustion products
   H.   Energy storage
   I.   Cogeneration and regeneration/reheat

VI. Fluid Mechanics and Fluid Machinery   **15%**
   A. Fluid statics
   B. Incompressible flow
   C. Fluid transport systems (e.g., pipes, ducts, series/parallel operations)
   D. Fluid machines: incompressible (e.g., turbines, pumps, hydraulic motors)
   E. Compressible flow
   F. Fluid machines: compressible (e.g., turbines, compressors, fans)
   G. Operating characteristics (e.g., fan laws, performance curves, efficiencies, work/power equations)
   H. Lift/drag
   I. Impulse/momentum

VII. Heat Transfer   **10%**
   A. Conduction
   B. Convection
   C. Radiation
   D. Composite walls and insulation
   E. Transient and periodic processes
   F. Heat exchangers
   G. Boiling and condensation heat transfer

VIII. Refrigeration and HVAC   **10%**
   A. Cycles
   B. Heating and cooling loads (e.g., degree day data, sensible heat, latent heat)
   C. Psychrometric charts
   D. Coefficient of performance
   E. Components (e.g., compressors, condensers, evaporators, expansion valve)

AFTERNOON SESSION IN OTHER/GENERAL ENGINEERING
(60 questions in 9 topic areas)

	Topic Area	Approximate Percentage of Test Content
I.	**Advanced Engineering Mathematics**	**10%**
	A. Differential equations	
	B. Partial differential calculus	
	C. Numerical solutions (e.g., differential equations, algebraic equations)	
	D. Linear algebra	
	E. Vector analysis	
II.	**Engineering Probability and Statistics**	**9%**
	A. Sample distributions and sizes	
	B. Design of experiments	
	C. Hypothesis testing	
	D. Goodness of fit (coefficient of correlation, chi square)	
	E. Estimation (e.g., point, confidence intervals) for two means	
III.	**Biology**	**5%**
	A. Cellular biology (e.g., structure, growth, cell organization)	
	B. Toxicology (e.g., human, environmental)	
	C. Industrial hygiene [e.g., personnel protection equipment (PPE), carcinogens]	
	D. Bioprocessing (e.g., fermentation, waste treatment, digestion)	
IV.	**Engineering Economics**	**10%**
	A. Cost estimating	
	B. Project selection	
	C. Lease/buy/make	
	D. Replacement analysis (e.g., optimal economic life)	
V.	**Application of Engineering Mechanics**	**13%**
	A. Stability analysis of beams, trusses, and frames	
	B. Deflection analysis	
	C. Failure theory (e.g., static and dynamic)	
	D. Failure analysis (e.g., creep, fatigue, fracture, buckling)	
VI.	**Engineering of Materials**	**11%**
	A. Material properties of:	
	1. metals	
	2. plastics	
	3. composites	
	4. concrete	
VII.	**Fluids**	**15%**
	A. Basic hydraulics (e.g., Manning equation, Bernoulli theorem, open-channel flow, pipe flow)	
	B. Laminar and turbulent flow	
	C. Friction losses (e.g., pipes, valves, fittings)	
	D. Flow measurement	
	E. Dimensionless numbers (e.g., Reynolds number)	
	F. Fluid transport systems (e.g., pipes, ducts, series/parallel operations)	
	G. Pumps, turbines, and compressors	
	H. Lift/drag	

VIII. Electricity and Magnetism   **12%**
    A.   Equivalent circuits (Norton, Thevenin)
    B.   AC circuits (frequency domain)
    C.   Network analysis (Kirchhoff laws)
    D.   RLC circuits
    E.   Sensors and instrumentation
    F.   Electrical machines

IX. Thermodynamics and Heat Transfer   **15%**
    A.   Thermodynamic properties (e.g., entropy, enthalpy, heat capacity)
    B.   Thermodynamic processes (e.g., isothermal, adiabatic, reversible, irreversible)
    C.   Equations of state (ideal and real gases)
    D.   Conduction, convection, and radiation heat transfer
    E.   Mass and energy balances
    F.   Property and phase diagrams (e.g., T-s, h-P)
    G.   Tables of thermodynamic properties
    H.   Cyclic processes and efficiency (e.g., refrigeration, power)
    I.   Phase equilibrium and phase change
    J.   Thermodynamic equilibrium
    K.   Combustion and combustion products (e.g., CO, CO_2, NO_X, ash, particulates)
    L.   Psychrometrics (e.g., humidity)

Do not write in this book or remove any pages.

Do all scratch work in your exam booklet.

UNITS

The FE exam and this handbook use both the metric system of units and the U.S. Customary System (USCS). In the USCS system of units, both force and mass are called pounds. Therefore, one must distinguish the pound-force (lbf) from the pound-mass (lbm).

The pound-force is that force which accelerates one pound-mass at 32.174 ft/sec². Thus, 1 lbf = 32.174 lbm-ft/sec². The expression 32.174 lbm-ft/(lbf-sec²) is designated as g_c and is used to resolve expressions involving both mass and force expressed as pounds. For instance, in writing Newton's second law, the equation would be written as $F = ma/g_c$, where F is in lbf, m in lbm, and a is in ft/sec².

Similar expressions exist for other quantities. Kinetic Energy, $KE = mv^2/2g_c$, with KE in (ft-lbf); Potential Energy, $PE = mgh/g_c$, with PE in (ft-lbf); Fluid Pressure, $p = \rho gh/g_c$, with p in (lbf/ft²); Specific Weight, $SW = \rho g/g_c$, in (lbf/ft³); Shear Stress, $\tau = (\mu/g_c)(dv/dy)$, with shear stress in (lbf/ft²). In all these examples, g_c should be regarded as a unit conversion factor. It is frequently not written explicitly in engineering equations. However, its use is required to produce a consistent set of units.

Note that the conversion factor g_c [lbm-ft/(lbf-sec²)] should not be confused with the local acceleration of gravity g, which has different units (m/s² or ft/sec²) and may be either its standard value (9.807 m/s² or 32.174 ft/sec²) or some other local value.

If the problem is presented in USCS units, it may be necessary to use the constant g_c in the equation to have a consistent set of units.

METRIC PREFIXES		
Multiple	Prefix	Symbol
10^{-18}	atto	a
10^{-15}	femto	f
10^{-12}	pico	p
10^{-9}	nano	n
10^{-6}	micro	μ
10^{-3}	milli	m
10^{-2}	centi	c
10^{-1}	deci	d
10^{1}	deka	da
10^{2}	hecto	h
10^{3}	kilo	k
10^{6}	mega	M
10^{9}	giga	G
10^{12}	tera	T
10^{15}	peta	P
10^{18}	exa	E

COMMONLY USED EQUIVALENTS	
1 gallon of water weighs	8.34 lbf
1 cubic foot of water weighs	62.4 lbf
1 cubic inch of mercury weighs	0.491 lbf
The mass of 1 cubic meter of water is	1,000 kilograms

TEMPERATURE CONVERSIONS
°F = 1.8 (°C) + 32
°C = (°F − 32)/1.8
°R = °F + 459.69
K = °C + 273.15

FUNDAMENTAL CONSTANTS

Quantity		Symbol	Value	Units
electron charge		e	1.6022×10^{-19}	C (coulombs)
Faraday constant		F	96,485	coulombs/(mol)
gas constant	metric	$\overline{R}$	8,314	J/(kmol•K)
gas constant	metric	$\overline{R}$	8.314	kPa·m³/(kmol•K)
gas constant	USCS	$\overline{R}$	1,545	ft-lbf/(lb mole-°R)
		$\overline{R}$	0.08206	L-atm/(mole-K)
gravitation - newtonian constant		G	6.673×10^{-11}	m³/(kg•s²)
gravitation - newtonian constant		G	6.673×10^{-11}	N•m²/kg²
gravity acceleration (standard)	metric	g	9.807	m/s²
gravity acceleration (standard)	USCS	g	32.174	ft/sec²
molar volume (ideal gas), $T = 273.15K$, $p = 101.3$ kPa		V_m	22,414	L/kmol
speed of light in vacuum		c	299,792,000	m/s
Stephan-Boltzmann constant		σ	5.67×10^{-8}	W/(m²•K⁴)

CONVERSION FACTORS

Multiply	By	To Obtain	Multiply	By	To Obtain
acre	43,560	square feet (ft^2)	joule (J)	9.478×10^{-4}	Btu
ampere-hr (A-hr)	3,600	coulomb (C)	J	0.7376	ft-lbf
ångström (Å)	1×10^{-10}	meter (m)	J	1	newton•m (N•m)
atmosphere (atm)	76.0	cm, mercury (Hg)	J/s	1	watt (W)
atm, std	29.92	in, mercury (Hg)			
atm, std	14.70	lbf/in^2 abs (psia)	kilogram (kg)	2.205	pound (lbm)
atm, std	33.90	ft, water	kgf	9.8066	newton (N)
atm, std	1.013×10^5	pascal (Pa)	kilometer (km)	3,281	feet (ft)
			km/hr	0.621	mph
bar	1×10^5	Pa	kilopascal (kPa)	0.145	lbf/in^2 (psi)
barrels–oil	42	gallons–oil	kilowatt (kW)	1.341	horsepower (hp)
Btu	1,055	joule (J)	kW	3,413	Btu/hr
Btu	2.928×10^{-4}	kilowatt-hr (kWh)	kW	737.6	(ft-lbf)/sec
Btu	778	ft-lbf	kW-hour (kWh)	3,413	Btu
Btu/hr	3.930×10^{-4}	horsepower (hp)	kWh	1.341	hp-hr
Btu/hr	0.293	watt (W)	kWh	3.6×10^6	joule (J)
Btu/hr	0.216	ft-lbf/sec	kip (K)	1,000	lbf
			K	4,448	newton (N)
calorie (g-cal)	3.968×10^{-3}	Btu			
cal	1.560×10^{-6}	hp-hr	liter (L)	61.02	in^3
cal	4.186	joule (J)	L	0.264	gal (US Liq)
cal/sec	4.184	watt (W)	L	10^{-3}	m^3
centimeter (cm)	3.281×10^{-2}	foot (ft)	L/second (L/s)	2.119	ft^3/min (cfm)
cm	0.394	inch (in)	L/s	15.85	gal (US)/min (gpm)
centipoise (cP)	0.001	pascal•sec (Pa•s)			
centipoise (cP)	1	g/(m•s)			
centistokes (cSt)	1×10^{-6}	m^2/sec (m^2/s)	meter (m)	3.281	feet (ft)
cubic feet/second (cfs)	0.646317	million gallons/day (mgd)	m	1.094	yard
cubic foot (ft^3)	7.481	gallon	metric ton	1,000	kilogram (kg)
cubic meters (m^3)	1,000	Liters	m/second (m/s)	196.8	feet/min (ft/min)
electronvolt (eV)	1.602×10^{-19}	joule (J)	mile (statute)	5,280	feet (ft)
			mile (statute)	1.609	kilometer (km)
foot (ft)	30.48	cm	mile/hour (mph)	88.0	ft/min (fpm)
ft	0.3048	meter (m)	mph	1.609	km/h
ft-pound (ft-lbf)	1.285×10^{-3}	Btu	mm of Hg	1.316×10^{-3}	atm
ft-lbf	3.766×10^{-7}	kilowatt-hr (kWh)	mm of H$_2$O	9.678×10^{-5}	atm
ft-lbf	0.324	calorie (g-cal)			
ft-lbf	1.356	joule (J)	newton (N)	0.225	lbf
			newton (N)	1	kg•m/s^2
ft-lbf/sec	1.818×10^{-3}	horsepower (hp)	N•m	0.7376	ft-lbf
			N•m	1	joule (J)
gallon (US Liq)	3.785	liter (L)			
gallon (US Liq)	0.134	ft^3	pascal (Pa)	9.869×10^{-6}	atmosphere (atm)
gallons of water	8.3453	pounds of water	Pa	1	newton/m^2 (N/m^2)
gamma (γ, Γ)	1×10^{-9}	tesla (T)	Pa•sec (Pa•s)	10	poise (P)
gauss	1×10^{-4}	T	pound (lbm, avdp)	0.454	kilogram (kg)
gram (g)	2.205×10^{-3}	pound (lbm)	lbf	4.448	N
			lbf-ft	1.356	N•m
hectare	1×10^4	square meters (m^2)	lbf/in^2 (psi)	0.068	atm
hectare	2.47104	acres	psi	2.307	ft of H$_2$O
horsepower (hp)	42.4	Btu/min	psi	2.036	in. of Hg
hp	745.7	watt (W)	psi	6,895	Pa
hp	33,000	(ft-lbf)/min			
hp	550	(ft-lbf)/sec	radian	180/π	degree
hp-hr	2,545	Btu			
hp-hr	1.98×10^6	ft-lbf	stokes	1×10^{-4}	m^2/s
hp-hr	2.68×10^6	joule (J)			
hp-hr	0.746	kWh	therm	1×10^5	Btu
			ton	2,000	pounds (lb)
inch (in)	2.540	centimeter (cm)	watt (W)	3.413	Btu/hr
in of Hg	0.0334	atm	W	1.341×10^{-3}	horsepower (hp)
in of Hg	13.60	in of H$_2$O	W	1	joule/s (J/s)
in of H$_2$O	0.0361	lbf/in^2 (psi)	weber/m^2 (Wb/m^2)	10,000	gauss
in of H$_2$O	0.002458	atm			

MATHEMATICS

STRAIGHT LINE

The general form of the equation is

$$Ax + By + C = 0$$

The standard form of the equation is

$$y = mx + b,$$

which is also known as the *slope-intercept* form.

The *point-slope* form is $\quad y - y_1 = m(x - x_1)$

Given two points: slope, $\quad m = (y_2 - y_1)/(x_2 - x_1)$

The angle between lines with slopes m_1 and m_2 is

$$\alpha = \arctan\left[(m_2 - m_1)/(1 + m_2 \cdot m_1)\right]$$

Two lines are perpendicular if $m_1 = -1/m_2$

The distance between two points is

$$d = \sqrt{(y_2 - y_1)^2 + (x_2 - x_1)^2}$$

QUADRATIC EQUATION

$$ax^2 + bx + c = 0$$

$$x = \text{Roots} = \frac{-b \pm \sqrt{b^2 - 4ac}}{2a}$$

CONIC SECTIONS

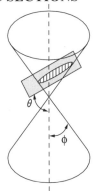

$$e = \text{eccentricity} = \cos\theta/(\cos\phi)$$

[Note: X' and Y', in the following cases, are translated axes.]

Case 1. Parabola $e = 1$:

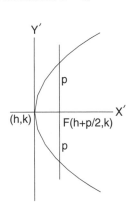

$(y - k)^2 = 2p(x - h)$; Center at (h, k)
is the standard form of the equation. When $h = k = 0$,
Focus: $(p/2, 0)$; Directrix: $x = -p/2$

Case 2. Ellipse $e < 1$:

♦

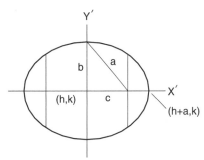

is the standard form of the equation. When $h = k = 0$,

Eccentricity: $\quad e = \sqrt{1 - (b^2/a^2)} = c/a$

$b = a\sqrt{1 - e^2}$;

Focus: $(\pm ae, 0)$; Directrix: $x = \pm a/e$

Case 3. Hyperbola $e > 1$:

♦

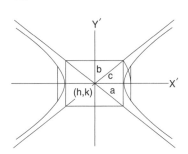

$$\frac{(x - h)^2}{a^2} - \frac{(y - k)^2}{b^2} = 1; \text{ Center at } (h, k)$$

is the standard form of the equation. When $h = k = 0$,

Eccentricity: $e = \sqrt{1 + (b^2/a^2)} = c/a$

$b = a\sqrt{e^2 - 1}$;

Focus: $(\pm ae, 0)$; Directrix: $x = \pm a/e$

♦ Brink, R.W., *A First Year of College Mathematics*, D. Appleton-Century Co., Inc., 1937.

Case 4. Circle $e = 0$:

$(x - h)^2 + (y - k)^2 = r^2$; Center at (h, k) is the general form of the equation with radius

$$r = \sqrt{(x - h)^2 + (y - k)^2}$$

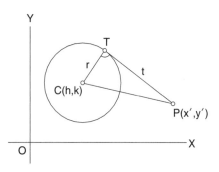

Length of the tangent from a point. Using the general form of the equation of a circle, the length of the tangent is found from

$$t^2 = (x' - h)^2 + (y' - k)^2 - r^2$$

by substituting the coordinates of a point $P(x',y')$ and the coordinates of the center of the circle into the equation and computing.

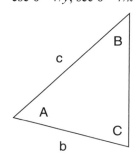

Conic Section Equation

The general form of the conic section equation is

$$Ax^2 + Bxy + Cy^2 + Dx + Ey + F = 0$$

where not both A and C are zero.

If $B^2 - AC < 0$, an *ellipse* is defined.

If $B^2 - AC > 0$, a *hyperbola* is defined.

If $B^2 - AC = 0$, the conic is a *parabola*.

If $A = C$ and $B = 0$, a *circle* is defined.

If $A = B = C = 0$, a *straight line* is defined.

$$x^2 + y^2 + 2ax + 2by + c = 0$$

is the normal form of the conic section equation, if that conic section has a principal axis parallel to a coordinate axis.

$h = -a; \, k = -b$

$r = \sqrt{a^2 + b^2 - c}$

If $a^2 + b^2 - c$ is positive, a *circle*, center $(-a, -b)$.

If $a^2 + b^2 - c$ equals zero, a *point* at $(-a, -b)$.

If $a^2 + b^2 - c$ is negative, locus is *imaginary*.

QUADRIC SURFACE (SPHERE)

The general form of the equation is

$$(x - h)^2 + (y - k)^2 + (z - m)^2 = r^2$$

with center at (h, k, m).

In a three-dimensional space, the distance between two points is

$$d = \sqrt{(x_2 - x_1)^2 + (y_2 - y_1)^2 + (z_2 - z_1)^2}$$

LOGARITHMS

The logarithm of x to the Base b is defined by

$\log_b (x) = c$, where $b^c = x$

Special definitions for $b = e$ or $b = 10$ are:

$\ln x$, Base $= e$

$\log x$, Base $= 10$

To change from one Base to another:

$\log_b x = (\log_a x)/(\log_a b)$

e.g., $\ln x = (\log_{10} x)/(\log_{10} e) = 2.302585 \, (\log_{10} x)$

Identities

$\log_b b^n = n$

$\log x^c = c \log x; \, x^c = \text{antilog} \, (c \log x)$

$\log xy = \log x + \log y$

$\log_b b = 1; \, \log 1 = 0$

$\log x/y = \log x - \log y$

TRIGONOMETRY

Trigonometric functions are defined using a right triangle.

$\sin \theta = y/r, \, \cos \theta = x/r$

$\tan \theta = y/x, \, \cot \theta = x/y$

$\csc \theta = r/y, \, \sec \theta = r/x$

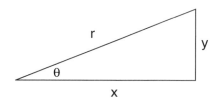

Law of Sines

$$\frac{a}{\sin A} = \frac{b}{\sin B} = \frac{c}{\sin C}$$

Law of Cosines

$a^2 = b^2 + c^2 - 2bc \cos A$

$b^2 = a^2 + c^2 - 2ac \cos B$

$c^2 = a^2 + b^2 - 2ab \cos C$

♦ Brink, R.W., *A First Year of College Mathematics*, D. Appleton-Century Co., Inc., Englewood Cliffs, NJ, 1937.

Identities

$\csc\theta = 1/\sin\theta$

$\sec\theta = 1/\cos\theta$

$\tan\theta = \sin\theta/\cos\theta$

$\cot\theta = 1/\tan\theta$

$\sin^2\theta + \cos^2\theta = 1$

$\tan^2\theta + 1 = \sec^2\theta$

$\cot^2\theta + 1 = \csc^2\theta$

$\sin(\alpha + \beta) = \sin\alpha\cos\beta + \cos\alpha\sin\beta$

$\cos(\alpha + \beta) = \cos\alpha\cos\beta - \sin\alpha\sin\beta$

$\sin 2\alpha = 2\sin\alpha\cos\alpha$

$\cos 2\alpha = \cos^2\alpha - \sin^2\alpha = 1 - 2\sin^2\alpha = 2\cos^2\alpha - 1$

$\tan 2\alpha = (2\tan\alpha)/(1 - \tan^2\alpha)$

$\cot 2\alpha = (\cot^2\alpha - 1)/(2\cot\alpha)$

$\tan(\alpha + \beta) = (\tan\alpha + \tan\beta)/(1 - \tan\alpha\tan\beta)$

$\cot(\alpha + \beta) = (\cot\alpha\cot\beta - 1)/(\cot\alpha + \cot\beta)$

$\sin(\alpha - \beta) = \sin\alpha\cos\beta - \cos\alpha\sin\beta$

$\cos(\alpha - \beta) = \cos\alpha\cos\beta + \sin\alpha\sin\beta$

$\tan(\alpha - \beta) = (\tan\alpha - \tan\beta)/(1 + \tan\alpha\tan\beta)$

$\cot(\alpha - \beta) = (\cot\alpha\cot\beta + 1)/(\cot\beta - \cot\alpha)$

$\sin(\alpha/2) = \pm\sqrt{(1 - \cos\alpha)/2}$

$\cos(\alpha/2) = \pm\sqrt{(1 + \cos)/2}$

$\tan(\alpha/2) = \pm\sqrt{(1 - \cos\alpha)/(1 + \cos\alpha)}$

$\cot(\alpha/2) = \pm\sqrt{(1 + \cos\alpha)/(1 - \cos\alpha)}$

$\sin\alpha\sin\beta = (1/2)[\cos(\alpha - \beta) - \cos(\alpha + \beta)]$

$\cos\alpha\cos\beta = (1/2)[\cos(\alpha - \beta) + \cos(\alpha + \beta)]$

$\sin\alpha\cos\beta = (1/2)[\sin(\alpha + \beta) + \sin(\alpha - \beta)]$

$\sin\alpha + \sin\beta = 2\sin(1/2)(\alpha + \beta)\cos(1/2)(\alpha - \beta)$

$\sin\alpha - \sin\beta = 2\cos(1/2)(\alpha + \beta)\sin(1/2)(\alpha - \beta)$

$\cos\alpha + \cos\beta = 2\cos(1/2)(\alpha + \beta)\cos(1/2)(\alpha - \beta)$

$\cos\alpha - \cos\beta = -2\sin(1/2)(\alpha + \beta)\sin(1/2)(\alpha - \beta)$

COMPLEX NUMBERS

Definition $i = \sqrt{-1}$

$(a + ib) + (c + id) = (a + c) + i(b + d)$

$(a + ib) - (c + id) = (a - c) + i(b - d)$

$(a + ib)(c + id) = (ac - bd) + i(ad + bc)$

$\dfrac{a + ib}{c + id} = \dfrac{(a + ib)(c - id)}{(c + id)(c - id)} = \dfrac{(ac + bd) + i(bc - ad)}{c^2 + d^2}$

Polar Coordinates

$x = r\cos\theta;\ y = r\sin\theta;\ \theta = \arctan(y/x)$

$r = |x + iy| = \sqrt{x^2 + y^2}$

$x + iy = r(\cos\theta + i\sin\theta) = re^{i\theta}$

$[r_1(\cos\theta_1 + i\sin\theta_1)][r_2(\cos\theta_2 + i\sin\theta_2)] =$
$$r_1 r_2[\cos(\theta_1 + \theta_2) + i\sin(\theta_1 + \theta_2)]$$

$(x + iy)^n = [r(\cos\theta + i\sin\theta)]^n$
$$= r^n(\cos n\theta + i\sin n\theta)$$

$$\frac{r_1(\cos\theta_1 + i\sin\theta_1)}{r_2(\cos\theta_2 + i\sin\theta_2)} = \frac{r_1}{r_2}\left[\cos(\theta_1 - \theta_2) + \sin(\theta_2)\right]$$

Euler's Identity

$e^{i\theta} = \cos\theta + i\sin\theta$

$e^{-i\theta} = \cos\theta - i\sin\theta$

$$\cos\theta = \frac{e^{i\theta} + e^{-i\theta}}{2},\ \sin\theta = \frac{e^{i\theta} - e^{-i\theta}}{2i}$$

Roots

If k is any positive integer, any complex number (other than zero) has k distinct roots. The k roots of r $(\cos\theta + i\sin\theta)$ can be found by substituting successively $n = 0, 1, 2, ..., (k - 1)$ in the formula

$$w = \sqrt[k]{r}\left[\cos\left(\frac{\theta}{k} + n\frac{360°}{k}\right) + i\sin\left(\frac{\theta}{k} + n\frac{360°}{k}\right)\right]$$

Also, see **Algebra of Complex Numbers** in the **ELECTRICAL AND COMPUTER ENGINEERING** section.

MATRICES

A matrix is an ordered rectangular array of numbers with m rows and n columns. The element a_{ij} refers to row i and column j.

Multiplication

If $A = (a_{ik})$ is an $m \times n$ matrix and $B = (b_{kj})$ is an $n \times s$ matrix, the matrix product AB is an $m \times s$ matrix

$$C = (c_{ij}) = \left(\sum_{l=1}^{n} a_{il}b_{lj}\right)$$

where n is the common integer representing the number of columns of A and the number of rows of B (l and $k = 1, 2, ..., n$).

Addition

If $A = (a_{ij})$ and $B = (b_{ij})$ are two matrices of the same size $m \times n$, the sum $A + B$ is the $m \times n$ matrix $C = (c_{ij})$ where $c_{ij} = a_{ij} + b_{ij}$.

Identity

The matrix $\mathbf{I} = (a_{ij})$ is a square $n \times n$ identity matrix where $a_{ii} = 1$ for $i = 1, 2, ..., n$ and $a_{ij} = 0$ for $i \neq j$.

Transpose

The matrix B is the transpose of the matrix A if each entry b_{ji} in B is the same as the entry a_{ij} in A and conversely. In equation form, the transpose is $B = A^T$.

Inverse

The inverse B of a square $n \times n$ matrix A is

$$B = A^{-1} = \frac{\text{adj}(A)}{|A|}, \text{ where}$$

adj(A) = adjoint of A (obtained by replacing A^T elements with their cofactors, see **DETERMINANTS**) and
$|A|$ = determinant of A.

Also, $\mathbf{AA^{-1}} = \mathbf{A^{-1}A} = \mathbf{I}$ where $\mathbf{I}$ is the identity matrix.

DETERMINANTS

A *determinant of order n* consists of n^2 numbers, called the *elements* of the determinant, arranged in n rows and n columns and enclosed by two vertical lines.

In any determinant, the *minor* of a given element is the determinant that remains after all of the elements are struck out that lie in the same row and in the same column as the given element. Consider an element which lies in the *j*th column and the *i*th row. The *cofactor* of this element is the value of the minor of the element (if $i + j$ is *even*), and it is the negative of the value of the minor of the element (if $i + j$ is *odd*).

If n is greater than 1, the *value* of a determinant of order n is the sum of the n products formed by multiplying each element of some specified row (or column) by its cofactor. This sum is called the *expansion of the determinant* [according to the elements of the specified row (or column)]. For a second-order determinant:

$$\begin{vmatrix} a_1 & a_2 \\ b_1 & b_2 \end{vmatrix} = a_1 b_2 - a_2 b_1$$

For a third-order determinant:

$$\begin{vmatrix} a_1 & a_2 & a_3 \\ b_1 & b_2 & b_3 \\ c_1 & c_2 & c_3 \end{vmatrix} = a_1 b_2 c_3 + a_2 b_3 c_1 + a_3 b_1 c_2 - a_3 b_2 c_1 - a_2 b_1 c_3 - a_1 b_3 c_2$$

VECTORS

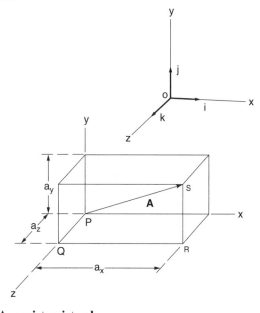

$$\mathbf{A} = a_x \mathbf{i} + a_y \mathbf{j} + a_z \mathbf{k}$$

Addition and *subtraction:*

$$\mathbf{A} + \mathbf{B} = (a_x + b_x)\mathbf{i} + (a_y + b_y)\mathbf{j} + (a_z + b_z)\mathbf{k}$$
$$\mathbf{A} - \mathbf{B} = (a_x - b_x)\mathbf{i} + (a_y - b_y)\mathbf{j} + (a_z - b_z)\mathbf{k}$$

The *dot product* is a *scalar product* and represents the projection of $\mathbf{B}$ onto $\mathbf{A}$ times $|\mathbf{A}|$. It is given by

$$\mathbf{A \cdot B} = a_x b_x + a_y b_y + a_z b_z$$
$$= |\mathbf{A}||\mathbf{B}|\cos\theta = \mathbf{B \cdot A}$$

The *cross product* is a *vector product* of magnitude $|\mathbf{B}|\,|\mathbf{A}|\sin\theta$ which is perpendicular to the plane containing $\mathbf{A}$ and $\mathbf{B}$. The product is

$$\mathbf{A \times B} = \begin{vmatrix} \mathbf{i} & \mathbf{j} & \mathbf{k} \\ a_x & a_y & a_z \\ b_x & b_y & b_z \end{vmatrix} = -\mathbf{B \times A}$$

The sense of $\mathbf{A \times B}$ is determined by the right-hand rule.

$$\mathbf{A \times B} = |\mathbf{A}||\mathbf{B}|\mathbf{n}\sin\theta, \text{ where}$$

$\mathbf{n}$ = unit vector perpendicular to the plane of $\mathbf{A}$ and $\mathbf{B}$.

Gradient, Divergence, and Curl

$$\nabla\phi = \left(\frac{\partial}{\partial x}\mathbf{i} + \frac{\partial}{\partial y}\mathbf{j} + \frac{\partial}{\partial z}\mathbf{k}\right)\phi$$

$$\nabla \cdot \mathbf{V} = \left(\frac{\partial}{\partial x}\mathbf{i} + \frac{\partial}{\partial y}\mathbf{j} + \frac{\partial}{\partial z}\mathbf{k}\right) \cdot (V_1\mathbf{i} + V_2\mathbf{j} + V_3\mathbf{k})$$

$$\nabla \times \mathbf{V} = \left(\frac{\partial}{\partial x}\mathbf{i} + \frac{\partial}{\partial y}\mathbf{j} + \frac{\partial}{\partial z}\mathbf{k}\right) \times (V_1\mathbf{i} + V_2\mathbf{j} + V_3\mathbf{k})$$

The Laplacian of a scalar function ϕ is

$$\nabla^2\phi = \frac{\partial^2\phi}{\partial x^2} + \frac{\partial^2\phi}{\partial y^2} + \frac{\partial^2\phi}{\partial z^2}$$

Identities

$\mathbf{A \cdot B} = \mathbf{B \cdot A};\ \mathbf{A \cdot (B + C)} = \mathbf{A \cdot B} + \mathbf{A \cdot C}$

$\mathbf{A \cdot A} = |\mathbf{A}|^2$

$\mathbf{i \cdot i} = \mathbf{j \cdot j} = \mathbf{k \cdot k} = 1$

$\mathbf{i \cdot j} = \mathbf{j \cdot k} = \mathbf{k \cdot i} = 0$

If $\mathbf{A \cdot B} = 0$, then either $\mathbf{A} = 0$, $\mathbf{B} = 0$, or $\mathbf{A}$ is perpendicular to $\mathbf{B}$.

$\mathbf{A \times B} = -\mathbf{B \times A}$

$\mathbf{A \times (B + C)} = (\mathbf{A \times B}) + (\mathbf{A \times C})$

$\mathbf{(B + C) \times A} = (\mathbf{B \times A}) + (\mathbf{C \times A})$

$\mathbf{i \times i} = \mathbf{j \times j} = \mathbf{k \times k} = \mathbf{0}$

$\mathbf{i \times j} = \mathbf{k} = -\mathbf{j \times i};\ \mathbf{j \times k} = \mathbf{i} = -\mathbf{k \times j}$

$\mathbf{k \times i} = \mathbf{j} = -\mathbf{i \times k}$

If $\mathbf{A \times B} = 0$, then either $\mathbf{A} = 0$, $\mathbf{B} = 0$, or $\mathbf{A}$ is parallel to $\mathbf{B}$.

$\nabla^2\phi = \nabla \cdot (\nabla\phi) = (\nabla \cdot \nabla)\phi$

$\nabla \times \nabla\phi = \mathbf{0}$

$\nabla \cdot (\nabla \times \mathbf{A}) = \mathbf{0}$

$\nabla \times (\nabla \times \mathbf{A}) = \nabla(\nabla \cdot \mathbf{A}) - \nabla^2\mathbf{A}$

PROGRESSIONS AND SERIES

Arithmetic Progression

To determine whether a given finite sequence of numbers is an arithmetic progression, subtract each number from the following number. If the differences are equal, the series is arithmetic.

1. The first term is a.
2. The common difference is d.
3. The number of terms is n.
4. The last or nth term is l.
5. The sum of n terms is S.

$$l = a + (n-1)d$$
$$S = n(a+l)/2 = n\,[2a + (n-1)\,d]/2$$

Geometric Progression

To determine whether a given finite sequence is a geometric progression (G.P.), divide each number after the first by the preceding number. If the quotients are equal, the series is geometric.

1. The first term is a.
2. The common ratio is r.
3. The number of terms is n.
4. The last or nth term is l.
5. The sum of n terms is S.

$$l = ar^{n-1}$$
$$S = a\,(1-r^n)/(1-r); \; r \neq 1$$
$$S = (a-rl)/(1-r); \; r \neq 1$$
$$\lim_{n\to\infty} S_n = a/(1-r); \; r < 1$$

A G.P. converges if $|r| < 1$ and it diverges if $|r| > 1$.

Properties of Series

$$\sum_{i=1}^{n} c = nc; \qquad c = \text{constant}$$

$$\sum_{i=1}^{n} cx_i = c \sum_{i=1}^{n} x_i$$

$$\sum_{i=1}^{n} \left(x_i + y_i - z_i \right) = \sum_{i=1}^{n} x_i + \sum_{i=1}^{n} y_i - \sum_{i=1}^{n} z_i$$

$$\sum_{x=1}^{n} x = \left(n + n^2 \right)/2$$

Power Series

$$\sum_{i=0}^{\infty} a_i (x-a)^i$$

1. A power series, which is convergent in the interval $-R < x < R$, defines a function of x that is continuous for all values of x within the interval and is said to represent the function in that interval.
2. A power series may be differentiated term by term within its interval of convergence. The resulting series has the same interval of convergence as the original series (except possibly at the end points of the series).
3. A power series may be integrated term by term provided the limits of integration are within the interval of convergence of the series.
4. Two power series may be added, subtracted, or multiplied, and the resulting series in each case is convergent, at least, in the interval common to the two series.
5. Using the process of long division (as for polynomials), two power series may be divided one by the other within their common interval of convergence.

Taylor's Series

$$f(x) = f(a) + \frac{f'(a)}{1!}(x-a) + \frac{f''(a)}{2!}(x-a)^2 + \dots + \frac{f^{(n)}(a)}{n!}(x-a)^n + \dots$$

is called *Taylor's series*, and the function $f(x)$ is said to be expanded about the point a in a Taylor's series.

If $a = 0$, the Taylor's series equation becomes a *Maclaurin's series*.

DIFFERENTIAL CALCULUS

The Derivative

For any function $y = f(x)$,
the derivative $= D_x y = dy/dx = y'$

$$y' = \lim_{\Delta x \to 0} \left[(\Delta y)/(\Delta x) \right]$$
$$= \lim_{\Delta x \to 0} \left\{ [f(x + \Delta x) = f(x)]/(\Delta x) \right\}$$
$$y' = \text{the slope of the curve } f(x).$$

<u>Test for a Maximum</u>

$y = f(x)$ is a maximum for
$x = a$, if $f'(a) = 0$ and $f''(a) < 0$.

<u>Test for a Minimum</u>

$y = f(x)$ is a minimum for
$x = a$, if $f'(a) = 0$ and $f''(a) > 0$.

<u>Test for a Point of Inflection</u>

$y = f(x)$ has a point of inflection at $x = a$,
if $f''(a) = 0$, and
if $f''(x)$ changes sign as x increases through
$x = a$.

The Partial Derivative

In a function of two independent variables x and y, a derivative with respect to one of the variables may be found if the other variable is *assumed* to remain constant. If y is *kept fixed*, the function

$$z = f(x, y)$$

becomes a function of the *single variable x*, and its derivative (if it exists) can be found. This derivative is called the *partial derivative of z with respect to x*. The partial derivative with respect to x is denoted as follows:

$$\frac{\partial z}{\partial x} = \frac{\partial f(x,y)}{\partial x}$$

The Curvature of Any Curve

♦

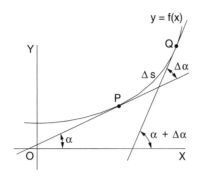

The curvature K of a curve at P is the limit of its average curvature for the arc PQ as Q approaches P. This is also expressed as: the curvature of a curve at a given point is the rate-of-change of its inclination with respect to its arc length.

$$K = \lim_{\Delta s \to 0} \frac{\Delta \alpha}{\Delta s} = \frac{d\alpha}{ds}$$

Curvature in Rectangular Coordinates

$$K = \frac{y''}{\left[1 + (y')^2\right]^{3/2}}$$

When it may be easier to differentiate the function with respect to y rather than x, the notation x' will be used for the derivative.

$$x' = dx/dy$$

$$K = \frac{-x''}{\left[1 + (x')^2\right]^{3/2}}$$

The Radius of Curvature

The *radius of curvature* R at any point on a curve is defined as the absolute value of the reciprocal of the curvature K at that point.

$$R = \frac{1}{|K|} \qquad (K \neq 0)$$

$$R = \left| \frac{\left[1 + (y')^2\right]^{3/2}}{|y''|} \right| \qquad (y'' \neq 0)$$

L'Hospital's Rule (L'Hôpital's Rule)

If the fractional function $f(x)/g(x)$ assumes one of the indeterminate forms $0/0$ or ∞/∞ (where α is finite or infinite), then

$$\lim_{x \to \alpha} f(x)/g(x)$$

is equal to the first of the expressions

$$\lim_{x \to \alpha} \frac{f'(x)}{g'(x)}, \lim_{x \to \alpha} \frac{f''(x)}{g''(x)}, \lim_{x \to \alpha} \frac{f'''(x)}{g'''(x)}$$

which is not indeterminate, provided such first indicated limit exists.

INTEGRAL CALCULUS

The definite integral is defined as:

$$\lim_{n \to \infty} \sum_{i=1}^{n} f(x_i) \Delta x_i = \int_a^b f(x)\, dx$$

Also, $\Delta x_i \to 0$ for all i.

A table of derivatives and integrals is available in the Derivatives and Indefinite Integrals section. The integral equations can be used along with the following methods of integration:

A. Integration by Parts (integral equation #6),
B. Integration by Substitution, and
C. Separation of Rational Fractions into Partial Fractions.

♦ Wade, Thomas L., *Calculus*, Ginn & Company/Simon & Schuster Publishers, 1953.

DERIVATIVES AND INDEFINITE INTEGRALS

In these formulas, u, v, and w represent functions of x. Also, a, c, and n represent constants. All arguments of the trigonometric functions are in radians. A constant of integration should be added to the integrals. To avoid terminology difficulty, the following definitions are followed: $\arcsin u = \sin^{-1} u$, $(\sin u)^{-1} = 1/\sin u$.

1. $dc/dx = 0$

2. $dx/dx = 1$

3. $d(cu)/dx = c\, du/dx$

4. $d(u + v - w)/dx = du/dx + dv/dx - dw/dx$

5. $d(uv)/dx = u\, dv/dx + v\, du/dx$

6. $d(uvw)/dx = uv\, dw/dx + uw\, dv/dx + vw\, du/dx$

7. $\dfrac{d(u/v)}{dx} = \dfrac{v\, du/dx - u\, dv/dx}{v^2}$

8. $d(u^n)/dx = nu^{n-1}\, du/dx$

9. $d[f(u)]/dx = \{d[f(u)]/du\}\, du/dx$

10. $du/dx = 1/(dx/du)$

11. $\dfrac{d(\log_a u)}{dx} = (\log_a e)\dfrac{1}{u}\dfrac{du}{dx}$

12. $\dfrac{d(\ln u)}{dx} = \dfrac{1}{u}\dfrac{du}{dx}$

13. $\dfrac{d(a^u)}{dx} = (\ln a)a^u\dfrac{du}{dx}$

14. $d(e^u)/dx = e^u\, du/dx$

15. $d(u^v)/dx = vu^{v-1}\, du/dx + (\ln u)\, u^v\, dv/dx$

16. $d(\sin u)/dx = \cos u\, du/dx$

17. $d(\cos u)/dx = -\sin u\, du/dx$

18. $d(\tan u)/dx = \sec^2 u\, du/dx$

19. $d(\cot u)/dx = -\csc^2 u\, du/dx$

20. $d(\sec u)/dx = \sec u \tan u\, du/dx$

21. $d(\csc u)/dx = -\csc u \cot u\, du/dx$

22. $\dfrac{d(\sin^{-1} u)}{dx} = \dfrac{1}{\sqrt{1 - u^2}}\dfrac{du}{dx}$ $(-\pi/2 \le \sin^{-1} u \le \pi/2)$

23. $\dfrac{d(\cos^{-1} u)}{dx} = -\dfrac{1}{\sqrt{1 - u^2}}\dfrac{du}{dx}$ $(0 \le \cos^{-1} u \le \pi)$

24. $\dfrac{d(\tan^{-1} u)}{dx} = \dfrac{1}{1 + u^2}\dfrac{du}{dx}$ $(-\pi/2 < \tan^{-1} u < \pi/2)$

25. $\dfrac{d(\cot^{-1} u)}{dx} = -\dfrac{1}{1 + u^2}\dfrac{du}{dx}$ $(0 < \cot^{-1} u < \pi)$

26. $\dfrac{d(\sec^{-1} u)}{dx} = \dfrac{1}{u\sqrt{u^2 - 1}}\dfrac{du}{dx}$

$(0 < \sec^{-1} u < \pi/2)(-\pi \le \sec^{-1} u < -\pi/2)$

27. $\dfrac{d(\csc^{-1} u)}{dx} = -\dfrac{1}{u\sqrt{u^2 - 1}}\dfrac{du}{dx}$

$(0 < \csc^{-1} u \le \pi/2)(-\pi < \csc^{-1} u \le -\pi/2)$

1. $\int df(x) = f(x)$

2. $\int dx = x$

3. $\int a\, f(x)\, dx = a \int f(x)\, dx$

4. $\int [u(x) \pm v(x)]\, dx = \int u(x)\, dx \pm \int v(x)\, dx$

5. $\int x^m dx = \dfrac{x^{m+1}}{m+1}$ $(m \ne -1)$

6. $\int u(x)\, dv(x) = u(x)\, v(x) - \int v(x)\, du(x)$

7. $\int \dfrac{dx}{ax + b} = \dfrac{1}{a}\ln|ax + b|$

8. $\int \dfrac{dx}{\sqrt{x}} = 2\sqrt{x}$

9. $\int a^x dx = \dfrac{a^x}{\ln a}$

10. $\int \sin x\, dx = -\cos x$

11. $\int \cos x\, dx = \sin x$

12. $\int \sin^2 x\, dx = \dfrac{x}{2} - \dfrac{\sin 2x}{4}$

13. $\int \cos^2 x\, dx = \dfrac{x}{2} + \dfrac{\sin 2x}{4}$

14. $\int x \sin x\, dx = \sin x - x \cos x$

15. $\int x \cos x\, dx = \cos x + x \sin x$

16. $\int \sin x \cos x\, dx = (\sin^2 x)/2$

17. $\int \sin ax \cos bx\, dx = -\dfrac{\cos(a - b)x}{2(a - b)} - \dfrac{\cos(a + b)x}{2(a + b)} \left(a^2 \ne b^2\right)$

18. $\int \tan x\, dx = -\ln|\cos x| = \ln|\sec x|$

19. $\int \cot x\, dx = -\ln|\csc x| = \ln|\sin x|$

20. $\int \tan^2 x\, dx = \tan x - x$

21. $\int \cot^2 x\, dx = -\cot x - x$

22. $\int e^{ax}\, dx = (1/a)\, e^{ax}$

23. $\int xe^{ax}\, dx = (e^{ax}/a^2)(ax - 1)$

24. $\int \ln x\, dx = x\,[\ln(x) - 1]$ $(x > 0)$

25. $\int \dfrac{dx}{a^2 + x^2} = \dfrac{1}{a}\tan^{-1}\dfrac{x}{a}$ $(a \ne 0)$

26. $\int \dfrac{dx}{ax^2 + c} = \dfrac{1}{\sqrt{ac}}\tan^{-1}\left(x\sqrt{\dfrac{a}{c}}\right)$, $(a > 0, c > 0)$

27a. $\int \dfrac{dx}{ax^2 + bx + c} = \dfrac{2}{\sqrt{4ac - b^2}}\tan^{-1}\dfrac{2ax + b}{\sqrt{4ac - b^2}}$

$(4ac - b^2 > 0)$

27b. $\int \dfrac{dx}{ax^2 + bx + c} = \dfrac{1}{\sqrt{b^2 - 4ac}}\ln\left|\dfrac{2ax + b - \sqrt{b^2 - 4ac}}{2ax + b + \sqrt{b^2 - 4ac}}\right|$

$(b^2 - 4ac > 0)$

27c. $\int \dfrac{dx}{ax^2 + bx + c} = -\dfrac{2}{2ax + b}$, $(b^2 - 4ac = 0)$

MENSURATION OF AREAS AND VOLUMES

Nomenclature

A = total surface area
P = perimeter
V = volume

Parabola

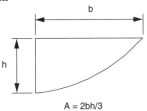

A = 2bh/3

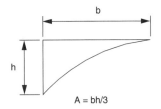

A = bh/3

Ellipse

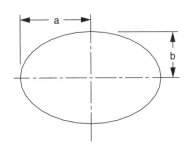

A = πab

$$P_{approx} = 2\pi\sqrt{(a^2 + b^2)/2}$$

$$P = \pi(a + b)\left[\begin{array}{l} 1 + \left(\tfrac{1}{2}\right)^2\lambda^2 + \left(\tfrac{1}{2} \times \tfrac{1}{4}\right)^2\lambda^4 \\ + \left(\tfrac{1}{2} \times \tfrac{1}{4} \times \tfrac{3}{6}\right)^2\lambda^6 + \left(\tfrac{1}{2} \times \tfrac{1}{4} \times \tfrac{3}{6} \times \tfrac{5}{8}\right)^2\lambda^8 \\ + \left(\tfrac{1}{2} \times \tfrac{1}{4} \times \tfrac{3}{6} \times \tfrac{5}{8} \times \tfrac{7}{10}\right)^2\lambda^{10} + ... \end{array}\right],$$

where

$$\lambda = (a - b)/(a + b)$$

Circular Segment

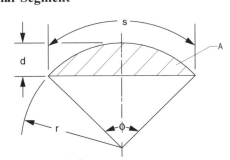

$$A = \left[r^2(\phi - \sin\phi)\right]/2$$

$$\phi = s/r = 2\left\{\arccos\left[(r - d)/r\right]\right\}$$

Circular Sector

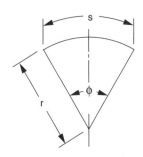

$$A = \phi r^2/2 = sr/2$$

$$\phi = s/r$$

Sphere

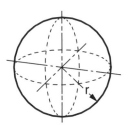

$$V = 4\pi r^3/3 = \pi d^3/6$$

$$A = 4\pi r^2 = \pi d^2$$

Parallelogram

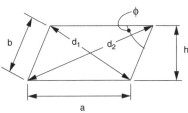

$$P = 2(a + b)$$

$$d_1 = \sqrt{a^2 + b^2 - 2ab(\cos\phi)}$$

$$d_2 = \sqrt{a^2 + b^2 + 2ab(\cos\phi)}$$

$$d_1^2 + d_2^2 = 2(a^2 + b^2)$$

$$A = ah = ab(\sin\phi)$$

If $a = b$, the parallelogram is a rhombus.

◆ Gieck, K. & Gieck R., *Engineering Formulas*, 6th ed., Gieck Publishing, 1967.

MENSURATION OF AREAS AND VOLUMES (continued)

Regular Polygon (*n* equal sides)

◆

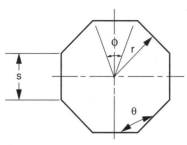

$$\phi = 2\pi/n$$

$$\theta = \left[\frac{\pi(n-2)}{n}\right] = \pi\left(1 - \frac{2}{n}\right)$$

$$P = ns$$

$$s = 2r\left[\tan(\phi/2)\right]$$

$$A = (nsr)/2$$

Prismoid

◆

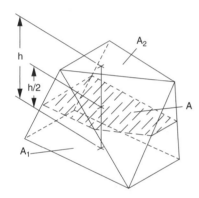

$$V = (h/6)\left(A_1 + A_2 + 4A\right)$$

Right Circular Cone

◆

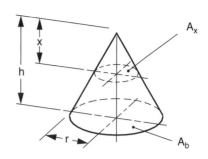

$$V = \left(\pi r^2 h\right)/3$$
$$A = \text{side area} + \text{base area}$$
$$= \pi r\left(r + \sqrt{r^2 + h^2}\right)$$
$$A_x : A_b = x^2 : h^2$$

Right Circular Cylinder

◆

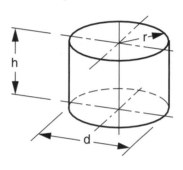

$$V = \pi r^2 h = \frac{\pi d^2 h}{4}$$
$$A = \text{side area} + \text{end areas} = 2\pi r(h + r)$$

Paraboloid of Revolution

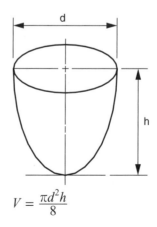

$$V = \frac{\pi d^2 h}{8}$$

◆Gieck, K. & R. Gieck, *Engineering Formulas*, 6th ed., Gieck Publishing, 1967.

CENTROIDS AND MOMENTS OF INERTIA

The *location of the centroid of an area*, bounded by the axes and the function $y = f(x)$, can be found by integration.

$$x_c = \frac{\int x\, dA}{A}$$

$$y_c = \frac{\int x\, dA}{A}$$

$$A = \int f(x)\, dx$$

$$dA = f(x)\, dx = g(y)\, dy$$

The *first moment of area* with respect to the *y*-axis and the *x*-axis, respectively, are:

$$M_y = \int x\, dA = x_c A$$
$$M_x = \int y\, dA = y_c A$$

The *moment of inertia (second moment of area)* with respect to the *y*-axis and the *x*-axis, respectively, are:

$$I_y = \int x^2\, dA$$
$$I_x = \int y^2\, dA$$

The moment of inertia taken with respect to an axis passing through the area's centroid is the *centroidal moment of inertia*. The *parallel axis theorem* for the moment of inertia with respect to another axis parallel with and located *d* units from the centroidal axis is expressed by

$$I_{\text{parallel axis}} = I_c + Ad^2$$

In a plane, $J = \int r^2\, dA = I_x + I_y$

Values for standard shapes are presented in tables in the **STATICS** and **DYNAMICS** sections.

DIFFERENTIAL EQUATIONS

A common class of ordinary linear differential equations is

$$b_n \frac{d^n y(x)}{dx^n} + \ldots + b_1 \frac{dy(x)}{dx} + b_0 y(x) = f(x)$$

where $b_n, \ldots, b_i, \ldots, b_1, b_0$ are constants.

When the equation is a homogeneous differential equation, $f(x) = 0$, the solution is

$$y_h(x) = C_1 e^{r_1 x} + C_2 e^{r_2 x} + \ldots + C_i e^{r_i x} + \ldots + C_n e^{r_n x}$$

where r_n is the *n*th distinct root of the characteristic polynomial $P(x)$ with

$$P(r) = b_n r^n + b_{n-1} r^{n-1} + \ldots + b_1 r + b_0$$

If the root $r_1 = r_2$, then $C_2 e^{r_2 x}$ is replaced with $C_2 x e^{r_1 x}$.

Higher orders of multiplicity imply higher powers of *x*. The complete solution for the differential equation is

$$y(x) = y_h(x) + y_p(x),$$

where $y_p(x)$ is any solution with $f(x)$ present. If $f(x)$ has $e^{r_n x}$ terms, then resonance is manifested. Furthermore, specific $f(x)$ forms result in specific $y_p(x)$ forms, some of which are:

$f(x)$	$y_p(x)$
A	B
$Ae^{\alpha x}$	$Be^{\alpha x}$, $\alpha \neq r_n$
$A_1 \sin \omega x + A_2 \cos \omega x$	$B_1 \sin \omega x + B_2 \cos \omega x$

If the independent variable is time *t*, then transient dynamic solutions are implied.

First-Order Linear Homogeneous Differential Equations with Constant Coefficients

$y' + ay = 0$, where *a* is a real constant:
Solution, $y = Ce^{-at}$

where $C = $ a constant that satisfies the initial conditions.

First-Order Linear Nonhomogeneous Differential Equations

$$\tau \frac{dy}{dt} + y = Kx(t) \qquad x(t) = \begin{Bmatrix} A & t < 0 \\ B & t > 0 \end{Bmatrix}$$

$$y(0) = KA$$

τ is the time constant
K is the gain
The solution is

$$y(t) = KA + (KB - KA)\left(1 - \exp\left(\frac{-t}{\tau}\right)\right) \text{ or}$$

$$\frac{t}{\tau} = \ln\left[\frac{KB - KA}{KB - y}\right]$$

Second-Order Linear Homogeneous Differential Equations with Constant Coefficients

An equation of the form

$$y'' + ay' + by = 0$$

can be solved by the method of undetermined coefficients where a solution of the form $y = Ce^{rx}$ is sought. Substitution of this solution gives

$$(r^2 + ar + b)\, Ce^{rx} = 0$$

and since Ce^{rx} cannot be zero, the characteristic equation must vanish or

$$r^2 + ar + b = 0$$

The roots of the characteristic equation are

$$r_{1,2} = -\frac{a \pm \sqrt{a^2 - 4b}}{2}$$

and can be real and distinct for $a^2 > 4b$, real and equal for $a^2 = 4b$, and complex for $a^2 < 4b$.

If $a^2 > 4b$, the solution is of the form (overdamped)

$$y = C_1 e^{r_1 x} + C_2 e^{r_2 x}$$

If $a^2 = 4b$, the solution is of the form (critically damped)

$$y = (C_1 + C_2 x)e^{r_1 x}$$

If $a^2 < 4b$, the solution is of the form (underdamped)

$$y = e^{\alpha x}(C_1 \cos \beta x + C_2 \sin \beta x), \text{ where}$$

$$\alpha = -a/2$$

$$\beta = \frac{\sqrt{4b - a^2}}{2}$$

FOURIER TRANSFORM

The Fourier transform pair, one form of which is

$$F(\omega) = \int_{-\infty}^{\infty} f(t) e^{-j\omega t} dt$$

$$f(t) = [1/(2\pi)] \int_{-\infty}^{\infty} F(\omega) e^{j\omega t} d\omega$$

can be used to characterize a broad class of signal models in terms of their frequency or spectral content. Some useful transform pairs are:

$f(t)$	$F(\omega)$
$\delta(t)$	1
$u(t)$	$\pi\delta(\omega) + 1/j\omega$
$u\left(t + \dfrac{\tau}{2}\right) - u\left(t - \dfrac{\tau}{2}\right) = r_{rect}\dfrac{t}{\tau}$	$\tau\dfrac{\sin(\omega\tau/2)}{\omega\tau/2}$
$e^{j\omega_o t}$	$2\pi\delta(\omega - \omega_o)$

Some mathematical liberties are required to obtain the second and fourth form. Other Fourier transforms are derivable from the Laplace transform by replacing s with $j\omega$ provided

$$f(t) = 0, t < 0$$

$$\int_0^{\infty} |f(t)| dt < \infty$$

Also refer to Fourier Series and Laplace Transforms in the **ELECTRICAL AND COMPUTER ENGINEERING** section of this handbook.

DIFFERENCE EQUATIONS

Difference equations are used to model discrete systems. Systems which can be described by difference equations include computer program variables iteratively evaluated in a loop, sequential circuits, cash flows, recursive processes, systems with time-delay components, etc. Any system whose input $v(t)$ and output $y(t)$ are defined only at the equally spaced intervals $t = kT$ can be described by a difference equation.

First-Order Linear Difference Equation

The difference equation

$$P_k = P_{k-1}(1 + i) - A$$

represents the balance P of a loan after the kth payment A. If P_k is defined as $y(k)$, the model becomes

$$y(k) - (1 + i)y(k - 1) = -A$$

Second-Order Linear Difference Equation

The Fibonacci number sequence can be generated by

$$y(k) = y(k - 1) + y(k - 2)$$

where $y(-1) = 1$ and $y(-2) = 1$. An alternate form for this model is $f(k + 2) = f(k + 1) + f(k)$ with $f(0) = 1$ and $f(1) = 1$.

NUMERICAL METHODS

Newton's Method for Root Extraction

Given a function $f(x)$ which has a simple root of $f(x) = 0$ at $x = a$ an important computational task would be to find that root. If $f(x)$ has a continuous first derivative then the $(j + 1)$st estimate of the root is

$$a^{j+1} = a^j - \left.\frac{f(x)}{\dfrac{df(x)}{dx}}\right|_{x = a^j}$$

The initial estimate of the root a^0 must be near enough to the actual root to cause the algorithm to converge to the root.

Newton's Method of Minimization

Given a scalar value function

$$h(x) = h(x_1, x_2, \ldots, x_n)$$

find a vector $x^* \in R_n$ such that

$$h(x^*) \le h(x) \text{ for all } x$$

Newton's algorithm is

$$x_{k+1} = x_k - \left(\left.\frac{\partial^2 h}{\partial x^2}\right|_{x = x_k}\right)^{-1} \left.\frac{\partial h}{\partial x}\right|_{x = x_k}, \text{ where}$$

$$\frac{\partial h}{\partial x} = \begin{bmatrix} \dfrac{\partial h}{\partial x_1} \\ \dfrac{\partial h}{\partial x_2} \\ \ldots \\ \ldots \\ \dfrac{\partial h}{\partial x_n} \end{bmatrix}$$

and

$$\frac{\partial^2 h}{\partial x^2} = \begin{bmatrix} \dfrac{\partial^2 h}{\partial x_1^2} & \dfrac{\partial^2 h}{\partial x_1 \partial x_2} & \ldots & \ldots & \dfrac{\partial^2 h}{\partial x_1 \partial x_n} \\ \dfrac{\partial^2 h}{\partial x_1 \partial x_2} & \dfrac{\partial^2 h}{\partial x_2^2} & \ldots & \ldots & \dfrac{\partial^2 h}{\partial x_2 \partial x_n} \\ \ldots & \ldots & \ldots & \ldots & \ldots \\ \ldots & \ldots & \ldots & \ldots & \ldots \\ \dfrac{\partial^2 h}{\partial x_1 \partial x_n} & \dfrac{\partial^2 h}{\partial x_2 \partial x_n} & \ldots & \ldots & \dfrac{\partial^2 h}{\partial x_n^2} \end{bmatrix}$$

Numerical Integration

Three of the more common numerical integration algorithms used to evaluate the integral

$$\int_a^b f(x)\,dx$$

are:

Euler's or Forward Rectangular Rule

$$\int_a^b f(x)\,dx \approx \Delta x \sum_{k=0}^{n-1} f(a + k\Delta x)$$

Trapezoidal Rule

for $n = 1$

$$\int_a^b f(x)\,dx \approx \Delta x \left[\frac{f(a) + f(b)}{2} \right]$$

for $n > 1$

$$\int_a^b f(x)\,dx \approx \frac{\Delta x}{2} \left[f(a) + 2 \sum_{k=1}^{n-1} f(a + k\Delta x) + f(b) \right]$$

Simpson's Rule/Parabolic Rule (n must be an even integer)

for $n = 2$

$$\int_a^b f(x)\,dx \approx \left(\frac{b-a}{6} \right) \left[f(a) + 4f\left(\frac{a+b}{2} \right) + f(b) \right]$$

for $n \geq 4$

$$\int_a^b f(x)\,dx \approx \frac{\Delta x}{3} \left[\begin{array}{l} f(a) + 2 \sum_{k=2,4,6,\ldots}^{n-2} f(a + k\Delta x) \\[2mm] +4 \sum_{k=1,3,5,\ldots}^{n-1} f(a + k\Delta x) + f(b) \end{array} \right]$$

with $\Delta x = (b - a)/n$

  n = number of intervals between data points

Numerical Solution of Ordinary Differential Equations

<u>Euler's Approximation</u>

Given a differential equation

  $dx/dt = f(x, t)$ with $x(0) = x_o$

At some general time $k\Delta t$

  $x[(k + 1)\Delta t] \cong x(k\Delta t) + \Delta t f[x(k\Delta t), k\Delta t]$

which can be used with starting condition x_o to solve recursively for $x(\Delta t)$, $x(2\Delta t)$, …, $x(n\Delta t)$.

The method can be extended to nth order differential equations by recasting them as n first-order equations.

In particular, when $dx/dt = f(x)$

  $x[(k + 1)\Delta t] \cong x(k\Delta t) + \Delta t f[x(k\Delta t)]$

which can be expressed as the recursive equation

  $x_{k+1} = x_k + \Delta t\,(dx_k/dt)$

Refer to the **ELECTRICAL AND COMPUTER ENGINEERING** section for additional information on Laplace transforms and algebra of complex numbers.

MECHANICS OF MATERIALS

UNIAXIAL STRESS-STRAIN

Stress-Strain Curve for Mild Steel

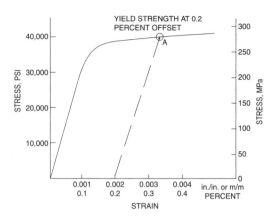

The slope of the linear portion of the curve equals the modulus of elasticity.

DEFINITIONS

Engineering Strain

$\varepsilon = \Delta L/L_o$, where

ε = engineering strain (units per unit),

ΔL = change in length (units) of member,

L_o = original length (units) of member.

Percent Elongation

$$\% \text{ Elongation} = \left(\frac{\Delta L}{L_o}\right) \times 100$$

Percent Reduction in Area (RA)

The % reduction in area from initial area, A_i, to final area, A_f, is:

$$\%RA = \left(\frac{A_i - A_f}{A_i}\right) \times 100$$

Shear Stress-Strain

$\gamma = \tau/G$, where

γ = shear strain,

τ = shear stress, and

G = *shear modulus* (constant in linear torsion-rotation relationship).

$$G = \frac{E}{2(1+v)}, \text{ where}$$

E = modulus of elasticity

v = *Poisson's ratio*, and

= – (lateral strain)/(longitudinal strain).

Uniaxial Loading and Deformation

$\sigma = P/A$, where

σ = stress on the cross section,

P = loading, and

A = cross-sectional area.

$\varepsilon = \delta/L$, where

δ = elastic longitudinal deformation and

L = length of member.

$$E = \sigma/\varepsilon = \frac{P/A}{\delta/L}$$

$$\delta = \frac{PL}{AE}$$

True stress is load divided by actual cross-sectional area whereas engineering stress is load divided by the initial area.

THERMAL DEFORMATIONS

$\delta_t = \alpha L(T - T_o)$, where

δ_t = deformation caused by a change in temperature,

α = temperature coefficient of expansion,

L = length of member,

T = final temperature, and

T_o = initial temperature.

CYLINDRICAL PRESSURE VESSEL

Cylindrical Pressure Vessel

For internal pressure only, the stresses at the inside wall are:

$$\sigma_t = P_i \frac{r_o^2 + r_i^2}{r_o^2 - r_i^2} \quad \text{and} \quad \sigma_r = -P_i$$

For external pressure only, the stresses at the outside wall are:

$$\sigma_t = -P_o \frac{r_o^2 + r_i^2}{r_o^2 - r_i^2} \quad \text{and} \quad \sigma_r = -P_o, \text{ where}$$

σ_t = tangential (hoop) stress,

σ_r = radial stress,

P_i = internal pressure,

P_o = external pressure,

r_i = inside radius, and

r_o = outside radius.

For vessels with end caps, the axial stress is:

$$\sigma_a = P_i \frac{r_i^2}{r_o^2 - r_i^2}$$

σ_t, σ_r, and σ_a are principal stresses.

♦ Flinn, Richard A. & Paul K. Trojan, *Engineering Materials & Their Applications*, 4th ed., Houghton Mifflin Co., Boston, 1990.

When the thickness of the cylinder wall is about one-tenth or less of inside radius, the cylinder can be considered as thin-walled. In which case, the internal pressure is resisted by the hoop stress and the axial stress.

$$\sigma_t = \frac{P_i r}{t} \quad \text{and} \quad \sigma_a = \frac{P_i r}{2t}$$

where t = wall thickness.

STRESS AND STRAIN

Principal Stresses

For the special case of a *two-dimensional* stress state, the equations for principal stress reduce to

$$\sigma_a, \sigma_b = \frac{\sigma_x + \sigma_y}{2} \pm \sqrt{\left(\frac{\sigma_x - \sigma_y}{2}\right)^2 + \tau_{xy}^2}$$

$$\sigma_c = 0$$

The two nonzero values calculated from this equation are temporarily labeled σ_a and σ_b and the third value σ_c is always zero in this case. Depending on their values, the three roots are then labeled according to the convention:
algebraically largest = σ_1, *algebraically smallest* = σ_3, *other* = σ_2. A typical 2D stress element is shown below with all indicated components shown in their positive sense.

♦

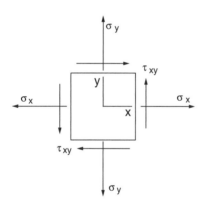

Mohr's Circle – Stress, 2D

To construct a Mohr's circle, the following sign conventions are used.
1. Tensile normal stress components are plotted on the horizontal axis and are considered positive. Compressive normal stress components are negative.
2. For constructing Mohr's circle only, shearing stresses are plotted above the normal stress axis when the pair of shearing stresses, acting on opposite and parallel faces of an element, forms a clockwise couple. Shearing stresses are plotted below the normal axis when the shear stresses form a counterclockwise couple.

The circle drawn with the center on the normal stress (horizontal) axis with center, C, and radius, R, where

$$C = \frac{\sigma_x + \sigma_y}{2}, \quad R = \sqrt{\left(\frac{\sigma_x - \sigma_y}{2}\right)^2 + \tau_{xy}^2}$$

The two nonzero principal stresses are then:

♦

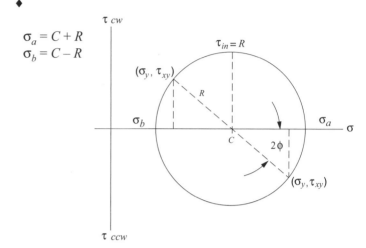

$$\sigma_a = C + R$$
$$\sigma_b = C - R$$

The maximum *inplane* shear stress is $\tau_{in} = R$. However, the maximum shear stress considering three dimensions is always

$$\tau_{max} = \frac{\sigma_1 - \sigma_3}{2}.$$

Hooke's Law

Three-dimensional case:

$$\varepsilon_x = (1/E)[\sigma_x - v(\sigma_y + \sigma_z)] \qquad \gamma_{xy} = \tau_{xy}/G$$
$$\varepsilon_y = (1/E)[\sigma_y - v(\sigma_z + \sigma_x)] \qquad \gamma_{yz} = \tau_{yz}/G$$
$$\varepsilon_z = (1/E)[\sigma_z - v(\sigma_x + \sigma_y)] \qquad \gamma_{zx} = \tau_{zx}/G$$

Plane stress case ($\sigma_z = 0$):
$$\varepsilon_x = (1/E)(\sigma_x - v\sigma_y)$$
$$\varepsilon_y = (1/E)(\sigma_y - v\sigma_x)$$
$$\varepsilon_z = -(1/E)(v\sigma_x + v\sigma_y)$$

$$\begin{Bmatrix} \sigma_x \\ \sigma_y \\ \tau_{xy} \end{Bmatrix} = \frac{E}{1 - v^2} \begin{bmatrix} 1 & v & 0 \\ v & 1 & 0 \\ 0 & 0 & \frac{1-v}{2} \end{bmatrix} \begin{Bmatrix} \varepsilon_x \\ \varepsilon_y \\ \gamma_{xy} \end{Bmatrix}$$

Uniaxial case ($\sigma_y = \sigma_z = 0$): $\qquad \sigma_x = E\varepsilon_x$ or $\sigma = E\varepsilon$, where

$\varepsilon_x, \varepsilon_y, \varepsilon_z$ = normal strain,

$\sigma_x, \sigma_y, \sigma_z$ = normal stress,

$\gamma_{xy}, \gamma_{yz}, \gamma_{zx}$ = shear strain,

$\tau_{xy}, \tau_{yz}, \tau_{zx}$ = shear stress,

E = modulus of elasticity,

G = shear modulus, and

v = Poisson's ratio.

♦ Crandall, S.H. and N.C. Dahl, *An Introduction to Mechanics of Solids*, McGraw-Hill, New York, 1959.

STATIC LOADING FAILURE THEORIES

See **MATERIALS SCIENCE/STRUCTURE OF MATTER** for Stress Concentration in Brittle Materials.

Brittle Materials

Maximum-Normal-Stress Theory

The maximum-normal-stress theory states that failure occurs when one of the three principal stresses equals the strength of the material. If $\sigma_1 \geq \sigma_2 \geq \sigma_3$, then the theory predicts that failure occurs whenever $\sigma_1 \geq S_{ut}$ or $\sigma_3 \leq -S_{uc}$ where S_{ut} and S_{uc} are the tensile and compressive strengths, respectively.

Coulomb-Mohr Theory

The Coulomb-Mohr theory is based upon the results of tensile and compression tests. On the σ, τ coordinate system, one circle is plotted for S_{ut} and one for S_{uc}. As shown in the figure, lines are then drawn tangent to these circles. The Coulomb-Mohr theory then states that fracture will occur for any stress situation that produces a circle that is either tangent to or crosses the envelope defined by the lines tangent to the S_{ut} and S_{uc} circles.

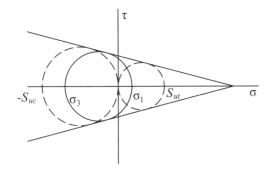

If $\sigma_1 \geq \sigma_2 \geq \sigma_3$ and $\sigma_3 < 0$, then the theory predicts that yielding will occur whenever

$$\frac{\sigma_1}{S_{ut}} - \frac{\sigma_3}{S_{uc}} \geq 1$$

Ductile Materials

Maximum-Shear-Stress Theory

The maximum-shear-stress theory states that yielding begins when the maximum shear stress equals the maximum shear stress in a tension-test specimen of the same material when that specimen begins to yield. If $\sigma_1 \geq \sigma_2 \geq \sigma_3$, then the theory predicts that yielding will occur whenever $\tau_{max} \geq S_y/2$ where S_y is the yield strength.

$$\tau_{max} = \frac{\sigma_1 - \sigma_3}{2}.$$

Distortion-Energy Theory

The distortion-energy theory states that yielding begins whenever the distortion energy in a unit volume equals the distortion energy in the same volume when uniaxially stressed to the yield strength. The theory predicts that yielding will occur whenever

$$\left[\frac{(\sigma_1 - \sigma_2)^2 + (\sigma_2 - \sigma_3)^2 + (\sigma_1 - \sigma_3)^2}{2} \right]^{1/2} \geq S_y$$

The term on the left side of the inequality is known as the effective or Von Mises stress. For a biaxial stress state the effective stress becomes

$$\sigma' = \left(\sigma_A^2 - \sigma_A \sigma_B + \sigma_B^2 \right)^{1/2}$$

or

$$\sigma' = \left(\sigma_x^2 - \sigma_x \sigma_y + \sigma_y^2 + 3\tau_{xy}^2 \right)^{1/2}$$

where σ_A and σ_B are the two nonzero principal stresses and σ_x, σ_y, and τ_{xy} are the stresses in orthogonal directions.

VARIABLE LOADING FAILURE THEORIES

<u>Modified Goodman Theory:</u> The modified Goodman criterion states that a fatigue failure will occur whenever

$$\frac{\sigma_a}{S_e} + \frac{\sigma_m}{S_{ut}} \geq 1 \quad \text{or} \quad \frac{\sigma_{max}}{S_y} \geq 1, \quad \sigma_m \geq 0,$$

where

S_e = fatigue strength,

S_{ut} = ultimate strength,

S_y = yield strength,

σ_a = alternating stress, and

σ_m = mean stress.

σ_{max} = $\sigma_m + \sigma_a$

<u>Soderberg Theory:</u> The Soderberg theory states that a fatigue failure will occur whenever

$$\frac{\sigma_a}{S_e} + \frac{\sigma_m}{S_y} \geq 1 \qquad \sigma_m \geq 0$$

<u>Endurance Limit for Steels:</u> When test data is unavailable, the endurance limit for steels may be estimated as

$$S'_e = \begin{Bmatrix} 0.5\,S_{ut},\, S_{ut} \leq 1,400 \text{ MPa} \\ 700 \text{ MPa},\, S_{ut} > 1,400 \text{ MPa} \end{Bmatrix}$$

Endurance Limit Modifying Factors: Endurance limit modifying factors are used to account for the differences between the endurance limit as determined from a rotating beam test, S'_e, and that which would result in the real part, S_e.

$$S_e = k_a k_b k_c k_d k_e S'_e$$

where

Surface Factor, $k_a = aS_{ut}^b$

Surface Finish	Factor a		Exponent b
	kpsi	MPa	
Ground	1.34	1.58	–0.085
Machined or CD	2.70	4.51	–0.265
Hot rolled	14.4	57.7	–0.718
As forged	39.9	272.0	–0.995

Size Factor, k_b:

For bending and torsion:

$d \leq 8$ mm;	$k_b = 1$
8 mm $\leq d \leq$ 250 mm;	$k_b = 1.189 d_{eff}^{-0.097}$
$d > 250$ mm;	$0.6 \leq k_b \leq 0.75$
For axial loading:	$k_b = 1$

Load Factor, k_c:

$k_c = 0.923$	axial loading, $S_{ut} \leq 1{,}520$ MPa
$k_c = 1$	axial loading, $S_{ut} > 1{,}520$ MPa
$k_c = 1$	bending
$k_c = 0.577$	torsion

Temperature Factor, k_d:

for T $\leq$ 450°C, $k_d = 1$

Miscellaneous Effects Factor, k_e: Used to account for strength reduction effects such as corrosion, plating, and residual stresses. In the absence of known effects, use $k_e = 1$.

TORSION

Torsion stress in circular solid or thick-walled (t > 0.1 r) shafts:

$$\tau = \frac{Tr}{J}$$

where J = polar moment of inertia (see table at end of **STATICS** section).

TORSIONAL STRAIN

$$\gamma_{\phi z} = \underset{\Delta z \to 0}{\text{limit}} \ r(\Delta \phi / \Delta z) = r(d\phi/dz)$$

The shear strain varies in direct proportion to the radius, from zero strain at the center to the greatest strain at the outside of the shaft. $d\phi/dz$ is the twist per unit length or the rate of twist.

$$\tau_{\phi z} = G\gamma_{\phi z} = Gr(d\phi/dz)$$

$$T = G(d\phi/dz) \int_A r^2 dA = GJ(d\phi/dz)$$

$$\phi = \int_o^L \frac{T}{GJ} dz = \frac{TL}{GJ}, \text{ where}$$

ϕ = total angle (radians) of twist,

T = torque, and

L = length of shaft.

T/ϕ gives the *twisting moment per radian of twist.* This is called the *torsional stiffness* and is often denoted by the symbol k or c.

For Hollow, Thin-Walled Shafts

$$\tau = \frac{T}{2A_m t}, \text{ where}$$

t = thickness of shaft wall and

A_m = the total mean area enclosed by the shaft measured to the midpoint of the wall.

BEAMS

Shearing Force and Bending Moment Sign Conventions

1. The bending moment is *positive* if it produces bending of the beam *concave upward* (compression in top fibers and tension in bottom fibers).

2. The shearing force is *positive* if the *right portion of the beam tends to shear downward with respect to the left.*

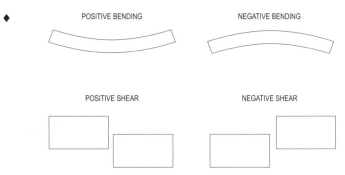

♦ Timoshenko, S. and Gleason H. MacCullough, *Elements of Strengths of Materials*, K. Van Nostrand Co./Wadsworth Publishing Co., 1949.

The relationship between the load (q), shear (V), and moment (M) equations are:

$$q(x) = -\frac{dV(x)}{dx}$$

$$V = \frac{dM(x)}{dx}$$

$$V_2 - V_1 = \int_{x_1}^{x_2}\left[-q(x)\right]dx$$

$$M_2 - M_1 = \int_{x_1}^{x_2}V(x)\,dx$$

Stresses in Beams

$$\varepsilon_x = -y/\rho, \text{ where}$$

ρ = the radius of curvature of the deflected axis of the beam, and

y = the distance from the neutral axis to the longitudinal fiber in question.

Using the stress-strain relationship $\sigma = E\varepsilon$,
Axial Stress: $\quad \sigma_x = -Ey/\rho$, where

σ_x = the normal stress of the fiber located y-distance from the neutral axis.

$$1/\rho = M/(EI), \text{ where}$$

M = the moment at the section and

I = the *moment of inertia* of the cross section.

$$\sigma_x = -My/I, \text{ where}$$

y = the distance from the neutral axis to the fiber location above or below the axis. Let $y = c$, where c = distance from the neutral axis to the outermost fiber of a symmetrical beam section.

$$\sigma_x = \pm Mc/I$$

Let $S = I/c$: then, $\sigma_x = \pm M/S$, where

S = the *elastic section modulus* of the beam member.

Transverse shear flow: $\quad q = VQ/I$ and

Transverse shear stress: $\quad \tau_{xy} = VQ/(Ib)$, where

q = shear flow,

τ_{xy} = shear stress on the surface,

V = shear force at the section,

b = width or thickness of the cross-section, and

Q = $A'\overline{y'}$, where

A' = area above the layer (or plane) upon which the desired transverse shear stress acts and

$\overline{y'}$ = distance from neutral axis to area centroid.

Deflection of Beams
Using $1/\rho = M/(EI)$,

$$EI\frac{d^2y}{dx^2} = M, \text{ differential equation of deflection curve}$$

$$EI\frac{d^3y}{dx^3} = dM(x)/dx = V$$

$$EI\frac{d^4y}{dx^4} = dV(x)/dx = -q$$

Determine the deflection curve equation by double integration (apply boundary conditions applicable to the deflection and/or slope).

$$EI\,(dy/dx) = \int M(x)\,dx$$

$$EIy = \int\left[\int M(x)\,dx\right]dx$$

The constants of integration can be determined from the physical geometry of the beam.

COLUMNS
For long columns with pinned ends:
Euler's Formula

$$P_{cr} = \frac{\pi^2 EI}{\ell^2}, \text{ where}$$

P_{cr} = critical axial loading,

ℓ = unbraced column length.

substitute $I = r^2 A$:

$$\frac{P_{cr}}{A} = \frac{\pi^2 E}{(\ell/r)^2}, \text{ where}$$

r = *radius of gyration* and

ℓ/r = *slenderness ratio* for the column.

For further column design theory, see the **CIVIL ENGINEERING** and **MECHANICAL ENGINEERING** sections.

ELASTIC STRAIN ENERGY

If the strain remains within the elastic limit, the work done during deflection (extension) of a member will be transformed into potential energy and can be recovered.

If the final load is P and the corresponding elongation of a tension member is δ, then the total energy U stored is equal to the work W done during loading.

$$U = W = P\delta/2$$

The strain energy per unit volume is

$$u = U/AL = \sigma^2/2E \qquad \text{(for tension)}$$

MATERIAL PROPERTIES

Material	Units	Steel	Aluminum	Cast Iron	Wood (Fir)
Modulus of Elasticity, E	Mpsi	29.0	10.0	14.5	1.6
	GPa	200.0	69.0	100.0	11.0
Modulus of Rigidity, G	Mpsi	11.5	3.8	6.0	0.6
	GPa	80.0	26.0	41.4	4.1
Poisson's Ratio, v		0.30	0.33	0.21	0.33
Coefficient of Thermal Expansion, α	$10^{-6}/°F$	6.5	13.1	6.7	1.7
	$10^{-6}/°C$	11.7	23.6	12.1	3.0

Beam Deflection Formulas – Special Cases
(δ is positive downward)

Diagram	δ	δ_{max}	Slope
P at distance a, cantilever; δ_{max}, ϕ_{max}	$\delta = \dfrac{Pa^2}{6EI}(3x-a),\text{ for } x>a$ $\delta = \dfrac{Px^2}{6EI}(-x+3a),\text{ for } x\le a$	$\delta_{max}=\dfrac{Pa^2}{6EI}(3L-a)$	$\phi_{max}=\dfrac{Pa^2}{2EI}$
w (LOAD PER UNIT LENGTH), cantilever; δ_{max}, ϕ_{max}	$\delta=\dfrac{wx^2}{24EI}\left(x^2+6L^2-4Lx\right)$	$\delta_{max}=\dfrac{wL^4}{8EI}$	$\phi_{max}=\dfrac{wL^3}{6EI}$
M, cantilever; δ_{max}, ϕ_{max}	$\delta=\dfrac{Mx^2}{2EI}$	$\delta_{max}=\dfrac{ML^2}{2EI}$	$\phi_{max}=\dfrac{ML}{EI}$
P at a,b; simply supported; $R_1=Pb/L$, $R_2=Pa/L$	$\delta=\dfrac{Pb}{6LEI}\left[\dfrac{L}{b}(x-a)^3-x^3+(L^2-b^2)x\right],\text{ for } x>a$ $\delta=\dfrac{Pb}{6LEI}\left[-x^3+(L^2-b^2)x\right],\text{ for } x\le a$	$\delta_{max}=\dfrac{Pb\left(L^2-b^2\right)^{3/2}}{9\sqrt{3}LEI}$ at $x=\sqrt{\dfrac{L^2-b^2}{3}}$	$\phi_1=\dfrac{Pab(2L-a)}{6LEI}$ $\phi_2=\dfrac{Pab(2L-b)}{6LEI}$
w (LOAD PER UNIT LENGTH); simply supported; $R_1=wL/2$, $R_2=wL/2$	$\delta=\dfrac{wx}{24EI}\left(L^3-2Lx^2+x^3\right)$	$\delta_{max}=\dfrac{5wL^4}{384EI}$	$\phi_1=\phi_2=\dfrac{wL^3}{24EI}$
w (LOAD PER UNIT LENGTH); fixed–fixed; $R_1=R_2=\dfrac{wL}{2}$ and $M_1=M_2=\dfrac{wL^2}{12}$	$\delta(x)=\dfrac{wx^2}{24EI}\left(L^2-Lx+x^2\right)$	$\left|\delta_{max}\right|=\dfrac{wL^4}{384EI}\text{ at } x=\dfrac{L}{2}$	$\left|\phi_{max}\right|=0.008\dfrac{wL^3}{24EI}$ at $x=\dfrac{1}{2}\pm\dfrac{L}{\sqrt{12}}$

Crandall, S.H. and N.C. Dahl, *An Introduction to Mechanics of Solids*, McGraw-Hill, New York, 1959.

ENGINEERING PROBABILITY AND STATISTICS

DISPERSION, MEAN, MEDIAN, AND MODE VALUES

If $X_1, X_2, \ldots, X_n$ represent the values of a random sample of n items or observations, the *arithmetic mean* of these items or observations, denoted $\overline{X}$, is defined as

$$\overline{X} = (1/n)(X_1 + X_2 + \ldots + X_n) = (1/n) \sum_{i=1}^{n} X_i$$

$\overline{X} \to \mu$ for sufficiently large values of n.

The *weighted arithmetic mean* is

$$\overline{X}_w = \frac{\sum w_i X_i}{\sum w_i}, \text{ where}$$

X_i = the value of the ith observation, and
w_i = the weight applied to X_i.

The *variance* of the population is the *arithmetic mean* of the *squared deviations from the population mean*. If μ is the arithmetic mean of a discrete population of size N, the *population variance* is defined by

$$\sigma^2 = (1/N)\left[(X_1 - \mu)^2 + (X_2 - \mu)^2 + \ldots + (X_N - \mu)^2\right]$$

$$= (1/N) \sum_{i=1}^{N} (X_i - \mu)^2$$

The *standard deviation* of the population is
$$\sigma = \sqrt{(1/N) \sum (X_i - \mu)^2}$$

The *sample variance* is
$$s^2 = [1/(n-1)] \sum_{i=1}^{n} (X_i - \overline{X})^2$$

The *sample standard deviation* is
$$s = \sqrt{[1/(n-1)] \sum_{i=1}^{n} (X_i - \overline{X})^2}$$

The *sample coefficient of variation* = $CV = s/\overline{X}$

The *sample geometric mean* = $\sqrt[n]{X_1 X_2 X_3 \ldots X_n}$

The *sample root-mean-square value* = $\sqrt{(1/n) \sum X_i^2}$

When the discrete data are rearranged in increasing order and n is odd, the median is the value of the $\left(\dfrac{n+1}{2}\right)^{\text{th}}$ item

When n is even, the median is the average of the $\left(\dfrac{n}{2}\right)^{\text{th}}$ and $\left(\dfrac{n}{2} + 1\right)^{\text{th}}$ items.

The *mode* of a set of data is the value that occurs with greatest frequency.

The *sample range* R is the largest sample value minus the smallest sample value.

PERMUTATIONS AND COMBINATIONS

A *permutation* is a particular sequence of a given set of objects. A *combination* is the set itself without reference to order.

1. The number of different *permutations* of n distinct objects *taken r at a time* is
    $$P(n,r) = \frac{n!}{(n-r)!}$$

    nPr is an alternative notation for $P(n,r)$

2. The number of different *combinations* of n distinct objects *taken r at a time* is
    $$C(n,r) = \frac{P(n,r)}{r!} = \frac{n!}{[r!(n-r)!]}$$

    nCr and $\begin{pmatrix} n \\ r \end{pmatrix}$ are alternative notations for $C(n,r)$

3. The number of different *permutations* of n objects *taken n at a time*, given that n_i are of type i, where $i = 1, 2, \ldots, k$ and $\sum n_i = n$, is
    $$P(n; n_1, n_2, \ldots, n_k) = \frac{n!}{n_1! n_2! \ldots n_k!}$$

SETS

DeMorgan's Law

$$\overline{A \cup B} = \overline{A} \cap \overline{B}$$
$$\overline{A \cap B} = \overline{A} \cup \overline{B}$$

Associative Law

$$A \cup (B \cup C) = (A \cup B) \cup C$$
$$A \cap (B \cap C) = (A \cap B) \cap C$$

Distributive Law

$$A \cup (B \cap C) = (A \cup B) \cap (A \cup C)$$
$$A \cap (B \cup C) = (A \cap B) \cup (A \cap C)$$

LAWS OF PROBABILITY

Property 1. General Character of Probability

The probability $P(E)$ of an event E is a real number in the range of 0 to 1. The probability of an impossible event is 0 and that of an event certain to occur is 1.

Property 2. Law of Total Probability

$$P(A + B) = P(A) + P(B) - P(A, B), \text{ where}$$

$P(A + B)$ = the probability that either A or B occur alone or that both occur together,

$P(A)$ = the probability that A occurs,

$P(B)$ = the probability that B occurs, and

$P(A, B)$ = the probability that both A and B occur simultaneously.

Property 3. Law of Compound or Joint Probability

If neither $P(A)$ nor $P(B)$ is zero,

$$P(A, B) = P(A)P(B \mid A) = P(B)P(A \mid B),\text{ where}$$

$P(B \mid A) =$ the probability that B occurs given the fact that A has occurred, and

$P(A \mid B) =$ the probability that A occurs given the fact that B has occurred.

If either $P(A)$ or $P(B)$ is zero, then $P(A, B) = 0$.

Bayes Theorem

$$P(B_j \mid A) = \frac{P(B_j)P(A \mid B_j)}{\sum\limits_{i=1}^{n} P(A \mid B_i)P(B_i)}$$

where $P(A_j)$ is the probability of event A_j within the population of A

$P(B_j)$ is the probability of event B_j within the population of B

PROBABILITY FUNCTIONS

A random variable X has a probability associated with each of its possible values. The probability is termed a discrete probability if X can assume only discrete values, or

$$X = x_1, x_2, x_3, \ldots, x_n$$

The *discrete probability* of any single event, $X = x_i$, occurring is defined as $P(x_i)$ while the *probability mass function* of the random variable X is defined by

$$f(x_k) = P(X = x_k), k = 1, 2, \ldots, n$$

Probability Density Function

If X is continuous, the *probability density function, f*, is defined such that

$$P(a \le X \le b) = \int_a^b f(x)\,dx$$

Cumulative Distribution Functions

The *cumulative distribution function, F*, of a discrete random variable X that has a probability distribution described by $P(x_i)$ is defined as

$$F(x_m) = \sum\limits_{k=1}^{m} P(x_k) = P(X \le x_m), m = 1, 2, \ldots, n$$

If X is continuous, the *cumulative distribution function, F*, is defined by

$$F(x) = \int_{-\infty}^{x} f(t)\,dt$$

which implies that $F(a)$ is the probability that $X \le a$.

Expected Values

Let X be a discrete random variable having a probability mass function

$$f(x_k), k = 1, 2, \ldots, n$$

The expected value of X is defined as

$$\mu = E[X] = \sum\limits_{k=1}^{n} x_k\, f(x_k)$$

The variance of X is defined as

$$\sigma^2 = V[X] = \sum\limits_{k=1}^{n} (x_k - \mu)^2 f(x_k)$$

Let X be a continuous random variable having a density function $f(X)$ and let $Y = g(X)$ be some general function. The expected value of Y is:

$$E[Y] = E[g(X)] = \int_{-\infty}^{\infty} g(x)f(x)\,dx$$

The mean or expected value of the random variable X is now defined as

$$\mu = E[X] = \int_{-\infty}^{\infty} xf(x)\,dx$$

while the variance is given by

$$\sigma^2 = V[X] = E[(X - \mu)^2] = \int_{-\infty}^{\infty} (x - \mu)^2 f(x)\,dx$$

The standard deviation is given by

$$\sigma = \sqrt{V[X]}$$

The coefficient of variation is defined as σ/μ.

Sums of Random Variables

$$Y = a_1 X_1 + a_2 X_2 + \ldots + a_n X_n$$

The expected value of Y is:

$$\mu_y = E(Y) = a_1 E(X_1) + a_2 E(X_2) + \ldots + a_n E(X_n)$$

If the random variables are statistically *independent*, then the variance of Y is:

$$\sigma_y^2 = V(Y) = a_1^2 V(X_1) + a_2^2 V(X_2) + \ldots + a_n^2 V(X_n)$$
$$= a_1^2 \sigma_1^2 + a_2^2 \sigma_2^2 + \ldots + a_n^2 \sigma_n^2$$

Also, the standard deviation of Y is:

$$\sigma_y = \sqrt{\sigma_y^2}$$

Binomial Distribution

$P(x)$ is the probability that x successes will occur in n trials. If p = probability of success and q = probability of failure = $1 - p$, then

$$P_n(x) = C(n,x)p^x q^{n-x} = \frac{n!}{x!(n-x)!}p^x q^{n-x},$$

where

x $= 0, 1, 2, ..., n,$

$C(n, x)$ = the number of combinations, and

n, p = parameters.

Normal Distribution (Gaussian Distribution)

This is a unimodal distribution, the mode being $x = \mu$, with two points of inflection (each located at a distance σ to either side of the mode). The averages of n observations tend to become normally distributed as n increases. The variate x is said to be normally distributed if its density function $f(x)$ is given by an expression of the form

$$f(x) = \frac{1}{\sigma\sqrt{2\pi}}e^{-\frac{1}{2}\left(\frac{x-\mu}{\sigma}\right)^2}, \text{ where}$$

μ = the population mean,

σ = the standard deviation of the population, and

$-\infty \le x \le \infty$

When $\mu = 0$ and $\sigma^2 = \sigma = 1$, the distribution is called a *standardized* or *unit normal* distribution. Then

$$f(x) = \frac{1}{\sqrt{2\pi}}e^{-x^2/2}, \text{ where } -\infty \le x \le \infty.$$

It is noted that $Z = \dfrac{x-\mu}{\sigma}$ follows a standardized normal distribution function.

A unit normal distribution table is included at the end of this section. In the table, the following notations are utilized:

$F(x)$ = the area under the curve from $-\infty$ to x,

$R(x)$ = the area under the curve from x to ∞, and

$W(x)$ = the area under the curve between $-x$ and x.

The Central Limit Theorem

Let $X_1, X_2, ..., X_n$ be a sequence of independent and identically distributed random variables each having mean μ and variance σ^2. Then for large n, the Central Limit Theorem asserts that the sum

$Y = X_1 + X_2 + ... X_n$ is approximately normal.

$$\mu_{\bar{y}} = \mu$$

and the standard deviation

$$\sigma_{\bar{y}} = \frac{\sigma}{\sqrt{n}}$$

t-Distribution

The variate t is defined as the quotient of two independent variates x and r where x is *unit normal* and r is *the root mean square* of n other independent *unit normal variates*; that is, $t = x/r$. The following is the *t*-distribution with n degrees of freedom:

$$f(t) = \frac{\Gamma[(n+1)]/2}{\Gamma(n/2)\sqrt{n\pi}} \frac{1}{\left(1 + t^2/n\right)^{(n+1)/2}}$$

where $-\infty \le t \le \infty$.

A table at the end of this section gives the values of $t_{\alpha, n}$ for values of α and n. Note that in view of the symmetry of the t-distribution, $t_{1-\alpha,n} = -t_{\alpha,n}$.

The function for α follows:

$$\alpha = \int_{t_{\alpha,n}}^{\infty} f(t)\,dt$$

χ^2 - Distribution

If $Z_1, Z_2, ..., Z_n$ are independent unit normal random variables, then

$$\chi^2 = Z_1^2 + Z_2^2 + ... + Z_n^2$$

is said to have a chi-square distribution with n degrees of freedom. The density function is shown as follows:

$$f(\chi^2) = \frac{\frac{1}{2}e^{-\frac{x^2}{2}}\left(\frac{x^2}{2}\right)^{\frac{n}{2}-1}}{\Gamma\left(\frac{n}{2}\right)}, x^2 > 0$$

A table at the end of this section gives values of $\chi^2_{\alpha, n}$ for selected values of α and n.

Gamma Function

$\Gamma(n) = \int_0^{\infty} t^{n-1}e^{-t}dt, \ n > 0$

LINEAR REGRESSION

Least Squares

$$y = \hat{a} + \hat{b}x, \text{ where}$$

$$y-\text{intercept: } \hat{a} = \bar{y} - \hat{b}\bar{x},$$

$$\text{and slope: } \hat{b} = S_{xy}/S_{xx},$$

$$S_{xy} = \sum_{i=1}^{n}x_i y_i - (1/n)\left(\sum_{i=1}^{n}x_i\right)\left(\sum_{i=1}^{n}y_i\right),$$

$$S_{xx} = \sum_{i=1}^{n}x_i^2 - (1/n)\left(\sum_{i=1}^{n}x_i\right)^2,$$

$$n = \text{sample size},$$

$$\bar{y} = (1/n)\left(\sum_{i=1}^{n}y_i\right), \text{ and}$$

$$\bar{x} = (1/n)\left(\sum_{i=1}^{n}x_i\right).$$

Standard Error of Estimate

$$S_e^2 = \frac{S_{xx}S_{yy} - S_{xy}^2}{S_{xx}(n-2)} = MSE, \text{ where}$$

$$S_{yy} = \sum_{i=1}^{n} y_i^2 - (1/n)\left(\sum_{i=1}^{n} y_i\right)^2$$

Confidence Interval for a

$$\hat{a} \pm t_{\alpha/2, n-2}\sqrt{\left(\frac{1}{n} + \frac{\bar{x}^2}{S_{xx}}\right)MSE}$$

Confidence Interval for b

$$\hat{b} \pm t_{\alpha/2, n-2}\sqrt{\frac{MSE}{S_{xx}}}$$

Sample Correlation Coefficient

$$R = \frac{S_{xy}}{\sqrt{S_{xx}S_{yy}}}$$

$$R^2 = \frac{S_{xy}^2}{S_{xx}S_{yy}}$$

HYPOTHESIS TESTING

Consider an unknown parameter θ of a statistical distribution. Let the null hypothesis be

$H_0: \theta = \theta_0$

and let the alternative hypothesis be

$H_1: \theta = \theta_1$

Rejecting H_0 when it is true is known as a type I error, while accepting H_0 when it is wrong is known as a type II error. Furthermore, the probabilities of type I and type II errors are usually represented by the symbols α and β, respectively:

α = probability (type I error)

β = probability (type II error)

The probability of a type I error is known as the level of significance of the test.

Assume that the values of α and β are given. The sample size can be obtained from the following relationships. In (A) and (B), μ_1 is the value assumed to be the true mean.

(A) $H_0: \mu = \mu_0$; $H_1: \mu \neq \mu_0$

$$\beta = \Phi\left(\frac{\mu_0 - \mu}{\sigma/\sqrt{n}} + Z_{\alpha/2}\right) - \Phi\left(\frac{\mu_0 - \mu}{\sigma/\sqrt{n}} - Z_{\alpha/2}\right)$$

An approximate result is

$$n \simeq \frac{(Z_{\alpha/2} + Z_\beta)^2 \sigma^2}{(\mu_1 - \mu_0)^2}$$

(B) $H_0: \mu = \mu_0$; $H_1: \mu > \mu_0$

$$\beta = \Phi\left(\frac{\mu_0 - \mu}{\sigma/\sqrt{n}} + Z_\alpha\right)$$

$$n = \frac{(Z_\alpha + Z_\beta)^2 \sigma^2}{(\mu_1 - \mu_0)^2}$$

Refer to the Hypothesis Testing table in the **INDUSTRIAL ENGINEERING** section of this handbook.

CONFIDENCE INTERVALS

Confidence Interval for the Mean μ of a Normal Distribution

(A) Standard deviation σ is known

$$\overline{X} - Z_{\alpha/2}\frac{\sigma}{\sqrt{n}} \leq \mu \leq \overline{X} + Z_{\alpha/2}\frac{\sigma}{\sqrt{n}}$$

(B) Standard deviation σ is not known

$$\overline{X} - t_{\alpha/2}\frac{s}{\sqrt{n}} \leq \mu \leq \overline{X} + t_{\alpha/2}\frac{s}{\sqrt{n}}$$

where $t_{\alpha/2}$ corresponds to $n - 1$ degrees of freedom.

Confidence Interval for the Difference Between Two Means μ_1 and μ_2

(A) Standard deviations σ_1 and σ_2 known

$$\overline{X_1} - \overline{X_2} - Z_{\alpha/2}\sqrt{\frac{\sigma_1^2}{n_1} + \frac{\sigma_2^2}{n_2}} \leq \mu_1 - \mu_2 \leq \overline{X_1} - \overline{X_2} + Z_{\alpha/2}\sqrt{\frac{\sigma_1^2}{n_1} + \frac{\sigma_2^2}{n_2}}$$

(B) Standard deviations σ_1 and σ_2 are not known

$$\overline{X_1} - \overline{X_2} - t_{\alpha/2}\sqrt{\frac{\left(\frac{1}{n_1} + \frac{1}{n_2}\right)\left[(n_1 - 1)S_1^2 + (n_2 - 1)S_2^2\right]}{n_1 + n_2 - 2}} \leq \mu_1 - \mu_2 \leq \overline{X_1} - \overline{X_2} + t_{\alpha/2}\sqrt{\frac{\left(\frac{1}{n_1} + \frac{1}{n_2}\right)\left[(n_1 - 1)S_1^2 + (n_2 - 1)S_2^2\right]}{n_1 + n_2 - 2}}$$

where $t_{\alpha/2}$ corresponds to $n_1 + n_2 - 2$ degrees of freedom.

Confidence Intervals for the Variance σ^2 of a Normal Distribution

$$\frac{(n - 1)s^2}{x_{\alpha/2,\, n - 1}^2} \leq \sigma^2 \leq \frac{(n - 1)s^2}{x_{1 - \alpha/2,\, n - 1}^2}$$

Sample Size

$$z = \frac{\overline{X} - \mu}{\sigma/\sqrt{n}} \qquad n = \left[\frac{z_{\alpha/2}\,\sigma}{\overline{x} - \mu}\right]^2$$

UNIT NORMAL DISTRIBUTION

x	f(x)	F(x)	R(x)	2R(x)	W(x)
0.0	0.3989	0.5000	0.5000	1.0000	0.0000
0.1	0.3970	0.5398	0.4602	0.9203	0.0797
0.2	0.3910	0.5793	0.4207	0.8415	0.1585
0.3	0.3814	0.6179	0.3821	0.7642	0.2358
0.4	0.3683	0.6554	0.3446	0.6892	0.3108
0.5	0.3521	0.6915	0.3085	0.6171	0.3829
0.6	0.3332	0.7257	0.2743	0.5485	0.4515
0.7	0.3123	0.7580	0.2420	0.4839	0.5161
0.8	0.2897	0.7881	0.2119	0.4237	0.5763
0.9	0.2661	0.8159	0.1841	0.3681	0.6319
1.0	0.2420	0.8413	0.1587	0.3173	0.6827
1.1	0.2179	0.8643	0.1357	0.2713	0.7287
1.2	0.1942	0.8849	0.1151	0.2301	0.7699
1.3	0.1714	0.9032	0.0968	0.1936	0.8064
1.4	0.1497	0.9192	0.0808	0.1615	0.8385
1.5	0.1295	0.9332	0.0668	0.1336	0.8664
1.6	0.1109	0.9452	0.0548	0.1096	0.8904
1.7	0.0940	0.9554	0.0446	0.0891	0.9109
1.8	0.0790	0.9641	0.0359	0.0719	0.9281
1.9	0.0656	0.9713	0.0287	0.0574	0.9426
2.0	0.0540	0.9772	0.0228	0.0455	0.9545
2.1	0.0440	0.9821	0.0179	0.0357	0.9643
2.2	0.0355	0.9861	0.0139	0.0278	0.9722
2.3	0.0283	0.9893	0.0107	0.0214	0.9786
2.4	0.0224	0.9918	0.0082	0.0164	0.9836
2.5	0.0175	0.9938	0.0062	0.0124	0.9876
2.6	0.0136	0.9953	0.0047	0.0093	0.9907
2.7	0.0104	0.9965	0.0035	0.0069	0.9931
2.8	0.0079	0.9974	0.0026	0.0051	0.9949
2.9	0.0060	0.9981	0.0019	0.0037	0.9963
3.0	0.0044	0.9987	0.0013	0.0027	0.9973
Fractiles					
1.2816	0.1755	0.9000	0.1000	0.2000	0.8000
1.6449	0.1031	0.9500	0.0500	0.1000	0.9000
1.9600	0.0584	0.9750	0.0250	0.0500	0.9500
2.0537	0.0484	0.9800	0.0200	0.0400	0.9600
2.3263	0.0267	0.9900	0.0100	0.0200	0.9800
2.5758	0.0145	0.9950	0.0050	0.0100	0.9900

STUDENT'S *t*-DISTRIBUTION

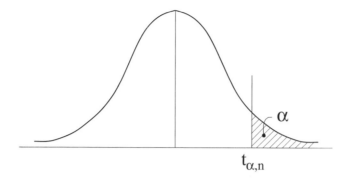

VALUES OF $t_{\alpha,n}$

df	0.25	0.20	0.15	0.10	0.05	0.025	0.01	0.005	df
1	1.000	1.376	1.963	3.078	6.314	12.706	31.821	63.657	1
2	0.816	1.061	1.386	1.886	2.920	4.303	6.965	9.925	2
3	0.765	0.978	1.350	1.638	2.353	3.182	4.541	5.841	3
4	0.741	0.941	1.190	1.533	2.132	2.776	3.747	4.604	4
5	**0.727**	**0.920**	**1.156**	**1.476**	**2.015**	**2.571**	**3.365**	**4.032**	**5**
6	0.718	0.906	1.134	1.440	1.943	2.447	3.143	3.707	6
7	0.711	0.896	1.119	1.415	1.895	2.365	2.998	3.499	7
8	0.706	0.889	1.108	1.397	1.860	2.306	2.896	3.355	8
9	0.703	0.883	1.100	1.383	1.833	2.262	2.821	3.250	9
10	**0.700**	**0.879**	**1.093**	**1.372**	**1.812**	**2.228**	**2.764**	**3.169**	**10**
11	0.697	0.876	1.088	1.363	1.796	2.201	2.718	3.106	11
12	0.695	0.873	1.083	1.356	1.782	2.179	2.681	3.055	12
13	0.694	0.870	1.079	1.350	1.771	2.160	2.650	3.012	13
14	0.692	0.868	1.076	1.345	1.761	2.145	2.624	2.977	14
15	**0.691**	**0.866**	**1.074**	**1.341**	**1.753**	**2.131**	**2.602**	**2.947**	**15**
16	0.690	0.865	1.071	1.337	1.746	2.120	2.583	2.921	16
17	0.689	0.863	1.069	1.333	1.740	2.110	2.567	2.898	17
18	0.688	0.862	1.067	1.330	1.734	2.101	2.552	2.878	18
19	0.688	0.861	1.066	1.328	1.729	2.093	2.539	2.861	19
20	**0.687**	**0.860**	**1.064**	**1.325**	**1.725**	**2.086**	**2.528**	**2.845**	**20**
21	0.686	0.859	1.063	1.323	1.721	2.080	2.518	2.831	21
22	0.686	0.858	1.061	1.321	1.717	2.074	2.508	2.819	22
23	0.685	0.858	1.060	1.319	1.714	2.069	2.500	2.807	23
24	0.685	0.857	1.059	1.318	1.711	2.064	2.492	2.797	24
25	**0.684**	**0.856**	**1.058**	**1.316**	**1.708**	**2.060**	**2.485**	**2.787**	**25**
26	0.684	0.856	1.058	1.315	1.706	2.056	2.479	2.779	26
27	0.684	0.855	1.057	1.314	1.703	2.052	2.473	2.771	27
28	0.683	0.855	1.056	1.313	1.701	2.048	2.467	2.763	28
29	0.683	0.854	1.055	1.311	1.699	2.045	2.462	2.756	29
30	**0.683**	**0.854**	**1.055**	**1.310**	**1.697**	**2.042**	**2.457**	**2.750**	**30**
∞	0.674	0.842	1.036	1.282	1.645	1.960	2.326	2.576	∞

CRITICAL VALUES OF THE F DISTRIBUTION – TABLE

For a particular combination of numerator and denominator degrees of freedom, entry represents the critical values of F corresponding to a specified upper tail area (α).

$\alpha = 0.05$

$F(\alpha, df_1, df_2)$

Denominator df_2	Numerator df_1																		
	1	2	3	4	5	6	7	8	9	10	12	15	20	24	30	40	60	120	∞
1	161.4	199.5	215.7	224.6	230.2	234.0	236.8	238.9	240.5	241.9	243.9	245.9	248.0	249.1	250.1	251.1	252.2	253.3	254.3
2	18.51	19.00	19.16	19.25	19.30	19.33	19.35	19.37	19.38	19.40	19.41	19.43	19.45	19.45	19.46	19.47	19.48	19.49	19.50
3	10.13	9.55	9.28	9.12	9.01	8.94	8.89	8.85	8.81	8.79	8.74	8.70	8.66	8.64	8.62	8.59	8.57	8.55	8.53
4	7.71	6.94	6.59	6.39	6.26	6.16	6.09	6.04	6.00	5.96	5.91	5.86	5.80	5.77	5.75	5.72	5.69	5.66	5.63
5	6.61	5.79	5.41	5.19	5.05	4.95	4.88	4.82	4.77	4.74	4.68	4.62	4.56	4.53	4.50	4.46	4.43	4.40	4.36
6	5.99	5.14	4.76	4.53	4.39	4.28	4.21	4.15	4.10	4.06	4.00	3.94	3.87	3.84	3.81	3.77	3.74	3.70	3.67
7	5.59	4.74	4.35	4.12	3.97	3.87	3.79	3.73	3.68	3.64	3.57	3.51	3.44	3.41	3.38	3.34	3.30	3.27	3.23
8	5.32	4.46	4.07	3.84	3.69	3.58	3.50	3.44	3.39	3.35	3.28	3.22	3.15	3.12	3.08	3.04	3.01	2.97	2.93
9	5.12	4.26	3.86	3.63	3.48	3.37	3.29	3.23	3.18	3.14	3.07	3.01	2.94	2.90	2.86	2.83	2.79	2.75	2.71
10	4.96	4.10	3.71	3.48	3.33	3.22	3.14	3.07	3.02	2.98	2.91	2.85	2.77	2.74	2.70	2.66	2.62	2.58	2.54
11	4.84	3.98	3.59	3.36	3.20	3.09	3.01	2.95	2.90	2.85	2.79	2.72	2.65	2.61	2.57	2.53	2.49	2.45	2.40
12	4.75	3.89	3.49	3.26	3.11	3.00	2.91	2.85	2.80	2.75	2.69	2.62	2.54	2.51	2.47	2.43	2.38	2.34	2.30
13	4.67	3.81	3.41	3.18	3.03	2.92	2.83	2.77	2.71	2.67	2.60	2.53	2.46	2.42	2.38	2.34	2.30	2.25	2.21
14	4.60	3.74	3.34	3.11	2.96	2.85	2.76	2.70	2.65	2.60	2.53	2.46	2.39	2.35	2.31	2.27	2.22	2.18	2.13
15	4.54	3.68	3.29	3.06	2.90	2.79	2.71	2.64	2.59	2.54	2.48	2.40	2.33	2.29	2.25	2.20	2.16	2.11	2.07
16	4.49	3.63	3.24	3.01	2.85	2.74	2.66	2.59	2.54	2.49	2.42	2.35	2.28	2.24	2.19	2.15	2.11	2.06	2.01
17	4.45	3.59	3.20	2.96	2.81	2.70	2.61	2.55	2.49	2.45	2.38	2.31	2.23	2.19	2.15	2.10	2.06	2.01	1.96
18	4.41	3.55	3.16	2.93	2.77	2.66	2.58	2.51	2.46	2.41	2.34	2.27	2.19	2.15	2.11	2.06	2.02	1.97	1.92
19	4.38	3.52	3.13	2.90	2.74	2.63	2.54	2.48	2.42	2.38	2.31	2.23	2.16	2.11	2.07	2.03	1.98	1.93	1.88
20	4.35	3.49	3.10	2.87	2.71	2.60	2.51	2.45	2.39	2.35	2.28	2.20	2.12	2.08	2.04	1.99	1.95	1.90	1.84
21	4.32	3.47	3.07	2.84	2.68	2.57	2.49	2.42	2.37	2.32	2.25	2.18	2.10	2.05	2.01	1.96	1.92	1.87	1.81
22	4.30	3.44	3.05	2.82	2.66	2.55	2.46	2.40	2.34	2.30	2.23	2.15	2.07	2.03	1.98	1.94	1.89	1.84	1.78
23	4.28	3.42	3.03	2.80	2.64	2.53	2.44	2.37	2.32	2.27	2.20	2.13	2.05	2.01	1.96	1.91	1.86	1.81	1.76
24	4.26	3.40	3.01	2.78	2.62	2.51	2.42	2.36	2.30	2.25	2.18	2.11	2.03	1.98	1.94	1.89	1.84	1.79	1.73
25	4.24	3.39	2.99	2.76	2.60	2.49	2.40	2.34	2.28	2.24	2.16	2.09	2.01	1.96	1.92	1.87	1.82	1.77	1.71
26	4.23	3.37	2.98	2.74	2.59	2.47	2.39	2.32	2.27	2.22	2.15	2.07	1.99	1.95	1.90	1.85	1.80	1.75	1.69
27	4.21	3.35	2.96	2.73	2.57	2.46	2.37	2.31	2.25	2.20	2.13	2.06	1.97	1.93	1.88	1.84	1.79	1.73	1.67
28	4.20	3.34	2.95	2.71	2.56	2.45	2.36	2.29	2.24	2.19	2.12	2.04	1.96	1.91	1.87	1.82	1.77	1.71	1.65
29	4.18	3.33	2.93	2.70	2.55	2.43	2.35	2.28	2.22	2.18	2.10	2.03	1.94	1.90	1.85	1.81	1.75	1.70	1.64
30	4.17	3.32	2.92	2.69	2.53	2.42	2.33	2.27	2.21	2.16	2.09	2.01	1.93	1.89	1.84	1.79	1.74	1.68	1.62
40	4.08	3.23	2.84	2.61	2.45	2.34	2.25	2.18	2.12	2.08	2.00	1.92	1.84	1.79	1.74	1.69	1.64	1.58	1.51
60	4.00	3.15	2.76	2.53	2.37	2.25	2.17	2.10	2.04	1.99	1.92	1.84	1.75	1.70	1.65	1.59	1.53	1.47	1.39
120	3.92	3.07	2.68	2.45	2.29	2.17	2.09	2.02	1.96	1.91	1.83	1.75	1.66	1.61	1.55	1.50	1.43	1.35	1.25
∞	3.84	3.00	2.60	2.37	2.21	2.10	2.01	1.94	1.88	1.83	1.75	1.67	1.57	1.52	1.46	1.39	1.32	1.22	1.00

CRITICAL VALUES OF X² DISTRIBUTION

Degrees of Freedom	$X^2_{.995}$	$X^2_{.990}$	$X^2_{.975}$	$X^2_{.950}$	$X^2_{.900}$	$X^2_{.100}$	$X^2_{.050}$	$X^2_{.025}$	$X^2_{.010}$	$X^2_{.005}$
1	0.0000393	0.0001571	0.0009821	0.0039321	0.0157908	2.70554	3.84146	5.02389	6.63490	7.87944
2	0.0100251	0.0201007	0.0506356	0.102587	0.210720	4.60517	5.99147	7.37776	9.21034	10.5966
3	0.0717212	0.114832	0.215795	0.351846	0.584375	6.25139	7.81473	9.34840	11.3449	12.8381
4	0.206990	0.297110	0.484419	0.710721	1.063623	7.77944	9.48773	11.1433	13.2767	14.8602
5	0.411740	0.554300	0.831211	1.145476	1.61031	9.23635	11.0705	12.8325	15.0863	16.7496
6	0.675727	0.872085	1.237347	1.63539	2.20413	10.6446	12.5916	14.4494	16.8119	18.5476
7	0.989265	1.239043	1.68987	2.16735	2.83311	12.0170	14.0671	16.0128	18.4753	20.2777
8	1.344419	1.646482	2.17973	2.73264	3.48954	13.3616	15.5073	17.5346	20.0902	21.9550
9	1.734926	2.087912	2.70039	3.32511	4.16816	14.6837	16.9190	19.0228	21.6660	23.5893
10	2.15585	2.55821	3.24697	3.94030	4.86518	15.9871	18.3070	20.4831	23.2093	25.1882
11	2.60321	3.05347	3.81575	4.57481	5.57779	17.2750	19.6751	21.9200	24.7250	26.7569
12	3.07382	3.57056	4.40379	5.22603	6.30380	18.5494	21.0261	23.3367	26.2170	28.2995
13	3.56503	4.10691	5.00874	5.89186	7.04150	19.8119	22.3621	24.7356	27.6883	29.8194
14	4.07468	4.66043	5.62872	6.57063	7.78953	21.0642	23.6848	26.1190	29.1413	31.3193
15	4.60094	5.22935	6.26214	7.26094	8.54675	22.3072	24.9958	27.4884	30.5779	32.8013
16	5.14224	5.81221	6.90766	7.96164	9.31223	23.5418	26.2962	28.8454	31.9999	34.2672
17	5.69724	6.40776	7.56418	8.67176	10.0852	24.7690	27.5871	30.1910	33.4087	35.7185
18	6.26481	7.01491	8.23075	9.39046	10.8649	25.9894	28.8693	31.5264	34.8053	37.1564
19	6.84398	7.63273	8.90655	10.1170	11.6509	27.2036	30.1435	32.8523	36.1908	38.5822
20	7.43386	8.26040	9.59083	10.8508	12.4426	28.4120	31.4104	34.1696	37.5662	39.9968
21	8.03366	8.89720	10.28293	11.5913	13.2396	29.6151	32.6705	35.4789	38.9321	41.4010
22	8.64272	9.54249	10.9823	12.3380	14.0415	30.8133	33.9244	36.7807	40.2894	42.7956
23	9.26042	10.19567	11.6885	13.0905	14.8479	32.0069	35.1725	38.0757	41.6384	44.1813
24	9.88623	10.8564	12.4011	13.8484	15.6587	33.1963	36.4151	39.3641	42.9798	45.5585
25	10.5197	11.5240	13.1197	14.6114	16.4734	34.3816	37.6525	40.6465	44.3141	46.9278
26	11.1603	12.1981	13.8439	15.3791	17.2919	35.5631	38.8852	41.9232	45.6417	48.2899
27	11.8076	12.8786	14.5733	16.1513	18.1138	36.7412	40.1133	43.1944	46.9630	49.6449
28	12.4613	13.5648	15.3079	16.9279	18.9392	37.9159	41.3372	44.4607	48.2782	50.9933
29	13.1211	14.2565	16.0471	17.7083	19.7677	39.0875	42.5569	45.7222	49.5879	52.3356
30	13.7867	14.9535	16.7908	18.4926	20.5992	40.2560	43.7729	46.9792	50.8922	53.6720
40	20.7065	22.1643	24.4331	26.5093	29.0505	51.8050	55.7585	59.3417	63.6907	66.7659
50	27.9907	29.7067	32.3574	34.7642	37.6886	63.1671	67.5048	71.4202	76.1539	79.4900
60	35.5346	37.4848	40.4817	43.1879	46.4589	74.3970	79.0819	83.2976	88.3794	91.9517
70	43.2752	45.4418	48.7576	51.7393	55.3290	85.5271	90.5312	95.0231	100.425	104.215
80	51.1720	53.5400	57.1532	60.3915	64.2778	96.5782	101.879	106.629	112.329	116.321
90	59.1963	61.7541	65.6466	69.1260	73.2912	107.565	113.145	118.136	124.116	128.299
100	67.3276	70.0648	74.2219	77.9295	82.3581	118.498	124.342	129.561	135.807	140.169

Source: Thompson, C. M., "Tables of the Percentage Points of the X²–Distribution," *Biometrika*, ©1941, 32, 188–189. Reproduced by permission of the *Biometrika* Trustees.

STATICS

FORCE
A *force* is a *vector* quantity. It is defined when its **(1)** magnitude, **(2)** point of application, and **(3)** direction are known.

RESULTANT (TWO DIMENSIONS)
The *resultant*, F, of n forces with components $F_{x,i}$ and $F_{y,i}$ has the magnitude of

$$F = \left[\left(\sum_{i=1}^{n} F_{x,i} \right)^2 + \left(\sum_{i=1}^{n} F_{y,i} \right)^2 \right]^{1/2}$$

The resultant direction with respect to the x-axis using fourquadrant angle functions is

$$\theta = \arctan \left(\sum_{i=1}^{n} F_{y,i} \middle/ \sum_{i=1}^{n} F_{x,i} \right)$$

The vector form of a force is
$$\boldsymbol{F} = F_x \mathbf{i} + F_y \mathbf{j}$$

RESOLUTION OF A FORCE
$F_x = F \cos \theta_x$; $F_y = F \cos \theta_y$; $F_z = F \cos \theta_z$

$\cos \theta_x = F_x / F$; $\cos \theta_y = F_y / F$; $\cos \theta_z = F_z / F$

Separating a force into components when the geometry of force is known and $R = \sqrt{x^2 + y^2 + z^2}$

$F_x = (x/R)F$; $\qquad F_y = (y/R)F$; $\qquad F_z = (z/R)F$

MOMENTS (COUPLES)
A system of two forces that are equal in magnitude, opposite in direction, and parallel to each other is called a *couple*. A *moment* $\boldsymbol{M}$ is defined as the cross product of the *radius vector* $\boldsymbol{r}$ and the *force* $\boldsymbol{F}$ from a point to the line of action of the force.

$\boldsymbol{M} = \boldsymbol{r} \times \boldsymbol{F}$; $\qquad M_x = yF_z - zF_y$,
$\qquad\qquad\qquad\quad M_y = zF_x - xF_z$, and
$\qquad\qquad\qquad\quad M_z = xF_y - yF_x$.

SYSTEMS OF FORCES
$\boldsymbol{F} = \Sigma \, \boldsymbol{F}_n$

$\boldsymbol{M} = \Sigma \, (\boldsymbol{r}_n \times \boldsymbol{F}_n)$

Equilibrium Requirements
$\Sigma \, \boldsymbol{F}_n = 0$

$\Sigma \, \boldsymbol{M}_n = 0$

CENTROIDS OF MASSES, AREAS, LENGTHS, AND VOLUMES
Formulas for centroids, moments of inertia, and first moment of areas are presented in the **MATHEMATICS** section for continuous functions. The following discrete formulas are for defined regular masses, areas, lengths, and volumes:

$$\boldsymbol{r}_c = \Sigma \, m_n \boldsymbol{r}_n / \Sigma \, m_n, \text{ where}$$

m_n = the *mass of each particle* making up the system,

$\boldsymbol{r}_n$ = the *radius vector* to each particle from a selected reference point, and

$\boldsymbol{r}_c$ = the *radius vector* to the *center of the total mass* from the selected reference point.

The *moment of area* (M_a) is defined as
$$M_{ay} = \Sigma \, x_n a_n$$
$$M_{ax} = \Sigma \, y_n a_n$$
$$M_{az} = \Sigma \, z_n a_n$$

The *centroid of area* is defined as
$$\left. \begin{array}{l} x_{ac} = M_{ay}/A \\ y_{ac} = M_{ax}/A \\ z_{ac} = M_{az}/A \end{array} \right\} \begin{array}{l} \text{with respect to center of} \\ \text{the coordinate system} \end{array}$$
where $\quad A = \Sigma \, a_n$

The *centroid of a line* is defined as
$$x_{lc} = (\Sigma \, x_n l_n)/L, \text{ where } L = \Sigma \, l_n$$
$$y_{lc} = (\Sigma \, y_n l_n)/L$$
$$z_{lc} = (\Sigma \, z_n l_n)/L$$

The *centroid of volume* is defined as
$$x_{vc} = (\Sigma \, x_n v_n)/V, \text{ where } V = \Sigma \, v_n$$
$$y_{vc} = (\Sigma \, y_n v_n)/V$$
$$z_{vc} = (\Sigma \, z_n v_n)/V$$

MOMENT OF INERTIA
The *moment of inertia*, or the second moment of area, is defined as
$$I_y = \int x^2 \, dA$$
$$I_x = \int y^2 \, dA$$

The *polar moment of inertia* J of an area about a point is equal to the sum of the moments of inertia of the area about any two perpendicular axes in the area and passing through the same point.
$$I_z = J = I_y + I_x = \int (x^2 + y^2) \, dA$$
$$= r_p^2 A, \text{ where}$$

r_p = the *radius of gyration* (see the **DYNAMICS** section and the next page of this section).

Moment of Inertia Parallel Axis Theorem

The moment of inertia of an area about any axis is defined as the moment of inertia of the area about a parallel centroidal axis plus a term equal to the area multiplied by the square of the perpendicular distance d from the centroidal axis to the axis in question.

$$I'_x = I_{x_c} + d_x^2 A$$

$$I'_y = I_{y_c} + d_y^2 A, \text{ where}$$

d_x, d_y = distance between the two axes in question,

I_{x_c}, I_{y_c} = the moment of inertia about the centroidal axis, and

I'_x, I'_y = the moment of inertia about the new axis.

Radius of Gyration

The *radius of gyration* r_p, r_x, r_y is the distance from a reference axis at which all of the area can be considered to be concentrated to produce the moment of inertia.

$$r_x = \sqrt{I_x / A}; \quad r_y = \sqrt{I_y / A}; \quad r_p = \sqrt{J / A}$$

Product of Inertia

The *product of inertia* (I_{xy}, etc.) is defined as:

$I_{xy} = \int xy dA$, with respect to the *xy*-coordinate system,

$I_{xz} = \int xz dA$, with respect to the *xz*-coordinate system, and

$I_{yz} = \int yz dA$, with respect to the *yz*-coordinate system.

The *parallel-axis theorem* also applies:

$I'_{xy} = I_{x_c y_c} + d_x d_y A$ for the *xy*-coordinate system, etc.

where

d_x = x-axis distance between the two axes in question, and

d_y = y-axis distance between the two axes in question.

FRICTION

The largest frictional force is called the *limiting friction*. Any further increase in applied forces will cause motion.

$$F \leq \mu_s N, \text{ where}$$

F = friction force,

μ_s = *coefficient of static friction*, and

N = normal force between surfaces in contact.

SCREW THREAD (also see MECHANICAL ENGINEERING section)

For a *screw-jack, square thread*,

$$M = Pr \tan (\alpha \pm \phi), \text{ where}$$

+ is for screw tightening,

– is for screw loosening,

M = external moment applied to axis of screw,

P = load on jack applied along and on the line of the axis,

r = the mean thread radius,

α = the *pitch angle* of the thread, and

μ = $\tan \phi$ = the appropriate coefficient of friction.

BELT FRICTION

$$F_1 = F_2 e^{\mu\theta}, \text{ where}$$

F_1 = force being applied in the direction of impending motion,

F_2 = force applied to resist impending motion,

μ = coefficient of static friction, and

θ = the total *angle of contact* between the surfaces expressed in radians.

STATICALLY DETERMINATE TRUSS

Plane Truss

A plane truss is a rigid framework satisfying the following conditions:

1. The members of the truss lie in the same plane.
2. The members are connected at their ends by frictionless pins.
3. All of the external loads lie in the plane of the truss and are applied at the joints only.
4. The truss reactions and member forces can be determined using the equations of equilibrium.

   $\Sigma F = 0; \Sigma M = 0$
5. A truss is statically indeterminate if the reactions and member forces cannot be solved with the equations of equilibrium.

Plane Truss: Method of Joints

The method consists of solving for the forces in the members by writing the two equilibrium equations for each joint of the truss.

$$\Sigma F_V = 0 \text{ and } \Sigma F_H = 0, \text{ where}$$

F_H = horizontal forces and member components and

F_V = vertical forces and member components.

Plane Truss: Method of Sections

The method consists of drawing a free-body diagram of a portion of the truss in such a way that the unknown truss member force is exposed as an external force.

CONCURRENT FORCES

A concurrent-force system is one in which the lines of action of the applied forces all meet at one point. A *two-force* body in static equilibrium has two applied forces that are equal in magnitude, opposite in direction, and collinear. A *three-force* body in static equilibrium has three applied forces whose lines of action meet at a point. As a consequence, if the direction and magnitude of two of the three forces are known, the direction and magnitude of the third can be determined.

Figure	Area & Centroid	Area Moment of Inertia	(Radius of Gyration)^2	Product of Inertia
	$A = bh/2$ $x_c = 2b/3$ $y_c = h/3$	$I_{x_c} = bh^3/36$ $I_{y_c} = b^3h/36$ $I_x = bh^3/12$ $I_y = b^3h/4$	$r_{x_c}^2 = h^2/18$ $r_{y_c}^2 = b^2/18$ $r_x^2 = h^2/6$ $r_y^2 = b^2/2$	$I_{x_c y_c} = Abh/36 = b^2h^2/72$ $I_{xy} = Abh/4 = b^2h^2/8$
	$A = bh/2$ $x_c = b/3$ $y_c = h/3$	$I_{x_c} = bh^3/36$ $I_{y_c} = b^3h/36$ $I_x = bh^3/12$ $I_y = b^3h/12$	$r_{x_c}^2 = h^2/18$ $r_{y_c}^2 = b^2/18$ $r_x^2 = h^2/6$ $r_y^2 = b^2/6$	$I_{x_c y_c} = -Abh/36 = -b^2h^2/72$ $I_{xy} = Abh/12 = b^2h^2/24$
	$A = bh/2$ $x_c = (a+b)/3$ $y_c = h/3$	$I_{x_c} = bh^3/36$ $I_{y_c} = [bh(b^2 - ab + a^2)]/36$ $I_x = bh^3/12$ $I_y = [bh(b^2 + ab + a^2)]/12$	$r_{x_c}^2 = h^2/18$ $r_{y_c}^2 = (b^2 - ab + a^2)/18$ $r_x^2 = h^2/6$ $r_y^2 = (b^2 + ab + a^2)/6$	$I_{x_c y_c} = [Ah(2a - b)]/36$ $= [bh^2(2a - b)]/72$ $I_{xy} = [Ah(2a + b)]/12$ $= [bh^2(2a + b)]/24$
	$A = bh$ $x_c = b/2$ $y_c = h/2$	$I_{x_c} = bh^3/12$ $I_{y_c} = b^3h/12$ $I_x = bh^3/3$ $I_y = b^3h/3$ $J = [bh(b^2 + h^2)]/12$	$r_{x_c}^2 = h^2/12$ $r_{y_c}^2 = b^2/12$ $r_x^2 = h^2/3$ $r_y^2 = b^2/3$ $r_p^2 = (b^2 + h^2)/12$	$I_{x_c y_c} = 0$ $I_{xy} = Abh/4 = b^2h^2/4$
	$A = h(a+b)/2$ $y_c = \dfrac{h(2a+b)}{3(a+b)}$	$I_{x_c} = \dfrac{h^3(a^2 + 4ab + b^2)}{36(a+b)}$ $I_x = \dfrac{h^3(3a+b)}{12}$	$r_{x_c}^2 = \dfrac{h^2(a^2 + 4ab + b^2)}{18(a+b)}$ $r_x^2 = \dfrac{h^2(3a+b)}{6(a+b)}$	
	$A = ab\sin\theta$ $x_c = (b + a\cos\theta)/2$ $y_c = (a\sin\theta)/2$	$I_{x_c} = (a^3b\sin^3\theta)/12$ $I_{y_c} = [ab\sin\theta(b^2 + a^2\cos^2\theta)]/12$ $I_x = (a^3b\sin^3\theta)/3$ $I_y = [ab\sin\theta(b + a\cos\theta)^2]/3$ $\quad - (a^2b^2\sin\theta\cos\theta)/6$	$r_{x_c}^2 = (a\sin\theta)^2/12$ $r_{y_c}^2 = (b^2 + a^2\cos^2\theta)/12$ $r_x^2 = (a\sin\theta)^2/3$ $r_y^2 = (b + a\cos\theta)^2/3$ $\quad - (ab\cos\theta)/6$	$I_{x_c y_c} = (a^3b\sin^2\theta\cos\theta)/12$

Housner, George W. & Donald E. Hudson, *Applied Mechanics Dynamics*, D. Van Nostrand Company, Inc., Princeton, NJ, 1959. Table reprinted by permission of G.W. Housner & D.E. Hudson.

Figure	Area & Centroid	Area Moment of Inertia	(Radius of Gyration)²	Product of Inertia
	$A = \pi a^2$ $x_c = a$ $y_c = a$	$I_{x_c} = I_{y_c} = \pi a^4/4$ $I_x = I_y = 5\pi a^4/4$ $J = \pi a^4/2$	$r_{x_c}^2 = r_{y_c}^2 = a^2/4$ $r_x^2 = r_y^2 = 5a^2/4$ $r_p^2 = a^2/2$	$I_{x_c y_c} = 0$ $I_{xy} = Aa^2$
	$A = \pi(a^2 - b^2)$ $x_c = a$ $y_c = a$	$I_{x_c} = I_{y_c} = \pi(a^4 - b^4)/4$ $I_x = I_y = \dfrac{5\pi a^4}{4} - \pi a^2 b^2 - \dfrac{\pi b^4}{4}$ $J = \pi(a^4 - b^4)/2$	$r_{x_c}^2 = r_{y_c}^2 = (a^2 + b^2)/4$ $r_x^2 = r_y^2 = (5a^2 + b^2)/4$ $r_p^2 = (a^2 + b^2)/2$	$I_{x_c y_c} = 0$ $I_{xy} = Aa^2$ $= \pi a^2(a^2 - b^2)$
	$A = \pi a^2/2$ $x_c = a$ $y_c = 4a/(3\pi)$	$I_{x_c} = \dfrac{a^4(9\pi^2 - 64)}{72\pi}$ $I_{y_c} = \pi a^4/8$ $I_x = \pi a^4/8$ $I_y = 5\pi a^4/8$	$r_{x_c}^2 = \dfrac{a^2(9\pi^2 - 64)}{36\pi^2}$ $r_{y_c}^2 = a^2/4$ $r_x^2 = a^2/4$ $r_y^2 = 5a^2/4$	$I_{x_c y_c} = 0$ $I_{xy} = 2a^4/3$
  CIRCULAR SECTOR	$A = a^2\theta$ $x_c = \dfrac{2a}{3}\dfrac{\sin\theta}{\theta}$ $y_c = 0$	$I_x = a^4(\theta - \sin\theta\,\cos\theta)/4$ $I_y = a^4(\theta + \sin\theta\,\cos\theta)/4$	$r_x^2 = \dfrac{a^2}{4}\dfrac{(\theta - \sin\theta\,\cos\theta)}{\theta}$ $r_y^2 = \dfrac{a^2}{4}\dfrac{(\theta + \sin\theta\,\cos\theta)}{\theta}$	$I_{x_c y_c} = 0$ $I_{xy} = 0$
  CIRCULAR SEGMENT	$A = a^2\left[\theta - \dfrac{\sin 2\theta}{2}\right]$ $x_c = \dfrac{2a}{3}\dfrac{\sin^3\theta}{\theta - \sin\theta\cos\theta}$ $y_c = 0$	$I_x = \dfrac{Aa^2}{4}\left[1 - \dfrac{2\sin^3\theta\,\cos\theta}{3\theta - 3\sin\theta\,\cos\theta}\right]$ $I_y = \dfrac{Aa^2}{4}\left[1 + \dfrac{2\sin^3\theta\,\cos\theta}{\theta - \sin\theta\,\cos\theta}\right]$	$r_x^2 = \dfrac{a^2}{4}\left[1 - \dfrac{2\sin^3\theta\,\cos\theta}{3\theta - 3\sin\theta\,\cos\theta}\right]$ $r_y^2 = \dfrac{a^2}{4}\left[1 + \dfrac{2\sin^3\theta\,\cos\theta}{\theta - \sin\theta\,\cos\theta}\right]$	$I_{x_c y_c} = 0$ $I_{xy} = 0$

Housner, George W. & Donald E. Hudson, *Applied Mechanics Dynamics*, D. Van Nostrand Company, Inc., Princeton, NJ, 1959. Table reprinted by permission of G.W. Housner & D.E. Hudson.

Figure	Area & Centroid	Area Moment of Inertia	(Radius of Gyration)^2	Product of Inertia
 PARABOLA	$A = 4ab/3$ $x_c = 3a/5$ $y_c = 0$	$I_{x_c} = I_x = 4ab^3/15$ $I_{y_c} = 16a^3b/175$ $I_y = 4a^3b/7$	$r_{x_c}^2 = r_x^2 = b^2/5$ $r_{y_c}^2 = 12a^2/175$ $r_y^2 = 3a^2/7$	$I_{x_c y_c} = 0$ $I_{xy} = 0$
 HALF A PARABOLA	$A = 2ab/3$ $x_c = 3a/5$ $y_c = 3b/8$	$I_x = 2ab^3/15$ $I_y = 2ba^3/7$	$\dot{r}_x^2 = b^2/5$ $r_y^2 = 3a^2/7$	$I_{xy} = Aab/4 = a^2b^2$
 $y = (h/b^n)x^n$ n^{th} DEGREE PARABOLA	$A = bh/(n+1)$ $x_c = \dfrac{n+1}{n+2} b$ $y_c = \dfrac{h}{2}\dfrac{n+1}{2n+1}$	$I_x = \dfrac{bh^3}{3(3n+1)}$ $I_y = \dfrac{hb^3}{n+3}$	$r_x^2 = \dfrac{h^2(n+1)}{3(3n+1)}$ $r_y^2 = \dfrac{n+1}{n+3}b^2$	
 $y = (h/b^{1/n})x^{1/n}$ n^{th} DEGREE PARABOLA	$A = \dfrac{n}{n+1} bh$ $x_c = \dfrac{n+1}{2n+1} b$ $y_c = \dfrac{n+1}{2(n+2)}h$	$I_x = \dfrac{n}{3(n+3)}bh^3$ $I_y = \dfrac{n}{3n+1}b^3h$	$r_x^2 = \dfrac{n+1}{3(n+1)}h^2$ $r_y^2 = \dfrac{n+1}{3n+1}b^2$	

Housner, George W. & Donald E. Hudson, *Applied Mechanics Dynamics*, D. Van Nostrand Company, Inc., Princeton, NJ, 1959. Table reprinted by permission of G.W. Housner & D.E. Hudson.

DYNAMICS

KINEMATICS

Kinematics is the study of motion without consideration of the mass of, or the forces acting on, the system. For particle motion, let $r(t)$ be the position vector of the particle in an inertial reference frame. The velocity and acceleration of the particle are respectively defined as

$v = dr/dt$

$a = dv/dt$, where

v = the instantaneous velocity,

a = the instantaneous acceleration, and

t = time

Cartesian Coordinates

$r = xi + yj + zk$

$v = \dot{x}i + \dot{y}j + \dot{z}k$

$a = \ddot{x}i + \ddot{y}j + \ddot{z}k$, where

$\dot{x} = dx/dt = v_x$, etc.

$\ddot{x} = d^2x/dt^2 = a_x$, etc.

Radial and Transverse Components for Planar Motion

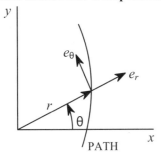

Unit vectors e_θ and e_r are, respectively, normal to and collinear with the position vector r. Thus:

$r = re_r$

$v = \dot{r}e_r + r\dot{\theta}e_\theta$

$a = (\ddot{r} - r\dot{\theta}^2)e_r + (r\ddot{\theta} + 2\dot{r}\dot{\theta})e_\theta$, where

r = the radial distance

θ = the angle between the x axis and e_r,

$\dot{r} = dr/dt$, etc.

$\ddot{r} = d^2r/dt^2$, etc.

Plane Circular Motion

A special case of transverse and radial components is for constant radius rotation about the origin, or plane circular motion.

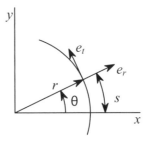

Here the vector quantities are defined as

$r = re_r$

$v = r\omega e_t$

$a = (-r\omega^2)e_r + r\alpha e_t$, where

r = the radius of the circle, and

θ = the angle between the x and e_r axes

The magnitudes of the angular velocity and acceleration, respectively, are defined as

$\omega = \dot{\theta}$, and

$\alpha = \dot{\omega} = \ddot{\theta}$

Arc length, tangential velocity, and tangential acceleration, respectively, are

$s = r\theta$,

$v_t = r\omega$

$a_t = r\alpha$

The normal acceleration is given by

$a_n = r\omega^2$

Normal and Tangential Components

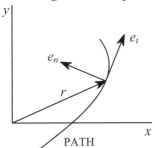

Unit vectors e_t and e_n are, respectively, tangent and normal to the path with e_n pointing to the center of curvature. Thus

$$\boldsymbol{v} = v(t)\boldsymbol{e_t}$$

$$\boldsymbol{a} = a(t)\boldsymbol{e_t} + \left(v_t^2/\rho\right)\boldsymbol{e_n}, \text{ where}$$

ρ = instantaneous radius of curvature

Constant Acceleration

The equations for the velocity and displacement when acceleration is a constant are given as

$$
\begin{aligned}
a(t) &= a_0 \\
v(t) &= a_0(t - t_0) + v_0 \\
s(t) &= a_0(t - t_0)^2/2 + v_0(t - t_0) + s_0, \text{ where}
\end{aligned}
$$

s = distance along the line of travel
s_0 = displacement at time t_0
v = velocity along the direction of travel
v_0 = velocity at time t_0
a_0 = constant acceleration
t = time, and
t_0 = some initial time

For a free-falling body, $a_0 = g$ (downward).
An additional equation for velocity as a function of position may be written as

$$v^2 = v_0^2 + 2a_0(s - s_0)$$

For constant angular acceleration, the equations for angular velocity and displacement are

$$
\begin{aligned}
\alpha(t) &= \alpha_0 \\
\omega(t) &= \alpha_0(t - t_0) + \omega_0 \\
\theta(t) &= \alpha_0(t - t_0)^2/2 + \omega_0(t - t_0) + \theta_0, \quad \text{where}
\end{aligned}
$$

θ = angular displacement
θ_0 = angular displacement at time t_0
ω = angular velocity
ω_0 = angular velocity at time t_0
α_0 = constant angular acceleration
t = time, and
t_0 = some initial time

An additional equation for angular velocity as a function of angular position may be written as

$$\omega^2 = \omega_0^2 + 2\alpha_0(\theta - \theta_0)$$

Non-constant Acceleration

When non-constant acceleration, $a(t)$, is considered, the equations for the velocity and displacement may be obtained from

$$v(t) = \int_{t_0}^{t} a(\tau)\,d\tau + v_{t_0}$$

$$s(t) = \int_{t_0}^{t} v(\tau)\,d\tau + s_{t_0}$$

For variable angular acceleration

$$\omega(t) = \int_{t_0}^{t} \alpha(\tau)\,d\tau + \omega_{t_0}$$

$$\theta(t) = \int_{t_0}^{t} \omega(\tau)\,d\tau + \theta_{t_0}$$

Projectile Motion

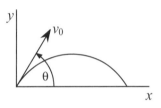

The equations for common projectile motion may be obtained from the constant acceleration equations as

$$
\begin{aligned}
a_x &= 0 \\
v_x &= v_0\cos(\theta) \\
x &= v_0\cos(\theta)t + x_0 \\
a_y &= -g \\
v_y &= -gt + v_0\sin(\theta) \\
y &= -gt^2/2 + v_0\sin(\theta)t + y_0
\end{aligned}
$$

CONCEPT OF WEIGHT

$W = mg$, where

W = weight, N (lbf),
m = mass, kg (lbf-sec^2/ft), and
g = local acceleration of gravity, m/sec^2 (ft/sec^2)

KINETICS

Newton's second law for a particle is

$\Sigma \boldsymbol{F} = d(m\boldsymbol{v})/dt$, where

$\Sigma \boldsymbol{F}$ = the sum of the applied forces acting on the particle,

m = the mass of the particle

$\boldsymbol{v}$ = the velocity of the particle

For constant mass,

$\Sigma \boldsymbol{F} = m\, d\boldsymbol{v}/dt = m\boldsymbol{a}$

One-Dimensional Motion of a Particle (Constant Mass)

When motion only exists in a single dimension then, without loss of generality, it may be assumed to be in the x direction, and

$a_x = F_x/m$, where

F_x = the resultant of the applied forces which in general can depend on t, x, and v_x.

If F_x only depends on t, then

$a_x(t) = F_x(t)/m,$

$v_x(t) = \int_{t_0}^{t} a_x(\tau)\,d\tau + v_{xt_0},$

$x(t) = \int_{t_0}^{t} v_x(\tau)\,d\tau + x_{t_0}$

If the force is constant (i.e. independent of time, displacement, and velocity) then

$a_x = F_x/m,$

$v_x = a_x(t - t_0) + v_{xt_0},$

$x = a_x(t - t_0)^2/2 + v_{xt_0}(t - t_0) + x_{t_0}$

Normal and Tangential Kinetics for Planar Problems

When working with normal and tangential directions, the scalar equations may be written as

$\Sigma F_t = ma_t = mdv_t/dt$ and

$\Sigma F_n = ma_n = m\left(v_t^2/\rho\right)$

Impulse and Momentum

Linear

Assuming constant mass, the equation of motion of a particle may be written as

$md\boldsymbol{v}/dt = \boldsymbol{F}$

$md\boldsymbol{v} = \boldsymbol{F}dt$

For a system of particles, by integrating and summing over the number of particles, this may be expanded to

$\Sigma m_i\left(\boldsymbol{v}_i\right)_{t_2} = \Sigma m_i\left(\boldsymbol{v}_i\right)_{t_1} + \Sigma \int_{t_1}^{t_2} \boldsymbol{F}_i\, dt$

The term on the left side of the equation is the linear momentum of a system of particles at time t_2. The first term on the right side of the equation is the linear momentum of a system of particles at time t_1. The second term on the right side of the equation is the impulse of the force $\boldsymbol{F}$ from time t_1 to t_2. It should be noted that the above equation is a vector equation. Component scalar equations may be obtained by considering the momentum and force in a set of orthogonal directions.

Angular Momentum or Moment of Momentum

The angular momentum or the moment of momentum about point 0 of a particle is defined as

$\mathbf{H}_0 = \mathbf{r} \times m\mathbf{v}$, or

$\mathbf{H}_0 = I_0\omega$

Taking the time derivative of the above, the equation of motion may be written as

$\dot{\mathbf{H}}_0 = d\left(I_0\omega\right)/dt = \mathbf{M}$, where

$\boldsymbol{M}$ is the moment applied to the particle. Now by integrating and summing over a system of any number of particles, this may be expanded to

$\Sigma\left(\mathbf{H}_{0i}\right)_{t_2} = \Sigma\left(\mathbf{H}_{0i}\right)_{t_1} + \Sigma \int_{t_1}^{t_2} \boldsymbol{M}_{0i}\,dt$

The term on the left side of the equation is the angular momentum of a system of particles at time t_2. The first term on the right side of the equation is the angular momentum of a system of particles at time t_1. The second term on the right side of the equation is the angular impulse of the moment $\boldsymbol{M}$ from time t_1 to t_2.

Work and Energy

Work W is defined as

$W = \int \boldsymbol{F} \cdot d\boldsymbol{r}$

(For particle flow, see **FLUID MECHANICS** section.)

Kinetic Energy

The kinetic energy of a particle is the work done by an external agent in accelerating the particle from rest to a velocity v. Thus

$T = mv^2/2$

In changing the velocity from $\boldsymbol{v}_1$ to $\boldsymbol{v}_2$, the change in kinetic energy is

$T_2 - T_1 = m\left(v_2^2 - v_1^2\right)/2$

Potential Energy
The work done by an external agent in the presence of a conservative field is termed the change in potential energy.

Potential Energy in Gravity Field
$$U = mgh, \text{ where}$$

h = the elevation above some specified datum.

Elastic Potential Energy
For a linear elastic spring with modulus, stiffness, or spring constant, the force in the spring is
$$F_s = k\,x, \text{ where}$$

x = the change in length of the spring from the undeformed length of the spring.

The potential energy stored in the spring when compressed or extended by an amount x is
$$U = k\,x^2/2$$

In changing the deformation in the spring from position x_1 to x_2, the change in the potential energy stored in the spring is
$$U_2 - U_1 = k\left(x_2^2 - x_1^2\right)/2$$

Principle of Work and Energy
If T_i and U_i are, respectively, the kinetic and potential energy of a particle at state i, then for conservative systems (no energy dissipation or gain), the law of conservation of energy is
$$T_2 + U_2 = T_1 + U_1$$

If nonconservative forces are present, then the work done by these forces must be accounted for. Hence
$$T_2 + U_2 = T_1 + U_1 + W_{1 \to 2}, \text{ where}$$

$W_{1 \to 2}$ = the work done by the nonconservative forces in moving between state 1 and state 2. Care must be exercised during computations to correctly compute the algebraic sign of the work term. If the forces serve to increase the energy of the system, $W_{1 \to 2}$ is positive. If the forces, such as friction, serve to dissipate energy, $W_{1 \to 2}$ is negative.

Impact
During an impact, momentum is conserved while energy may or may not be conserved. For direct central impact with no external forces
$$m_1 v_1 + m_2 v_2 = m_1 v_1' + m_2 v_2', \text{ where}$$

m_1, m_2 = the masses of the two bodies,

v_1, v_2 = the velocities of the bodies just before impact, and

v_1', v_2' = the velocities of the bodies just after impact.

For impacts, the relative velocity expression is
$$e = \frac{(v_2')_n - (v_1')_n}{(v_1)_n - (v_2)_n}, \text{ where}$$

e = coefficient of restitution,

$(v_i)_n$ = the velocity normal to the plane of impact just **before** impact, and

$(v_i')_n$ = the velocity normal to the plane of impact just **after** impact

The value of e is such that
$$0 \le e \le 1, \text{ with limiting values}$$
$e = 1$, perfectly elastic (energy conserved), and
$e = 0$, perfectly plastic (no rebound)

Knowing the value of e, the velocities after the impact are given as
$$(v_1')_n = \frac{m_2 (v_2)_n (1 + e) + (m_1 - em_2)(v_1)_n}{m_1 + m_2}$$
$$(v_2')_n = \frac{m_1 (v_1)_n (1 + e) - (em_1 - m_2)(v_2)_n}{m_1 + m_2}$$

Friction
The Laws of Friction are
1. The total friction force F that can be developed is independent of the magnitude of the area of contact.
2. The total friction force F that can be developed is proportional to the normal force N.
3. For low velocities of sliding, the total frictional force that can be developed is practically independent of the velocity, although experiments show that the force F necessary to initiate slip is greater than that necessary to maintain the motion.

The formula expressing the Laws of Friction is
$$F \le \mu N, \text{ where}$$

μ = the coefficient of friction.

In general
$$F < \mu_s N, \text{ no slip occuring,}$$
$$F = \mu_s N, \text{ at the point of impending slip, and}$$
$$F = \mu_k N, \text{ when slip is occuring.}$$

Here,

μ_s = the coefficient of static friction, and

μ_k = the coefficient of kinetic friction.

The coefficient of kinetic friction is often approximated as 75% of the coefficient of static friction.

Mass Moment of Inertia

The definitions for the mass moments of inertia are

$$I_x = \int \left(y^2 + z^2 \right) dm,$$

$$I_y = \int \left(x^2 + z^2 \right) dm, \text{ and}$$

$$I_z = \int \left(x^2 + y^2 \right) dm$$

A table listing moment of inertia formulas for some standard shapes is at the end of this section.

Parallel-Axis Theorem

The mass moments of inertia may be calculated about any axis through the application of the above definitions. However, once the moments of inertia have been determined about an axis passing through a body's mass center, it may be transformed to another parallel axis. The transformation equation is

$$I_{new} = I_c + md^2, \text{ where}$$

I_{new} = the mass moment of inertia about any specified axis

I_c = the mass moment of inertia about an axis that is parallel to the above specified axis but passes through the body's mass center

m = the mass of the body

d = the normal distance from the body's mass center to the above-specified axis

Radius of Gyration

The radius of gyration is defined as

$$r = \sqrt{I/m}$$

PLANE MOTION OF A RIGID BODY

Kinematics

Instantaneous Center of Rotation (Instant Centers)

An instantaneous center of rotation (instant center) is a point, common to two bodies, at which each has the same velocity (magnitude and direction) at a given instant. It is also a point on one body about which another body rotates, instantaneously.

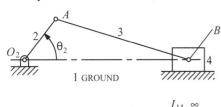

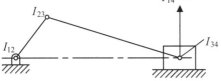

The figure shows a fourbar slider-crank. Link 2 (the crank) rotates about the fixed center, O_2. Link 3 couples the crank to the slider (link 4), which slides against ground (link 1). Using the definition of an instant center (IC), we see that the pins at O_2, A, and B are ICs that are designated I_{12}, I_{23}, and I_{34}. The easily observable IC is I_{14}, which is located at infinity with its direction perpendicular to the interface between links 1 and 4 (the direction of sliding). To locate the remaining two ICs (for a fourbar) we must make use of Kennedy's rule.

Kennedy's Rule: When three bodies move relative to one another they have three instantaneous centers, all of which lie on the same straight line.

To apply this rule to the slider-crank mechanism, consider links 1, 2, and 3 whose ICs are I_{12}, I_{23}, and I_{13}, all of which lie on a straight line. Consider also links 1, 3, and 4 whose ICs are I_{13}, I_{34}, and I_{14}, all of which lie on a straight line. Extending the line through I_{12} and I_{23} and the line through I_{34} and I_{14} to their intersection locates I_{13}, which is common to the two groups of links that were considered.

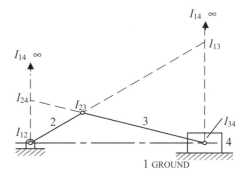

Similarly, if body groups 1, 2, 4 and 2, 3, 4 are considered, a line drawn through known ICs I_{12} and I_{14} to the intersection of a line drawn through known ICs I_{23} and I_{34} locates I_{24}.

The number of ICs, c, for a given mechanism is related to the number of links, n, by

$$c = \frac{n(n-1)}{2}$$

Relative Motion

The equations for the relative position, velocity, and acceleration may be written as

Translating Axis

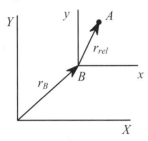

$$r_A = r_B + r_{rel}$$
$$v_A = v_B + \omega \times r_{rel}$$
$$a_A = a_B + \alpha \times r_{rel} + \omega \times (\omega \times r_{rel})$$

where, ω and α are, respectively, the angular velocity and angular acceleration of the relative position vector r_{rel}.

Rotating Axis

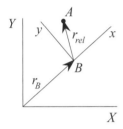

$$r_A = r_B + r_{rel}$$
$$v_A = v_B + \omega \times r_{rel} + v_{rel}$$
$$a_A = a_B + \alpha \times r_{rel} + \omega \times (\omega \times r_{rel}) + 2\omega \times v_{rel} + a_{rel}$$

where, ω and α are, respectively, the total angular velocity and acceleration of the relative position vector r_{rel}.

Rigid Body Rotation

For rigid body rotation θ
$$\omega = d\theta/dt,$$
$$\alpha = d\omega/dt, \text{ and}$$
$$\alpha d\theta = \omega d\omega$$

Kinetics

In general, Newton's second laws for a rigid body, with constant mass and mass moment of inertia, in plane motion may be written in vector form as

$$\Sigma F = ma_c$$
$$\Sigma M_c = I_c \alpha$$
$$\Sigma M_p = I_c \alpha + \rho_{pc} \times ma_c, \text{ where}$$

F are forces and a_c is the acceleration of the body's mass center both in the plane of motion, M_c are moments and α is the angular acceleration both about an axis normal to the plane of motion, I_c is the mass moment of inertia about the normal axis through the mass center, and ρ_{pc} is a vector from point p to point c.

Without loss of generality, the body may be assumed to be in the x-y plane. The scalar equations of motion may then be written as

$$\Sigma F_x = ma_{xc}$$
$$\Sigma F_y = ma_{yc}$$
$$\Sigma M_{zc} = I_{zc}\alpha, \text{ where}$$

zc indicates the z axis passing through the body's mass center, a_{xc} and a_{yc} are the acceleration of the body's mass center in the x and y directions, respectively, and α is the angular acceleration of the body about the z axis.

Rotation about an Arbitrary Fixed Axis

For rotation about some arbitrary fixed axis q

$$\Sigma M_q = I_q \alpha$$

If the applied moment acting about the fixed axis is constant then integrating with respect to time, from $t = 0$ yields

$$\alpha = M_q/I_q$$
$$\omega = \omega_0 + \alpha t$$
$$\theta = \theta_0 + \omega_0 t + \alpha t^2/2$$

where ω_0 and θ_0 are the values of angular velocity and angular displacement at time $t = 0$, respectively.

The change in kinetic energy is the work done in accelerating the rigid body from ω_0 to ω

$$I_q \omega^2/2 = I_q \omega_0^2/2 + \int_{\theta_0}^{\theta} M_q d\theta$$

Kinetic Energy

In general the kinetic energy for a rigid body may be written as

$$T = mv^2/2 + I_c \omega^2/2$$

For motion in the xy plane this reduces to

$$T = m\left(v_{cx}^2 + v_{cy}^2\right)/2 + I_c \omega_z^2/2$$

For motion about an instant center,
$$T = I_{IC}\omega^2/2$$

Free Vibration

The figure illustrates a single degree-of-freedom system.

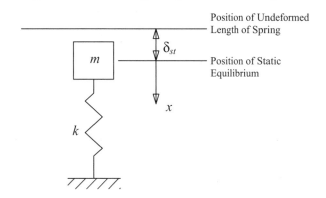

The equation of motion may be expressed as

$$m\ddot{x} = mg - k(x + \delta_{st})$$

where m is mass of the system, k is the spring constant of the system, δ_{st} is the static deflection of the system, and x is the displacement of the system from static equilibrium.

From statics it may be shown that

$$mg = k\delta_{st}$$

thus the equation of motion may be written as

$$m\ddot{x} + kx = 0, \text{ or }$$
$$\ddot{x} + (k/m)x = 0$$

The solution of this differential equation is

$$x(t) = C_1 \cos(\omega_n t) + C_2 \sin(\omega_n t)$$

where $\omega_n = \sqrt{k/m}$ is the undamped natural circular frequency and C_1 and C_2 are constants of integration whose values are determined from the initial conditions.

If the initial conditions are denoted as $x(0) = x_0$ and $\dot{x}(0) = v_0$, then

$$x(t) = x_0 \cos(\omega_n t) + (v_0/\omega_n)\sin(\omega_n t)$$

It may also be shown that the undamped natural frequency may be expressed in terms of the static deflection of the system as

$$\omega_n = \sqrt{g/\delta_{st}}$$

The undamped natural period of vibration may now be written as

$$\tau_n = 2\pi/\omega_n = 2\pi\sqrt{m/k} = 2\pi\sqrt{\delta_{st}/g}$$

Torsional Vibration

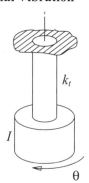

For torsional free vibrations it may be shown that the differential equation of motion is

$$\ddot{\theta} + (k_t/I)\theta = 0, \text{ where}$$

θ = the angular displacement of the system

k_t = the torsional stiffness of the massless rod

I = the mass moment of inertia of the end mass

The solution may now be written in terms of the initial conditions $\theta(0) = \theta_0$ and $\dot{\theta}(0) = \dot{\theta}_0$ as

$$\theta(t) = \theta_0 \cos(\omega_n t) + (\dot{\theta}_0/\omega_n)\sin(\omega_n t)$$

where the undamped natural circular frequency is given by

$$\omega_n = \sqrt{k_t/I}$$

The torsional stiffness of a solid round rod with associated polar moment-of-inertia J, length L, and shear modulus of elasticity G is given by

$$k_t = GJ/L$$

Thus the undamped circular natural frequency for a system with a solid round supporting rod may be written as

$$\omega_n = \sqrt{GJ/IL}$$

Similar to the linear vibration problem, the undamped natural period may be written as

$$\tau_n = 2\pi/\omega_n = 2\pi\sqrt{I/k_t} = 2\pi\sqrt{IL/GJ}$$

See second-order control-system models in the **MEASUREMENT AND CONTROLS** section for additional information on second-order systems.

Figure	Mass & Centroid	Mass Moment of Inertia	(Radius of Gyration)^2	Product of Inertia
	$M = \rho L A$ $x_c = L/2$ $y_c = 0$ $z_c = 0$ A = cross-sectional area of rod ρ = mass/vol.	$I_x = I_{x_c} = 0$ $I_{y_c} = I_{z_c} = ML^2/12$ $I_y = I_z = ML^2/3$	$r_x^2 = r_{x_c}^2 = 0$ $r_{y_c}^2 = r_{z_c}^2 = L^2/12$ $r_y^2 = r_z^2 = L^2/3$	$I_{x_c y_c}$, etc. = 0 I_{xy}, etc. = 0
	$M = 2\pi R \rho A$ $x_c = R$ = mean radius $y_c = R$ = mean radius $z_c = 0$ A = cross-sectional area of ring ρ = mass/vol.	$I_{x_c} = I_{y_c} = MR^2/2$ $I_{z_c} = MR^2$ $I_x = I_y = 3MR^2/2$ $I_z = 3MR^2$	$r_{x_c}^2 = r_{y_c}^2 = R^2/2$ $r_{z_c}^2 = R^2$ $r_x^2 = r_y^2 = 3R^2/2$ $r_z^2 = 3R^2$	$I_{x_c z_c}$, etc. = 0 $I_{z_c z_c} = MR^2$ $I_{xz} = I_{yz} = 0$
	$M = \pi R^2 \rho h$ $x_c = 0$ $y_c = h/2$ $z_c = 0$ ρ = mass/vol.	$I_{x_c} = I_{z_c} = M(3R^2 + h^2)/12$ $I_{y_c} = I_y = MR^2/2$ $I_x = I_z = M(3R^2 + 4h^2)/12$	$r_{x_c}^2 = r_{z_c}^2 = (3R^2 + h^2)/12$ $r_{y_c}^2 = r_y^2 = R^2/2$ $r_x^2 = r_z^2 = (3R^2 + 4h^2)/12$	$I_{x_c y_c}$, etc. = 0 I_{xy}, etc. = 0
	$M = \pi\left(R_1^2 - R_2^2\right)\rho h$ $x_c = 0$ $y_c = h/2$ $z_c = 0$ ρ = mass/vol.	$I_{x_c} = I_{z_c}$ $\quad = M\left(3R_1^2 + 3R_2^2 + h^2\right)/12$ $I_{y_c} = I_y = M\left(R_1^2 + R_2^2\right)/2$ $I_x = I_z$ $\quad = M\left(3R_1^2 + 3R_2^2 + 4h^2\right)/12$	$r_{x_c}^2 = r_{z_c}^2 = \left(3R_1^2 + 3R_2^2 + h^2\right)/12$ $r_{y_c}^2 = r_y^2 = \left(R_1^2 + R_2^2\right)/2$ $r_x^2 = r_z^2$ $\quad = \left(3R_1^2 + 3R_2^2 + 4h^2\right)/12$	$I_{x_c y_c}$, etc. = 0 I_{xy}, etc. = 0
	$M = \dfrac{4}{3}\pi R^3 \rho$ $x_c = 0$ $y_c = 0$ $z_c = 0$ ρ = mass/vol.	$I_{x_c} = I_x = 2MR^2/5$ $I_{y_c} = I_y = 2MR^2/5$ $I_{z_c} = I_z = 2MR^2/5$	$r_{x_c}^2 = r_x^2 = 2R^2/5$ $r_{y_c}^2 = r_y^2 = 2R^2/5$ $r_{z_c}^2 = r_z^2 = 2R^2/5$	$I_{x_c y_c}$, etc. = 0

Housner, George W. & Donald E. Hudson, *Applied Mechanics Dynamics*, D. Van Nostrand Company, Inc., Princeton, NJ, 1959. Table reprinted by permission of G.W. Housner & D.E. Hudson.

FLUID MECHANICS

DENSITY, SPECIFIC VOLUME, SPECIFIC WEIGHT, AND SPECIFIC GRAVITY

The definitions of density, specific volume, specific weight, and specific gravity follow:

$$\rho = \lim_{\Delta V \to 0} \Delta m / \Delta V$$

$$\gamma = \lim_{\Delta V \to 0} \Delta W / \Delta V$$

$$\gamma = \lim_{\Delta V \to 0} g \cdot \Delta m / \Delta V = \rho g$$

also $SG = \gamma / \gamma_w = \rho / \rho_w$, where

ρ = *density* (also *mass density*),

Δm = mass of infinitesimal volume,

ΔV = volume of infinitesimal object considered,

γ = *specific weight,*

    = ρg,

ΔW = weight of an infinitesimal volume,

SG = *specific gravity,*

ρ_w = mass density of water at standard conditions
    = 1,000 kg/m^3 (62.43 lbm/ft^3), and

γ_ω = specific weight of water at standard conditions,
    = 9,810 N/m^3 (62.4 lbf/ft^3), and
    = 9,810 kg/(m$^2 \cdot$s^2).

STRESS, PRESSURE, AND VISCOSITY

Stress is defined as

$$\tau(1) = \lim_{\Delta A \to 0} \Delta F / \Delta A, \text{ where}$$

$\tau(1)$ = surface stress vector at point 1,

ΔF = force acting on infinitesimal area ΔA, and

ΔA = infinitesimal area at point 1.

$$\tau_n = -P$$

$$\tau_t = \mu(d\mathrm{v}/dy) \text{ (one-dimensional; i.e., } y\text{), where}$$

τ_n and τ_t = the normal and tangential stress components at point 1,

P = the pressure at point 1,

μ = *absolute dynamic viscosity* of the fluid
    N•s/m^2 [lbm/(ft-sec)],

$d\mathrm{v}$ = differential velocity,

dy = differential distance, normal to boundary.

v = velocity at boundary condition, and

y = normal distance, measured from boundary.

$$\mathrm{v} = \mu/\rho, \text{ where}$$

υ = *kinematic viscosity*; m^2/s (ft^2/sec).

For a thin Newtonian fluid film and a linear velocity profile,

$$\mathrm{v}(y) = \mathrm{v}y/\delta; \ d\mathrm{v}/dy = \mathrm{v}/\delta, \text{ where}$$

v = velocity of plate on film and

δ = thickness of fluid film.

For a power law (non-Newtonian) fluid

$$\tau_t = K \, (d\mathrm{v}/dy)^n, \text{ where}$$

K = consistency index, and

n = power law index.

    $n < 1 \equiv$ pseudo plastic

    $n > 1 \equiv$ dilatant

SURFACE TENSION AND CAPILLARITY

Surface tension σ is the force per unit contact length

$$\sigma = F/L, \text{ where}$$

σ = surface tension, force/length,

F = surface force at the interface, and

L = length of interface.

The *capillary rise h* is approximated by

$$h = 4\sigma \cos \beta /(\gamma d), \text{ where}$$

h = the height of the liquid in the vertical tube,

σ = the surface tension,

β = the angle made by the liquid with the wetted tube wall,

γ = specific weight of the liquid, and

d = the diameter of the capillary tube.

THE PRESSURE FIELD IN A STATIC LIQUID

♦

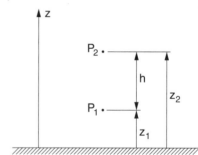

The difference in pressure between two different points is

$$P_2 - P_1 = -\gamma \, (z_2 - z_1) = -\gamma h = -\rho g h$$

For a simple manometer,

$$P_o = P_2 + \gamma_2 z_2 - \gamma_1 z_1$$

Absolute pressure = atmospheric pressure + gage pressure reading

Absolute pressure = atmospheric pressure − vacuum gage pressure reading

♦ Bober, W. & R.A. Kenyon, *Fluid Mechanics*, Wiley, New York, 1980. Diagrams reprinted by permission of William Bober & Richard A. Kenyon.

FORCES ON SUBMERGED SURFACES AND THE CENTER OF PRESSURE

◆

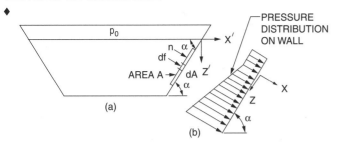

Forces on a submerged plane wall. (a) Submerged plane surface. (b) Pressure distribution.

The pressure on a point at a distance Z' below the surface is
$$p = p_o + \gamma Z', \text{ for } Z' \geq 0$$

If the tank were open to the atmosphere, the effects of p_o could be ignored.

The coordinates of the *center of pressure* (CP) are
$$y^* = (\gamma I_{y_c z_c} \sin\alpha)/(p_c A) \text{ and}$$

$$z^* = (\gamma I_{y_c} \sin\alpha)/(p_c A), \text{ where}$$

$y^* =$ the y-distance from the centroid (C) of area (A) to the center of pressure,

$z^* =$ the z-distance from the centroid (C) of area (A) to the center of pressure,

I_{y_c} and $I_{y_c z_c} =$ the moment and product of inertia of the area,

$p_c =$ the pressure at the centroid of area (A), and

$Z_c =$ the slant distance from the water surface to the centroid (C) of area (A).

◆

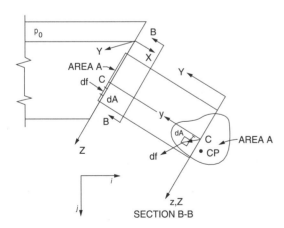

SECTION B-B

If the free surface is open to the atmosphere, then $p_o = 0$ and $p_c = \gamma Z_c \sin\alpha$.
$$y^* = I_{y_c z_c}/(AZ_c) \text{ and } z^* = I_{y_c}/(AZ_c)$$

The force on a rectangular plate can be computed as
$$\boldsymbol{F} = [p_1 A_v + (p_2 - p_1) A_v /2]\boldsymbol{i} + V_f \gamma_f \boldsymbol{j}, \text{ where}$$

$\boldsymbol{F} =$ force on the plate,

$p_1 =$ pressure at the top edge of the plate area,

$p_2 =$ pressure at the bottom edge of the plate area,

$A_v =$ vertical projection of the plate area,

$V_f =$ volume of column of fluid above plate, and

$\gamma_f =$ specific weight of the fluid.

ARCHIMEDES PRINCIPLE AND BUOYANCY

1. The buoyant force exerted on a submerged or floating body is equal to the weight of the fluid displaced by the body.

2. A floating body displaces a weight of fluid equal to its own weight; i.e., a floating body is in equilibrium.

The *center of buoyancy* is located at the centroid of the displaced fluid volume.

In the case of a body lying at the *interface of two immiscible fluids*, the buoyant force equals the sum of the weights of the fluids displaced by the body.

ONE-DIMENSIONAL FLOWS
The Continuity Equation
So long as the flow Q is continuous, the *continuity equation*, as applied to one-dimensional flows, states that the flow passing two points (1 and 2) in a stream is equal at each point, $A_1 v_1 = A_2 v_2$.
$$Q = A v$$

$$\dot{m} = \rho Q = \rho A v, \text{ where}$$

$Q =$ volumetric flow rate,

$\dot{m} =$ mass flow rate,

$A =$ cross section of area of flow,

$v =$ average flow velocity, and

$\rho =$ the fluid density.

For steady, one-dimensional flow, $\dot{m}$ is a constant. If, in addition, the density is constant, then Q is constant.

◆ Bober, W. & R.A. Kenyon, *Fluid Mechanics*, Wiley, New York, 1980. Diagrams reprinted by permission of William Bober & Richard A. Kenyon.

The Field Equation is derived when the energy equation is applied to one-dimensional flows. Assuming no friction losses and that no pump or turbine exists between sections 1 and 2 in the system,

$$\frac{P_2}{\gamma} + \frac{v_2^2}{2g} + z_2 = \frac{P_1}{\gamma} + \frac{v_1^2}{2g} + z_1 \ \text{ or}$$

$$\frac{P_2}{\rho} + \frac{v_2^2}{2} + z_2 g = \frac{P_1}{\rho} + \frac{v_1^2}{2} + z_1 g, \ \text{ where}$$

$P_1, P_2 =$ pressure at sections 1 and 2,

$v_1, v_2 =$ average velocity of the fluid at the sections,

$z_1, z_2 =$ the vertical distance from a datum to the sections (the potential energy),

$\gamma \quad\ =$ the specific weight of the fluid (ρg), and

$g \quad\ =$ the acceleration of gravity.

FLUID FLOW

The velocity distribution for *laminar flow* in circular tubes or between planes is

$$v(r) = v_{max}\left[1 - \left(\frac{r}{R}\right)^2\right], \text{where}$$

$r \quad =$ the distance (m) from the centerline,

$R \quad =$ the radius (m) of the tube or half the distance between the parallel planes,

$v \quad =$ the local velocity (m/s) at r, and

$v_{max} =$ the velocity (m/s) at the centerline of the duct.

$v_{max} =$ 1.18v, for fully turbulent flow

$v_{max} =$ 2v, for circular tubes in laminar flow and

$v_{max} =$ 1.5v, for parallel planes in laminar flow, where

$\overline{v} \quad =$ the average velocity (m/s) in the duct.

The shear stress distribution is

$$\frac{\tau}{\tau_w} = \frac{r}{R}, \ \text{ where}$$

τ and τ_w are the shear stresses at radii r and R respectively.

The *drag force* F_D on **objects immersed in a large body of flowing fluid or objects moving through a stagnant fluid** is

$$F_D = \frac{C_D \rho v^2 A}{2}, \ \text{ where}$$

$C_D \quad =$ the *drag coefficient*,

$v \quad =$ the velocity (m/s) of the flowing fluid or moving object, and

$A \quad =$ the *projected area* (m^2) of blunt objects such as spheres, ellipsoids, disks, and plates, cylinders, ellipses, and air foils with axes perpendicular to the flow.

For **flat plates placed parallel with the flow**

$C_D = 1.33/\text{Re}^{0.5} \ (10^4 < \text{Re} < 5 \times 10^5)$

$C_D = 0.031/\text{Re}^{1/7} \ (10^6 < \text{Re} < 10^9)$

The characteristic length in the Reynolds Number (Re) is the length of the plate parallel with the flow. For blunt objects, the characteristic length is the largest linear dimension (diameter of cylinder, sphere, disk, etc.) which is perpendicular to the flow.

AERODYNAMICS

Airfoil Theory

The lift force on an airfoil is given by

$$F_L = \frac{C_L \rho v^2 A_P}{2}$$

$C_L =$ the lift coefficient

$v \ =$ velocity (m/s) of the undisturbed fluid and

$A_P =$ the projected area of the airfoil as seen from above (plan area). This same area is used in defining the drag coefficient for an airfoil.

The lift coefficient can be approximated by the equation

$C_L \ = \ 2\pi k_1 \sin(\alpha + \beta)$ which is valid for small values of α and β.

$k_1 \ = \ $ a constant of proportionality

$\alpha \ = \ $ angle of attack (angle between chord of airfoil and direction of flow)

$\beta = $ negative of angle of attack for zero lift.

The drag coefficient may be approximated by

$$C_D = C_{D\infty} + \frac{C_L^2}{\pi AR}$$

$C_{D\infty} =$ infinite span drag coefficient

$$AR = \frac{b^2}{A_p} = \frac{A_p}{c^2}$$

The aerodynamic moment is given by

$$M = \frac{C_M \rho v^2 A_p c}{2}$$

where the moment is taken about the front quarter point of the airfoil.

$C_M = \quad$ moment coefficient

$A_p \ = \quad$ plan area

$c \ = \quad$ chord length

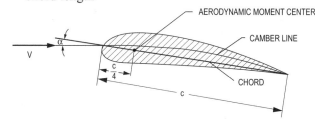

Reynolds Number

$$Re = vD\rho/\mu = vD/\upsilon$$

$$Re' = \frac{v^{(2-n)}D^n\rho}{K\left(\frac{3n+1}{4n}\right)^n 8^{(n-1)}}, \text{ where}$$

ρ = the mass density,

D = the diameter of the pipe, dimension of the fluid streamline, or characteristic length.

μ = the dynamic viscosity,

υ = the kinematic viscosity,

Re = the Reynolds number (Newtonian fluid),

Re' = the Reynolds number (Power law fluid), and

K and n are defined in the Stress, Pressure, and Viscosity section.

The critical Reynolds number $(Re)_c$ is defined to be the minimum Reynolds number at which a flow will turn turbulent.

Flow through a pipe is generally characterized as laminar for $Re < 2{,}100$ and fully turbulent for $Re > 10{,}000$, and transitional flow for $2{,}100 < Re < 10{,}000$.

Hydraulic Gradient (Grade Line)

The hydraulic gradient (grade line) is defined as an imaginary line above a pipe so that the vertical distance from the pipe axis to the line represents the *pressure head* at that point. If a row of piezometers were placed at intervals along the pipe, the grade line would join the water levels in the piezometer water columns.

Energy Line (Bernoulli Equation)

The Bernoulli equation states that the sum of the pressure, velocity, and elevation heads is constant. The energy line is this sum or the "total head line" above a horizontal datum. The difference between the hydraulic grade line and the energy line is the $v^2/2g$ term.

STEADY, INCOMPRESSIBLE FLOW IN CONDUITS AND PIPES

The energy equation for incompressible flow is

$$\frac{p_1}{\gamma} + z_1 + \frac{v_1^2}{2g} = \frac{p_2}{\gamma} + z_2 + \frac{v_2^2}{2g} + h_f \text{ or}$$

$$\frac{p_1}{\rho g} + z_1 + \frac{v_1^2}{2g} = \frac{p_2}{\rho g} + z_2 + \frac{v_2^2}{2g} + h_f$$

h_f = the head loss, considered a friction effect, and all remaining terms are defined above.

If the cross-sectional area and the elevation of the pipe are the same at both sections (1 and 2), then $z_1 = z_2$ and $v_1 = v_2$.

The pressure drop $p_1 - p_2$ is given by the following:

$$p_1 - p_2 = \gamma h_f = \rho g h_f$$

COMPRESSIBLE FLOW
See **MECHANICAL ENGINEERING** section.

The *Darcy-Weisbach equation* is

$$h_f = f\frac{L}{D}\frac{v^2}{2g}, \text{ where}$$

f = $f(Re, e/D)$, the Moody or Darcy friction factor,

D = diameter of the pipe,

L = length over which the pressure drop occurs,

e = roughness factor for the pipe, and all other symbols are defined as before.

An alternative formulation employed by chemical engineers is

$$h_f = \left(4f_{Fanning}\right)\frac{Lv^2}{D2g} = \frac{2f_{Fanning}Lv^2}{Dg}$$

Fanning friction factor, $f_{Fanning} = \frac{f}{4}$

A chart that gives f versus Re for various values of e/D, known as a *Moody* or *Stanton diagram*, is available at the end of this section.

Friction Factor for Laminar Flow

The equation for Q in terms of the pressure drop Δp_f is the Hagen-Poiseuille equation. This relation is valid only for flow in the laminar region.

$$Q = \frac{\pi R^4 \Delta p_f}{8\mu L} = \frac{\pi D^4 \Delta p_f}{128\mu L}$$

Flow in Noncircular Conduits

Analysis of flow in conduits having a noncircular cross section uses the *hydraulic diameter* D_H, or the *hydraulic radius* R_H, as follows

$$R_H = \frac{\text{cross-sectional area}}{\text{wetted perimeter}} = \frac{D_H}{4}$$

Minor Losses in Pipe Fittings, Contractions, and Expansions

Head losses also occur as the fluid flows through pipe fittings (i.e., elbows, valves, couplings, etc.) and sudden pipe contractions and expansions.

$$\frac{p_1}{\gamma} + z_1 + \frac{v_1^2}{2g} = \frac{p_2}{\gamma} + z_2 + \frac{v_2^2}{2g} + h_f + h_{f,\,fitting}$$

$$\frac{p_1}{\rho g} + z_1 + \frac{v_1^2}{2g} = \frac{p_2}{\rho g} + z_2 + \frac{v_2^2}{2g} + h_f + h_{f,\,fitting}, \text{ where}$$

$$h_{f,\,fitting} = C\frac{v^2}{2g}, \text{ and } \frac{v^2}{2g} = 1 \text{ velocity head}$$

Specific fittings have characteristic values of C, which will be provided in the problem statement. A generally accepted *nominal value* for head loss in *well-streamlined gradual contractions* is

$$h_{f,\,fitting} = 0.04 \, v^2/2g$$

The *head loss* at either an *entrance* or *exit* of a pipe from or to a reservoir is also given by the $h_{f, \text{fitting}}$ equation. Values for C for various cases are shown as follows.

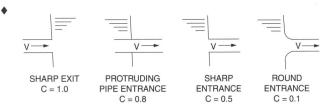

| SHARP EXIT $C = 1.0$ | PROTRUDING PIPE ENTRANCE $C = 0.8$ | SHARP ENTRANCE $C = 0.5$ | ROUND ENTRANCE $C = 0.1$ |

PUMP POWER EQUATION

$$\dot{W} = Q\gamma h/\eta = Q\rho g h/\eta, \text{ where}$$

Q = volumetric flow (m^3/s or cfs),

h = head (m or ft) the fluid has to be lifted,

η = efficiency, and

$\dot{W}$ = power (watts or ft-lbf/sec).

For additonal information on pumps refer to the **MECHANICAL ENGINEERING** section of this handbook.

COMPRESSIBLE FLOW

See the **MECHANICAL ENGINEERING** section for compressible flow and machinery associated with compressible flow (compressors, turbines, fans).

THE IMPULSE-MOMENTUM PRINCIPLE

The resultant force in a given direction acting on the fluid equals the rate of change of momentum of the fluid.

$$\Sigma F = Q_2\rho_2 v_2 - Q_1\rho_1 v_1, \text{ where}$$

ΣF = the resultant of all external forces acting on the control volume,

$Q_1\rho_1 v_1$ = the rate of momentum of the fluid flow entering the control volume in the same direction of the force, and

$Q_2\rho_2 v_2$ = the rate of momentum of the fluid flow leaving the control volume in the same direction of the force.

Pipe Bends, Enlargements, and Contractions

The force exerted by a flowing fluid on a bend, enlargement, or contraction in a pipe line may be computed using the impulse-momentum principle.

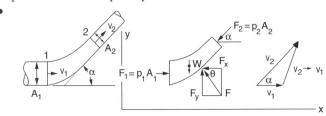

$$p_1 A_1 - p_2 A_2 \cos\alpha - F_x = Q\rho\,(v_2\cos\alpha - v_1)$$

$$F_y - W - p_2 A_2 \sin\alpha = Q\rho\,(v_2\sin\alpha - 0), \text{ where}$$

F = the force exerted by the bend on the fluid (the force exerted by the fluid on the bend is equal in magnitude and opposite in sign), F_x and F_y are the x-component and y-component of the force,

p = the internal pressure in the pipe line,

A = the cross-sectional area of the pipe line,

W = the weight of the fluid,

v = the velocity of the fluid flow,

α = the angle the pipe bend makes with the horizontal,

ρ = the density of the fluid, and

Q = the quantity of fluid flow.

Jet Propulsion

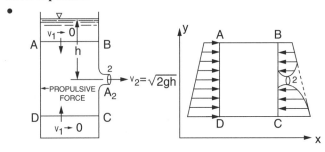

$$F = Q\rho(v_2 - 0)$$

$$F = 2\gamma h A_2, \text{ where}$$

F = the propulsive force,

γ = the specific weight of the fluid,

h = the height of the fluid above the outlet,

A_2 = the area of the nozzle tip,

Q = $A_2\sqrt{2gh}$, and

v_2 = $\sqrt{2gh}$.

Deflectors and Blades

Fixed Blade

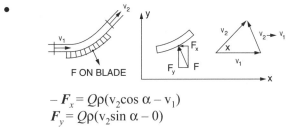

$$-F_x = Q\rho(v_2\cos\alpha - v_1)$$

$$F_y = Q\rho(v_2\sin\alpha - 0)$$

Moving Blade

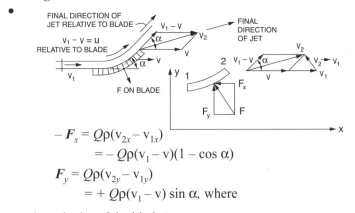

$$-F_x = Q\rho(v_{2x} - v_{1x})$$

$$= -Q\rho(v_1 - v)(1 - \cos\alpha)$$

$$F_y = Q\rho(v_{2y} - v_{1y})$$

$$= +Q\rho(v_1 - v)\sin\alpha, \text{ where}$$

v = the velocity of the blade.

♦ Bober, W. & R.A. Kenyon, *Fluid Mechanics*, Wiley, New York, 1980. Diagram reprinted by permission of William Bober & Richard A. Kenyon.

• Vennard, J.K., *Elementary Fluid Mechanics*, 6th ed., J.K. Vennard, 1954.

Impulse Turbine

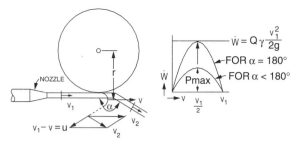

$$\dot{W} = Q\rho(v_1 - v)(1 - \cos\alpha)v, \text{ where}$$

$\dot{W}$ = power of the turbine.

$$\dot{W}_{max} = Q\rho(v_1^2/4)(1 - \cos\alpha)$$

When $\alpha = 180°$,

$$\dot{W}_{max} = (Q\rho v_1^2)/2 = (Q\gamma v_1^2)/2g$$

MULTIPATH PIPELINE PROBLEMS

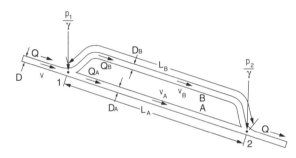

The same head loss occurs in each branch as in the combination of the two. The following equations may be solved simultaneously for v_A and v_B:

$$h_L = f_A \frac{L_A}{D_A}\frac{v_A^2}{2g} = f_B \frac{L_B}{D_B}\frac{v_B^2}{2g}$$

$$(\pi D^2/4)v = (\pi D_A^2/4)v_A + (\pi D_B^2/4)v_B$$

The flow Q can be divided into Q_A and Q_B when the pipe characteristics are known.

OPEN-CHANNEL FLOW AND/OR PIPE FLOW
Manning's Equation

$v = (k/n)R^{2/3}S^{1/2}$, where

k = 1 for SI units,

k = 1.486 for USCS units,

v = velocity (m/s, ft/sec),

n = roughness coefficient,

R = hydraulic radius (m, ft), and

S = slope of energy grade line (m/m, ft/ft).

Also see Hydraulic Elements Graph for Circular Sewers in the **CIVIL ENGINEERING** section.

Hazen-Williams Equation

$v = k_1 CR^{0.63}S^{0.54}$, where

C = roughness coefficient,

k_1 = 0.849 for SI units, and

k_1 = 1.318 for USCS units.

Other terms defined as above.

WEIR FORMULAS
See the **CIVIL ENGINEERING** section.

FLOW THROUGH A PACKED BED

A porous, fixed bed of solid particles can be characterized by

L = length of particle bed (m)

D_p = average particle diameter (m)

Φ_s = sphericity of particles, dimensionless (0–1)

ε = porosity or void fraction of the particle bed, dimensionless (0–1)

The Ergun equation can be used to estimate pressure loss through a packed bed under laminar and turbulent flow conditions.

$$\frac{\Delta p}{L} = \frac{150v_o\mu(1-\varepsilon)^2}{\Phi_s^2 D_p^2 \varepsilon^3} + \frac{1.75\rho v_o^2(1-\varepsilon)}{\Phi_s D_p \varepsilon^3}$$

Δp = pressure loss across packed bed (Pa)

v_o = superficial (flow through empty vessel) fluid velocity $\left(\dfrac{m}{s}\right)$

ρ = fluid density $\left(\dfrac{kg}{m^3}\right)$

μ = fluid viscosity $\left(\dfrac{kg}{m \cdot s}\right)$

FLUID MEASUREMENTS
The Pitot Tube – From the stagnation pressure equation for an *incompressible fluid,*

$$v = \sqrt{(2/\rho)(p_0 - p_s)} = \sqrt{2g(p_0 - p_s)/\gamma}, \text{ where}$$

v = the velocity of the fluid,

p_0 = the stagnation pressure, and

p_s = the static pressure of the fluid at the elevation where the measurement is taken.

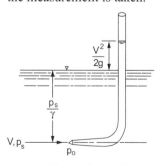

For a *compressible fluid,* use the above incompressible fluid equation if the Mach number ≤ 0.3.

- Vennard, J.K., *Elementary Fluid Mechanics*, 6th ed., J.K. Vennard, 1954.

MANOMETERS

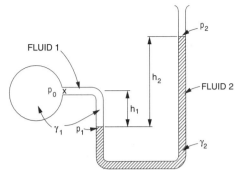

For a simple manometer,

$$p_0 = p_2 + \gamma_2 h_2 - \gamma_1 h_1 = p_2 + g\left(\rho_2 h_2 - \rho_1 h_1\right)$$

If $h_1 = h_2 = h$

$$p_0 = p_2 + (\gamma_2 - \gamma_1)h = p_2 + (\rho_2 - \rho_1)gh$$

Note that the difference between the two densities is used.

Another device that works on the same principle as the manometer is the simple barometer.

$$p_{atm} = p_A = p_v + \gamma h = p_B + \gamma h = p_B + \rho g h$$

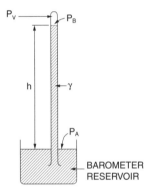

p_v = vapor pressure of the barometer fluid

Venturi Meters

$$Q = \frac{C_v A_2}{\sqrt{1 - \left(A_2/A_1\right)^2}} \sqrt{2g\left(\frac{p_1}{\gamma} + z_1 - \frac{p_2}{\gamma} - z_2\right)}, \text{ where}$$

C_v = the coefficient of velocity, and
γ = ρg.

The above equation is for *incompressible fluids*.

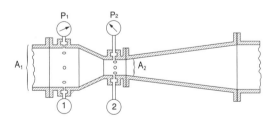

Orifices The cross-sectional area at the vena contracta A_2 is characterized by a *coefficient of contraction* C_c and given by $C_c A$.

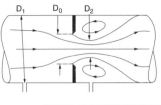

$$Q = CA_0 \sqrt{2g\left(\frac{p_1}{\gamma} + z_1 - \frac{p_2}{\gamma} - z_2\right)}$$

where C, the *coefficient of the meter (orifice coefficient)*, is given by

$$C = \frac{C_v C_c}{\sqrt{1 - C_c^2\left(A_0/A_1\right)^2}}$$

ORIFICES AND THEIR NOMINAL COEFFICIENTS			
SHARP EDGED	ROUNDED	SHORT TUBE	BORDA
C 0.61	0.98	0.80	0.51
C_c 0.62	1.00	1.00	0.52
C_v 0.98	0.98	0.80	0.98

For incompressible flow through a horizontal orifice meter installation

$$Q = CA_0 \sqrt{2\rho\left(p_1 - p_2\right)}$$

Submerged Orifice operating under steady-flow conditions:

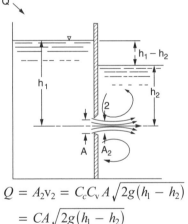

$$Q = A_2 v_2 = C_c C_v A \sqrt{2g\left(h_1 - h_2\right)}$$
$$= CA \sqrt{2g\left(h_1 - h_2\right)}$$

in which the product of C_c and C_v is defined as the *coefficient of discharge* of the orifice.

♦ Bober, W. & R.A. Kenyon, *Fluid Mechanics*, Wiley, New York, 1980. Diagram reprinted by permission of William Bober & Richard A. Kenyon.
• Vennard, J.K., *Elementary Fluid Mechanics*, 6th ed., J.K. Vennard, 1954.

Orifice Discharging Freely into Atmosphere

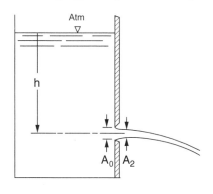

$$Q = CA_0 \sqrt{2gh}$$

in which h is measured from the liquid surface to the centroid of the orifice opening.

DIMENSIONAL HOMOGENEITY AND DIMENSIONAL ANALYSIS

Equations that are in a form that do not depend on the fundamental units of measurement are called *dimensionally homogeneous* equations. A special form of the dimensionally homogeneous equation is one that involves only *dimensionless groups* of terms.

Buckingham's Theorem: The *number of independent dimensionless groups* that may be employed to describe a phenomenon known to involve n variables is equal to the number $(n - \bar{r})$, where $\bar{r}$ is the number of basic dimensions (i.e., M, L, T) needed to express the variables dimensionally.

• Vennard, J.K., *Elementary Fluid Mechanics*, 6th ed., J.K. Vennard, 1954.

SIMILITUDE

In order to use a model to simulate the conditions of the prototype, the model must be *geometrically*, *kinematically*, and *dynamically similar* to the prototype system.

To obtain dynamic similarity between two flow pictures, all independent force ratios that can be written must be the same in both the model and the prototype. Thus, dynamic similarity between two flow pictures (when all possible forces are acting) is expressed in the five simultaneous equations below.

$$\left[\frac{F_I}{F_P}\right]_p = \left[\frac{F_I}{F_P}\right]_m = \left[\frac{\rho v^2}{p}\right]_p = \left[\frac{\rho v^2}{p}\right]_m$$

$$\left[\frac{F_I}{F_V}\right]_p = \left[\frac{F_I}{F_V}\right]_m = \left[\frac{vl\rho}{\mu}\right]_p = \left[\frac{vl\rho}{\mu}\right]_m = [Re]_p = [Re]_m$$

$$\left[\frac{F_I}{F_G}\right]_p = \left[\frac{F_I}{F_G}\right]_m = \left[\frac{v^2}{lg}\right]_p = \left[\frac{v^2}{lg}\right]_m = [Fr]_p = [Fr]_m$$

$$\left[\frac{F_I}{F_E}\right]_p = \left[\frac{F_I}{F_E}\right]_m = \left[\frac{\rho v^2}{E_v}\right]_p = \left[\frac{\rho v^2}{E_v}\right]_m = [Ca]_p = [Ca]_m$$

$$\left[\frac{F_I}{F_T}\right]_p = \left[\frac{F_I}{F_T}\right]_m = \left[\frac{\rho l v^2}{\sigma}\right]_p = \left[\frac{\rho l v^2}{\sigma}\right]_m = [We]_p = [We]_m$$

where
the subscripts p and m stand for *prototype* and *model* respectively, and

F_I = inertia force,

F_P = pressure force,

F_V = viscous force,

F_G = gravity force,

F_E = elastic force,

F_T = surface tension force,

Re = Reynolds number,

We = Weber number,

Ca = Cauchy number,

Fr = Froude number,

l = characteristic length,

v = velocity,

ρ = density,

σ = surface tension,

E_v = bulk modulus,

μ = dynamic viscosity,

p = pressure, and

g = acceleration of gravity.

PROPERTIES OF WATER[f] (SI METRIC UNITS)

Temperature °C	Specific Weight[a], γ, kN/m^3	Density[a], ρ, kg/m^3	Absolute Dynamic Viscosity[a], μ Pa·s	Kinematic Viscosity[a], υ m^2/s	Vapor Pressure[e], p_v, kPa
0	9.805	999.8	0.001781	0.000001785	0.61
5	9.807	1000.0	0.001518	0.000001518	0.87
10	9.804	999.7	0.001307	0.000001306	1.23
15	9.798	999.1	0.001139	0.000001139	1.70
20	9.789	998.2	0.001002	0.000001003	2.34
25	9.777	997.0	0.000890	0.000000893	3.17
30	9.764	995.7	0.000798	0.000000800	4.24
40	9.730	992.2	0.000653	0.000000658	7.38
50	9.689	988.0	0.000547	0.000000553	12.33
60	9.642	983.2	0.000466	0.000000474	19.92
70	9.589	977.8	0.000404	0.000000413	31.16
80	9.530	971.8	0.000354	0.000000364	47.34
90	9.466	965.3	0.000315	0.000000326	70.10
100	9.399	958.4	0.000282	0.000000294	101.33

PROPERTIES OF WATER (ENGLISH UNITS)

Temperature (°F)	Specific Weight γ (lb/ft^3)	Mass Density ρ (lb·sec^2/ft^4)	Absolute Dynamic Viscosity μ ($\times 10^{-5}$ lb·sec/ft^2)	Kinematic Viscosity υ ($\times 10^{-5}$ ft^2/sec)	Vapor Pressure p_v (psi)
32	62.42	1.940	3.746	1.931	0.09
40	62.43	1.940	3.229	1.664	0.12
50	62.41	1.940	2.735	1.410	0.18
60	62.37	1.938	2.359	1.217	0.26
70	62.30	1.936	2.050	1.059	0.36
80	62.22	1.934	1.799	0.930	0.51
90	62.11	1.931	1.595	0.826	0.70
100	62.00	1.927	1.424	0.739	0.95
110	61.86	1.923	1.284	0.667	1.24
120	61.71	1.918	1.168	0.609	1.69
130	61.55	1.913	1.069	0.558	2.22
140	61.38	1.908	0.981	0.514	2.89
150	61.20	1.902	0.905	0.476	3.72
160	61.00	1.896	0.838	0.442	4.74
170	60.80	1.890	0.780	0.413	5.99
180	60.58	1.883	0.726	0.385	7.51
190	60.36	1.876	0.678	0.362	9.34
200	60.12	1.868	0.637	0.341	11.52
212	59.83	1.860	0.593	0.319	14.70

♦ [a]From "Hydraulic Models,"*ASCE Manual of Engineering Practice*, No. 25, ASCE, 1942.
[e]From J.H. Keenan and F.G. Keyes, *Thermodynamic Properties of Steam*, John Wiley & Sons, 1936.
[f]Compiled from many sources including those indicated: *Handbook of Chemistry and Physics*, 54th ed.,
The CRC Press, 1973, and *Handbook of Tables for Applied Engineering Science*, The Chemical Rubber Co., 1970.
Vennard, J.K. and Robert L. Street, *Elementary Fluid Mechanics*, 6th ed., Wiley, New York, 1982.

MOODY (STANTON) DIAGRAM

Material	e (ft)	e (mm)
Riveted steel	10.003–0.03	0.9–9.0
Concrete	0.001–0.01	0.3–3.0
Cast iron	0.00085	0.25
Galvanized iron	0.0005	0.15
Commercial steel or wrought iron	0.00015	0.046
Drawn tubing	0.000005	0.0015

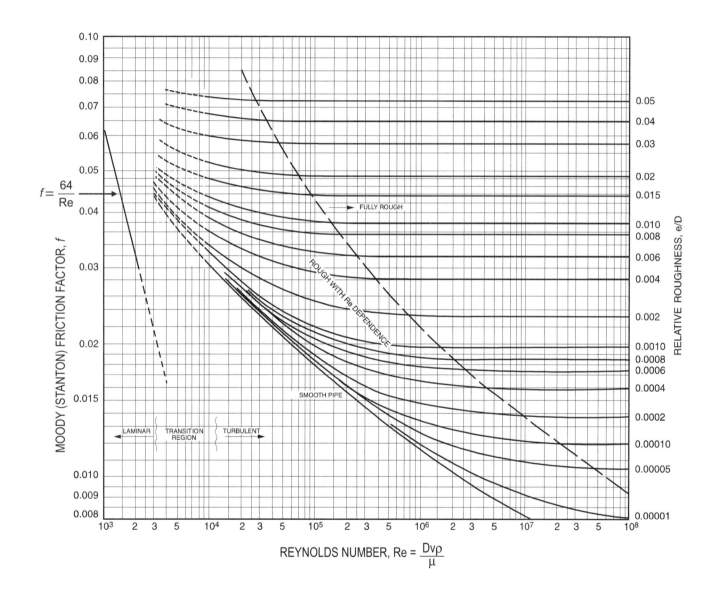

From ASHRAE (The American Society of Heating, Refrigerating and Air-Conditioning Engineers, Inc.)

DRAG COEFFICIENTS FOR SPHERES, DISKS, AND CYLINDERS

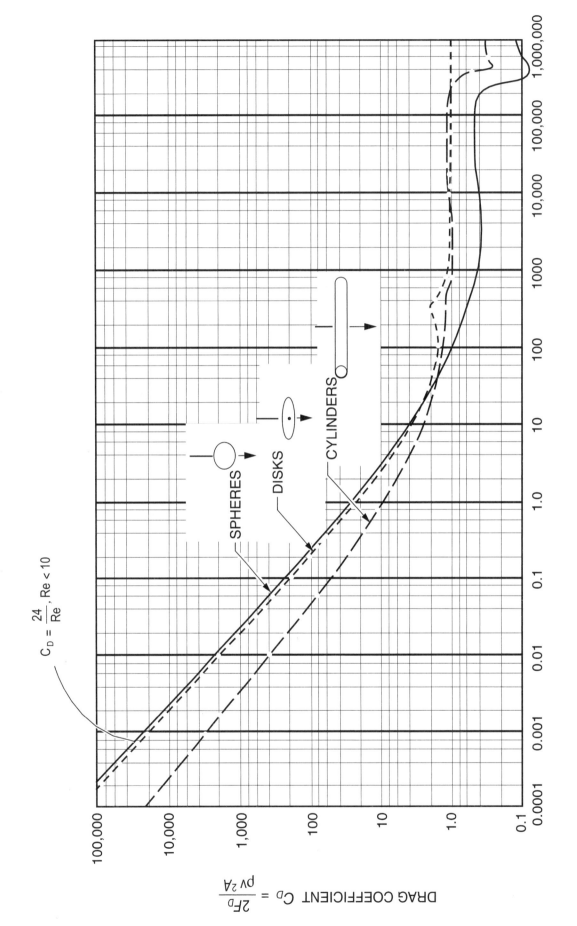

Note: Intermediate divisions are 2, 4, 6, and 8.

THERMODYNAMICS

PROPERTIES OF SINGLE-COMPONENT SYSTEMS

Nomenclature

1. Intensive properties are independent of mass.
2. Extensive properties are proportional to mass.
3. Specific properties are lowercase (extensive/mass).

State Functions (properties)

Absolute Pressure, P	(lbf/in^2 or Pa)
Absolute Temperature, T	(°R or K)
Volume, V	(ft^3 or m^3)
Specific Volume, $v = V/m$	(ft^3/lbm or m^3/kg)
Internal Energy, U	(Btu or kJ)
Specific Internal Energy, $u = U/m$	(usually in Btu/lbm or kJ/kg)
Enthalpy, H	(Btu or KJ)
Specific Enthalpy, $h = u + Pv = H/m$	(usually in Btu/lbm or kJ/kg)
Entropy, S	(Btu/°R or kJ/K)
Specific Entropy, $s = S/m$	[Btu/(lbm-°R) or kJ/(kg•K)]
Gibbs Free Energy, $g = h - Ts$	(usually in Btu/lbm or kJ/kg)
Helmholz Free Energy, $a = u - Ts$	(usually in Btu/lbm or kJ/kg)

Heat Capacity at Constant Pressure, $c_p = \left(\dfrac{\partial h}{\partial T}\right)_P$

Heat Capacity at Constant Volume, $c_v = \left(\dfrac{\partial u}{\partial T}\right)_v$

Quality x (applies to liquid-vapor systems at saturation) is defined as the mass fraction of the vapor phase:

$$x = m_g/(m_g + m_f), \text{ where}$$

m_g = mass of vapor, and

m_f = mass of liquid.

Specific volume of a two-phase system can be written:

$v = xv_g + (1-x)v_f$ or $v = v_f + xv_{fg}$, where

v_f = specific volume of saturated liquid,

v_g = specific volume of saturated vapor, and

v_{fg} = specific volume change upon vaporization.

$\quad = v_g - v_f$

Similar expressions exist for u, h, and s:

$u = xu_g + (1-x)u_f$ or $u = u_f + xu_{fg}$

$h = xh_g + (1-x)h_f$ or $h = h_f + xh_{fg}$

$s = xs_g + (1-x)s_f$ or $s = s_f + xs_{fg}$

For a simple substance, specification of any two intensive, independent properties is sufficient to fix all the rest.

For an ideal gas, $Pv = RT$ or $PV = mRT$, and

$$P_1v_1/T_1 = P_2v_2/T_2, \text{ where}$$

P = pressure,

v = specific volume,

m = mass of gas,

R = gas constant, and

T = absolute temperature.

V = volume

R is *specific to each gas* but can be found from

$$R = \frac{\overline{R}}{(mol.\ wt)}, \text{ where}$$

$\overline{R}$ = the universal gas constant

$\quad$ = 1,545 ft-lbf/(lbmol-°R) = 8,314 J/(kmol·K).

For *ideal gases*, $c_p - c_v = R$

Also, for *ideal gases*:

$$\left(\frac{\partial h}{\partial P}\right)_T = 0 \qquad \left(\frac{\partial u}{\partial v}\right)_T = 0$$

For cold air standard, *heat capacities are assumed to be constant* at their room temperature values. In that case, the following are true:

$$\Delta u = c_v \Delta T; \quad \Delta h = c_p \Delta T$$
$$\Delta s = c_p \ln(T_2/T_1) - R \ln(P_2/P_1); \text{ and}$$
$$\Delta s = c_v \ln(T_2/T_1) + R \ln(v_2/v_1).$$

For heat capacities that are temperature dependent, the value to be used in the above equations for Δh is known as the mean heat capacity $(\overline{c}_p)$ and is given by

$$\overline{c}_p = \frac{\int_{T_1}^{T_2} c_p dT}{T_2 - T_1}$$

Also, for *constant entropy* processes:

$$\frac{P_2}{P_1} = \left(\frac{v_1}{v_2}\right)^k; \qquad \frac{T_2}{T_1} = \left(\frac{P_2}{P_1}\right)^{\frac{k-1}{k}}$$

$$\frac{T_2}{T_1} = \left(\frac{v_1}{v_2}\right)^{k-1}, \text{ where } k = c_p/c_v$$

For real gases, several equations of state are available; one such equation is the van der Waals equation with constants based on the critical point:

$$\left(P + \frac{a}{v^2}\right)(v - b) = \overline{R}T$$

where $a = \left(\dfrac{27}{64}\right)\left(\dfrac{\overline{R}^2 T_c^2}{P_c}\right), \quad b = \dfrac{RT_c}{8P_c}$

where P_c and T_c are the pressure and temperature at the critical point, respectively.

FIRST LAW OF THERMODYNAMICS

The *First Law of Thermodynamics* is a statement of conservation of energy in a thermodynamic system. The net energy crossing the system boundary is equal to the change in energy inside the system.

Heat Q is *energy transferred* due to temperature difference and is considered positive if it is inward or added to the system.

Closed Thermodynamic System

No mass crosses system boundary

$$Q - W = \Delta U + \Delta KE + \Delta PE$$

where

ΔKE = change in kinetic energy, and

ΔPE = change in potential energy.

Energy can cross the boundary only in the form of heat or work. Work can be boundary work, w_b, or other work forms (electrical work, etc.)

Work $W \left(w = \dfrac{W}{m} \right)$ is considered *positive if it is outward* or *work done* by the system.

Reversible boundary work is given by $w_b = \int P\, dv$.

Special Cases of Closed Systems
Constant Pressure (***Charles' Law***):

$$w_b = P\Delta v$$
(ideal gas) T/v = constant

Constant Volume:

$$w_b = 0$$
(ideal gas) T/P = constant

Isentropic (ideal gas):

$$Pv^k = \text{constant}$$
$$w = (P_2 v_2 - P_1 v_1)/(1 - k)$$
$$= \overline{R}(T_2 - T_1)/(1 - k)$$

Constant Temperature (***Boyle's Law***):

(ideal gas) Pv = constant
$$w_b = \overline{R}T \ln (v_2 / v_1) = \overline{R}T \ln (P_1/P_2)$$

Polytropic (ideal gas):

$$Pv^n = \text{constant}$$
$$w = (P_2 v_2 - P_1 v_1)/(1 - n)$$

Open Thermodynamic System

Mass crosses the system boundary
There is flow work (Pv) done by mass entering the system. The reversible flow work is given by:

$$w_{rev} = -\int v\, dP + \Delta ke + \Delta pe$$

First Law applies whether or not processes are reversible.

FIRST LAW (energy balance)

$$\Sigma \dot{m}_i \left[h_i + V_i^2/2 + gZ_i \right] - \Sigma \dot{m}_e \left[h_e + V_e^2/2 + gZ_e \right]$$
$$+ \dot{Q}_{in} - \dot{W}_{net} = d(m_s u_s)/dt, \text{ where}$$

$\dot{W}_{net}$ = rate of net or shaft work transfer,

m_s = mass of fluid within the system,

u_s = specific internal energy of system, and

$\dot{Q}$ = rate of heat transfer (neglecting kinetic and potential energy of the system).

Special Cases of Open Systems
Constant Volume:

$$w_{rev} = -v (P_2 - P_1)$$

Constant Pressure:

$$w_{rev} = 0$$

Constant Temperature:

(ideal gas) Pv = constant
$$w_{rev} = \overline{R}T \ln (v_2 /v_1) = \overline{R}T \ln (P_1 /P_2)$$

Isentropic (ideal gas):

$$Pv^k = \text{constant}$$
$$w_{rev} = k (P_2 v_2 - P_1 v_1)/(1 - k)$$
$$= k\overline{R} (T_2 - T_1)/(1 - k)$$
$$w_{rev} = \frac{k}{k - 1} \overline{R}T_1 \left[1 - \left(\frac{P_2}{P_1} \right)^{(k - 1)/k} \right]$$

Polytropic:

$$Pv^n = \text{constant}$$
$$w_{rev} = n (P_2 v_2 - P_1 v_1)/(1 - n)$$

Steady-State Systems

The system does not change state with time. This assumption is valid for steady operation of turbines, pumps, compressors, throttling valves, nozzles, and heat exchangers, including boilers and condensers.

$$\Sigma \dot{m} \left(h_i + V_i^2/2 + gZ_i \right) - \Sigma \dot{m}_e \left(h_e + V_e^2/2 + gZ_e \right) + \dot{Q}_{in} - \dot{W}_{out} = 0$$

and

$$\Sigma \dot{m}_i = \Sigma \dot{m}_e$$

where

$\dot{m}$ = mass flow rate (subscripts i and e refer to inlet and exit states of system),

g = acceleration of gravity,

Z = elevation,

V = velocity, and

$\dot{W}$ = rate of work.

Special Cases of Steady-Flow Energy Equation
Nozzles, Diffusers: Velocity terms are significant. No elevation change, no heat transfer, and no work. Single mass stream.

$$h_i + V_i^2/2 = h_e + V_e^2/2$$

Isentropic Efficiency (nozzle) $= \dfrac{V_e^2 - V_i^2}{2 \left(h_i - h_{es} \right)}$, where

h_{es} = enthalpy at isentropic exit state.

Turbines, Pumps, Compressors: Often considered adiabatic (no heat transfer). Velocity terms usually can be ignored. There are significant work terms and a single mass stream.

$$h_i = h_e + w$$

Isentropic Efficiency (turbine) = $\dfrac{h_i - h_e}{h_i - h_{es}}$

Isentropic Efficiency (compressor, pump) = $\dfrac{h_{es} - h_i}{h_e - h_i}$

Throttling Valves and Throttling Processes: No work, no heat transfer, and single-mass stream. Velocity terms are often insignificant.

$$h_i = h_e$$

Boilers, Condensers, Evaporators, One Side in a Heat Exchanger: Heat transfer terms are significant. For a single-mass stream, the following applies:

$$h_i + q = h_e$$

Heat Exchangers: No heat or work. Two separate flow rates $\dot{m}_1$ and $\dot{m}_2$:

$$\dot{m}_1 \left(h_{1i} - h_{1e} \right) = \dot{m}_2 \left(h_{2e} - h_{2i} \right)$$

See **MECHANICAL ENGINEERING** section.

Mixers, Separators, Open or Closed Feedwater Heaters:

$$\Sigma \dot{m}_i h_i = \Sigma \dot{m}_e h_e \quad \text{and}$$
$$\Sigma \dot{m}_i = \Sigma \dot{m}_e$$

BASIC CYCLES

Heat engines take in heat Q_H at a high temperature T_H, produce a net amount of work W, and reject heat Q_L at a low temperature T_L. The efficiency η of a heat engine is given by:

$$\eta = W/Q_H = (Q_H - Q_L)/Q_H$$

The most efficient engine possible is the *Carnot Cycle*. Its efficiency is given by:

$$\eta_c = (T_H - T_L)/T_H, \text{ where}$$

T_H and T_L = absolute temperatures (Kelvin or Rankine).

The following heat-engine cycles are plotted on *P-v* and *T-s* diagrams (see later in this chapter):

Carnot, Otto, Rankine

Refrigeration cycles are the reverse of heat-engine cycles. Heat is moved from low to high temperature requiring work, *W*. Cycles can be used either for refrigeration or as heat pumps.

Coefficient of Performance (COP) is defined as:
COP = Q_H/W for heat pumps, and as
COP = Q_L/W for refrigerators and air conditioners.

Upper limit of COP is based on reversed Carnot Cycle:
$COP_c = T_H/(T_H - T_L)$ for heat pumps and
$COP_c = T_L/(T_H - T_L)$ for refrigeration.

1 ton refrigeration = 12,000 Btu/hr = 3,516 W

IDEAL GAS MIXTURES

$i = 1, 2, \ldots, n$ constituents. Each constituent is an ideal gas.
Mole Fraction:

$$x_i = N_i/N; N = \Sigma N_i; \Sigma x_i = 1$$

where N_i = number of moles of component i.

Mass Fraction: $y_i = m_i/m; m = \Sigma m_i; \Sigma y_i = 1$

Molecular Weight: $M = m/N = \Sigma x_i M_i$

Gas Constant: $R = \overline{R}/M$

To convert *mole fractions x_i* to *mass fractions y_i*:

$$y_i = \frac{x_i M_i}{\Sigma (x_i M_i)}$$

To convert *mass fractions* to *mole fractions*:

$$x_i = \frac{y_i/M_i}{\Sigma (y_i/M_i)}$$

Partial Pressures: $P = \Sigma P_i; P_i = \dfrac{m_i R_i T}{V}$

Partial Volumes: $V = \Sigma V_i; V_i = \dfrac{m_i R_i T}{P}$, where

P, V, T = the pressure, volume, and temperature of the mixture.

$$x_i = P_i/P = V_i/V$$

Other Properties:
$u = \Sigma (y_i u_i); h = \Sigma (y_i h_i); s = \Sigma (y_i s_i)$
u_i and h_i are evaluated at T, and
s_i is evaluated at T and P_i.

PSYCHROMETRICS
We deal here with a mixture of dry air (subscript *a*) and water vapor (subscript *v*):

$$P = P_a + P_v$$

Specific Humidity (absolute humidity, humidity ratio) ω:

$$\omega = m_v/m_a, \text{ where}$$
m_v = mass of water vapor and
m_a = mass of dry air.

$$\omega = 0.622 P_v/P_a = 0.622 P_v/(P - P_v)$$

Relative Humidity (rh) ϕ:

$$\phi = P_v/P_g, \text{ where}$$
P_g = saturation pressure at T.

Enthalpy *h*: $h = h_a + \omega h_v$

Dew-Point Temperature T_{dp}:
$$T_{dp} = T_{sat} \text{ at } P_g = P_v$$

Wet-bulb temperature T_{wb} is the temperature indicated by a thermometer covered by a wick saturated with liquid water and in contact with moving air.

Humid Volume: Volume of moist air/mass of dry air.

Psychrometric Chart
A plot of specific humidity as a function of dry-bulb temperature plotted for a value of atmospheric pressure. (See chart at end of section.)

PHASE RELATIONS
Clapeyron Equation for Phase Transitions:

$$\left(\frac{dP}{dT}\right)_{sat} = \frac{h_{fg}}{Tv_{fg}} = \frac{s_{fg}}{v_{fg}}, \text{ where}$$

h_{fg} = enthalpy change for phase transitions,

v_{fg} = volume change,

s_{fg} = entropy change,

T = absolute temperature, and

$(dP/dT)_{sat}$ = slope of phase transition (e.g.,vapor-liquid) saturation line.

Clausius-Clapeyron Equation
This equation results if it is assumed that (1) the volume change (v_{fg}) can be replaced with the vapor volume (v_g), (2) the latter can be replaced with $P/\overline{R}T$ from the ideal gas law, and (3) h_{fg} is independent of the temperature (T).

$$\ln_e\left(\frac{P_2}{P_1}\right) = \frac{h_{fg}}{\overline{R}} \cdot \frac{T_2 - T_1}{T_1 T_2}$$

Gibbs Phase Rule (*non-reacting systems*)
P + F = C + 2, where
P = number of phases making up a system
F = degrees of freedom, and
C = number of components in a system

COMBUSTION PROCESSES
First, the combustion equation should be written and balanced. For example, for the stoichiometric combustion of methane in oxygen:

$$CH_4 + 2\,O_2 \rightarrow CO_2 + 2\,H_2O$$

Combustion in Air
For each mole of oxygen, there will be 3.76 moles of nitrogen. For stoichiometric combustion of methane in air:

$$CH_4 + 2\,O_2 + 2(3.76)\,N_2 \rightarrow CO_2 + 2\,H_2O + 7.52\,N_2$$

Combustion in Excess Air
The excess oxygen appears as oxygen on the right side of the combustion equation.

Incomplete Combustion
Some carbon is burned to create carbon monoxide (CO).

Air-Fuel Ratio (*A/F*): $A/F = \dfrac{\text{mass of air}}{\text{mass of fuel}}$

Stoichiometric (theoretical) air-fuel ratio is the air-fuel ratio calculated from the stoichiometric combustion equation.

$$\text{Percent Theoretical Air} = \frac{(A/F)_{actual}}{(A/F)_{stoichiometric}} \times 100$$

$$\text{Percent Excess Air} = \frac{(A/F)_{actual} - (A/F)_{stoichiometric}}{(A/F)_{stoichiometric}} \times 100$$

SECOND LAW OF THERMODYNAMICS
Thermal Energy Reservoirs
$$\Delta S_{reservoir} = Q/T_{reservoir}, \text{ where}$$

Q is measured with respect to the reservoir.

Kelvin-Planck Statement of Second Law
No heat engine can operate in a cycle while transferring heat with a single heat reservoir.

COROLLARY to Kelvin-Planck: No heat engine can have a higher efficiency than a Carnot Cycle operating between the same reservoirs.

Clausius' Statement of Second Law
No refrigeration or heat pump cycle can operate without a net work input.

COROLLARY: No refrigerator or heat pump can have a higher COP than a Carnot Cycle refrigerator or heat pump.

VAPOR-LIQUID MIXTURES

Henry's Law at Constant Temperature
At equilibrium, the partial pressure of a gas is proportional to its concentration in a liquid. Henry's Law is valid for low concentrations; i.e., $x \approx 0$.

$P_i = Py_i = hx_i$, where
h = Henry's Law constant,
P_i = partial pressure of a gas in contact with a liquid,
x_i = mol fraction of the gas in the liquid,
y_i = mol fraction of the gas in the vapor, and
P = total pressure.

Raoult's Law for Vapor-Liquid Equilibrium
Valid for concentrations near 1; i.e., $x_i \approx 1$.

$P_i = x_i P_i^*$, where
P_i = partial pressure of component i,
x_i = mol fraction of component i in the liquid, and
P_i^* = vapor pressure of pure component i at the temperature of the mixture.

ENTROPY

$$ds = (1/T)\delta Q_{rev}$$

$$s_2 - s_1 = \int_1^2 (1/T)\delta Q_{rev}$$

Inequality of Clausius

$$\oint (1/T)\delta Q_{rev} \leq 0$$

$$\int_1^2 (1/T)\delta Q \leq s_2 - s_1$$

Isothermal, Reversible Process

$$\Delta s = s_2 - s_1 = Q/T$$

Isentropic Process

$$\Delta s = 0; \ ds = 0$$

A reversible adiabatic process is isentropic.

Adiabatic Process

$$\delta Q = 0; \ \Delta s \geq 0$$

Increase of Entropy Principle

$$\Delta s_{total} = \Delta s_{system} + \Delta s_{surroundings} \geq 0$$

$$\Delta \dot{s}_{total} = \Sigma \dot{m}_{out} s_{out} - \Sigma \dot{m}_{in} s_{in} - \Sigma \left(\dot{Q}_{external}/T_{external} \right) \geq 0$$

Temperature-Entropy (T-s) Diagram

$$Q_{rev} = \int_1^2 T \, ds$$

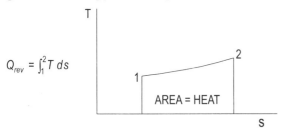

Entropy Change for Solids and Liquids

$$ds = c \, (dT/T)$$

$$s_2 - s_1 = \int c \, (dT/T) = c_{mean} \ln (T_2/T_1),$$

where c equals the heat capacity of the solid or liquid.

Irreversibility

$$I = w_{rev} - w_{actual}$$

EXERGY

Exergy is the portion of total energy available to do work.

Closed-System Exergy (Availability)

(no chemical reactions)

$$\phi = (u - u_o) - T_o (s - s_o) + p_o (v - v_o)$$

where the subscript o designates environmental conditions

$$w_{reversible} = \phi_1 - \phi_2$$

Open-System Exergy (Availability)

$$\psi = (h - h_o) - T_o (s - s_o) + V^2/2 + gz$$

$$w_{reversible} = \psi_1 - \psi_2$$

Gibbs Free Energy, ΔG

Energy released or absorbed in a reaction occurring reversibly at constant pressure and temperature.

Helmholtz Free Energy, ΔA

Energy released or absorbed in a reaction occurring reversibly at constant volume and temperature.

COMMON THERMODYNAMIC CYCLES

Carnot Cycle

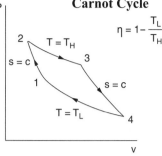

$$\eta = 1 - \frac{T_L}{T_H}$$

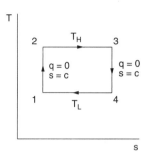

Reversed Carnot

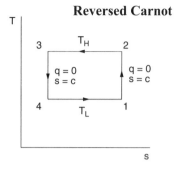

Otto Cycle
(Gasoline Engine)

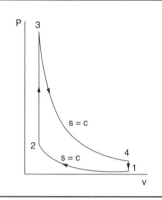

$$\eta = 1 - r^{1-k}$$

$$r = v_1/v_2$$

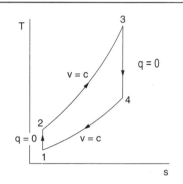

Rankine Cycle

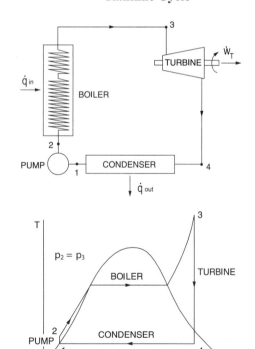

$$\eta = \frac{(h_3 - h_4) - (h_2 - h_1)}{h_3 - h_2}$$

Refrigeration
(Reversed Rankine Cycle)

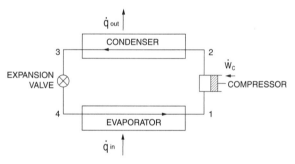

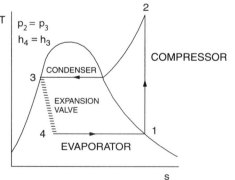

$$COP_{ref} = \frac{h_1 - h_4}{h_2 - h_1} \qquad COP_{HP} = \frac{h_2 - h_3}{h_2 - h_1}$$

STEAM TABLES
Saturated Water - Temperature Table

Temp. °C T	Sat. Press. kPa p_{sat}	Specific Volume m³/kg		Internal Energy kJ/kg			Enthalpy kJ/kg			Entropy kJ/(kg·K)		
		Sat. liquid v_f	Sat. vapor v_g	Sat. liquid u_f	Evap. u_{fg}	Sat. vapor u_g	Sat. liquid h_f	Evap. h_{fg}	Sat. vapor h_g	Sat. liquid s_f	Evap. s_{fg}	Sat. vapor s_g
0.01	0.6113	0.001 000	206.14	0.00	2375.3	2375.3	0.01	2501.3	2501.4	0.0000	9.1562	9.1562
5	0.8721	0.001 000	147.12	20.97	2361.3	2382.3	20.98	2489.6	2510.6	0.0761	8.9496	9.0257
10	1.2276	0.001 000	106.38	42.00	2347.2	2389.2	42.01	2477.7	2519.8	0.1510	8.7498	8.9008
15	1.7051	0.001 001	77.93	62.99	2333.1	2396.1	62.99	2465.9	2528.9	0.2245	8.5569	8.7814
20	2.339	0.001 002	57.79	83.95	2319.0	2402.9	83.96	2454.1	2538.1	0.2966	8.3706	8.6672
25	3.169	0.001 003	43.36	104.88	2304.9	2409.8	104.89	2442.3	2547.2	0.3674	8.1905	8.5580
30	4.246	0.001 004	32.89	125.78	2290.8	2416.6	125.79	2430.5	2556.3	0.4369	8.0164	8.4533
35	5.628	0.001 006	25.22	146.67	2276.7	2423.4	146.68	2418.6	2565.3	0.5053	7.8478	8.3531
40	7.384	0.001 008	19.52	167.56	2262.6	2430.1	167.57	2406.7	2574.3	0.5725	7.6845	8.2570
45	9.593	0.001 010	15.26	188.44	2248.4	2436.8	188.45	2394.8	2583.2	0.6387	7.5261	8.1648
50	12.349	0.001 012	12.03	209.32	2234.2	2443.5	209.33	2382.7	2592.1	0.7038	7.3725	8.0763
55	15.758	0.001 015	9.568	230.21	2219.9	2450.1	230.23	2370.7	2600.9	0.7679	7.2234	7.9913
60	19.940	0.001 017	7.671	251.11	2205.5	2456.6	251.13	2358.5	2609.6	0.8312	7.0784	7.9096
65	25.03	0.001 020	6.197	272.02	2191.1	2463.1	272.06	2346.2	2618.3	0.8935	6.9375	7.8310
70	31.19	0.001 023	5.042	292.95	2176.6	2569.6	292.98	2333.8	2626.8	0.9549	6.8004	7.7553
75	38.58	0.001 026	4.131	313.90	2162.0	2475.9	313.93	2321.4	2635.3	1.0155	6.6669	7.6824
80	47.39	0.001 029	3.407	334.86	2147.4	2482.2	334.91	2308.8	2643.7	1.0753	6.5369	7.6122
85	57.83	0.001 033	2.828	355.84	2132.6	2488.4	355.90	2296.0	2651.9	1.1343	6.4102	7.5445
90	70.14	0.001 036	2.361	376.85	2117.7	2494.5	376.92	2283.2	2660.1	1.1925	6.2866	7.4791
95	84.55	0.001 040	1.982	397.88	2102.7	2500.6	397.96	2270.2	2668.1	1.2500	6.1659	7.4159
	MPa											
100	0.101 35	0.001 044	1.6729	418.94	2087.6	2506.5	419.04	2257.0	2676.1	1.3069	6.0480	7.3549
105	0.120 82	0.001 048	1.4194	440.02	2072.3	2512.4	440.15	2243.7	2683.8	1.3630	5.9328	7.2958
110	0.143 27	0.001 052	1.2102	461.14	2057.0	2518.1	461.30	2230.2	2691.5	1.4185	5.8202	7.2387
115	0.169 06	0.001 056	1.0366	482.30	2041.4	2523.7	482.48	2216.5	2699.0	1.4734	5.7100	7.1833
120	0.198 53	0.001 060	0.8919	503.50	2025.8	2529.3	503.71	2202.6	2706.3	1.5276	5.6020	7.1296
125	0.2321	0.001 065	0.7706	524.74	2009.9	2534.6	524.99	2188.5	2713.5	1.5813	5.4962	7.0775
130	0.2701	0.001 070	0.6685	546.02	1993.9	2539.9	546.31	2174.2	2720.5	1.6344	5.3925	7.0269
135	0.3130	0.001 075	0.5822	567.35	1977.7	2545.0	567.69	2159.6	2727.3	1.6870	5.2907	6.9777
140	0.3613	0.001 080	0.5089	588.74	1961.3	2550.0	589.13	2144.7	2733.9	1.7391	5.1908	6.9299
145	0.4154	0.001 085	0.4463	610.18	1944.7	2554.9	610.63	2129.6	2740.3	1.7907	5.0926	6.8833
150	0.4758	0.001 091	0.3928	631.68	1927.9	2559.5	632.20	2114.3	2746.5	1.8418	4.9960	6.8379
155	0.5431	0.001 096	0.3468	653.24	1910.8	2564.1	653.84	2098.6	2752.4	1.8925	4.9010	6.7935
160	0.6178	0.001 102	0.3071	674.87	1893.5	2568.4	675.55	2082.6	2758.1	1.9427	4.8075	6.7502
165	0.7005	0.001 108	0.2727	696.56	1876.0	2572.5	697.34	2066.2	2763.5	1.9925	4.7153	6.7078
170	0.7917	0.001 114	0.2428	718.33	1858.1	2576.5	719.21	2049.5	2768.7	2.0419	4.6244	6.6663
175	0.8920	0.001 121	0.2168	740.17	1840.0	2580.2	741.17	2032.4	2773.6	2.0909	4.5347	6.6256
180	1.0021	0.001 127	0.194 05	762.09	1821.6	2583.7	763.22	2015.0	2778.2	2.1396	4.4461	6.5857
185	1.1227	0.001 134	0.174 09	784.10	1802.9	2587.0	785.37	1997.1	2782.4	2.1879	4.3586	6.5465
190	1.2544	0.001 141	0.156 54	806.19	1783.8	2590.0	807.62	1978.8	2786.4	2.2359	4.2720	6.5079
195	1.3978	0.001 149	0.141 05	828.37	1764.4	2592.8	829.98	1960.0	2790.0	2.2835	4.1863	6.4698
200	1.5538	0.001 157	0.127 36	850.65	1744.7	2595.3	852.45	1940.7	2793.2	2.3309	4.1014	6.4323
205	1.7230	0.001 164	0.115 21	873.04	1724.5	2597.5	875.04	1921.0	2796.0	2.3780	4.0172	6.3952
210	1.9062	0.001 173	0.104 41	895.53	1703.9	2599.5	897.76	1900.7	2798.5	2.4248	3.9337	6.3585
215	2.104	0.001 181	0.094 79	918.14	1682.9	2601.1	920.62	1879.9	2800.5	2.4714	3.8507	6.3221
220	2.318	0.001 190	0.086 19	940.87	1661.5	2602.4	943.62	1858.5	2802.1	2.5178	3.7683	6.2861
225	2.548	0.001 199	0.078 49	963.73	1639.6	2603.3	966.78	1836.5	2803.3	2.5639	3.6863	6.2503
230	2.795	0.001 209	0.071 58	986.74	1617.2	2603.9	990.12	1813.8	2804.0	2.6099	3.6047	6.2146
235	3.060	0.001 219	0.065 37	1009.89	1594.2	2604.1	1013.62	1790.5	2804.2	2.6558	3.5233	6.1791
240	3.344	0.001 229	0.059 76	1033.21	1570.8	2604.0	1037.32	1766.5	2803.8	2.7015	3.4422	6.1437
245	3.648	0.001 240	0.054 71	1056.71	1546.7	2603.4	1061.23	1741.7	2803.0	2.7472	3.3612	6.1083
250	3.973	0.001 251	0.050 13	1080.39	1522.0	2602.4	1085.36	1716.2	2801.5	2.7927	3.2802	6.0730
255	4.319	0.001 263	0.045 98	1104.28	1596.7	2600.9	1109.73	1689.8	2799.5	2.8383	3.1992	6.0375
260	4.688	0.001 276	0.042 21	1128.39	1470.6	2599.0	1134.37	1662.5	2796.9	2.8838	3.1181	6.0019
265	5.081	0.001 289	0.038 77	1152.74	1443.9	2596.6	1159.28	1634.4	2793.6	2.9294	3.0368	5.9662
270	5.499	0.001 302	0.035 64	1177.36	1416.3	2593.7	1184.51	1605.2	2789.7	2.9751	2.9551	5.9301
275	5.942	0.001 317	0.032 79	1202.25	1387.9	2590.2	1210.07	1574.9	2785.0	3.0208	2.8730	5.8938
280	6.412	0.001 332	0.030 17	1227.46	1358.7	2586.1	1235.99	1543.6	2779.6	3.0668	2.7903	5.8571
285	6.909	0.001 348	0.027 77	1253.00	1328.4	2581.4	1262.31	1511.0	2773.3	3.1130	2.7070	5.8199
290	7.436	0.001 366	0.025 57	1278.92	1297.1	2576.0	1289.07	1477.1	2766.2	3.1594	2.6227	5.7821
295	7.993	0.001 384	0.023 54	1305.2	1264.7	2569.9	1316.3	1441.8	2758.1	3.2062	2.5375	5.7437
300	8.581	0.001 404	0.021 67	1332.0	1231.0	2563.0	1344.0	1404.9	2749.0	3.2534	2.4511	5.7045
305	9.202	0.001 425	0.019 948	1359.3	1195.9	2555.2	1372.4	1366.4	2738.7	3.3010	2.3633	5.6643
310	9.856	0.001 447	0.018 350	1387.1	1159.4	2546.4	1401.3	1326.0	2727.3	3.3493	2.2737	5.6230
315	10.547	0.001 472	0.016 867	1415.5	1121.1	2536.6	1431.0	1283.5	2714.5	3.3982	2.1821	5.5804
320	11.274	0.001 499	0.015 488	1444.6	1080.9	2525.5	1461.5	1238.6	2700.1	3.4480	2.0882	5.5362
330	12.845	0.001 561	0.012 996	1505.3	993.7	2498.9	1525.3	1140.6	2665.9	3.5507	1.8909	5.4417
340	14.586	0.001 638	0.010 797	1570.3	894.3	2464.6	1594.2	1027.9	2622.0	3.6594	1.6763	5.3357
350	16.513	0.001 740	0.008 813	1641.9	776.6	2418.4	1670.6	893.4	2563.9	3.7777	1.4335	5.2112
360	18.651	0.001 893	0.006 945	1725.2	626.3	2351.5	1760.5	720.3	2481.0	3.9147	1.1379	5.0526
370	21.03	0.002 213	0.004 925	1844.0	384.5	2228.5	1890.5	441.6	2332.1	4.1106	0.6865	4.7971
374.14	22.09	0.003 155	0.003 155	2029.6	0	2029.6	2099.3	0	2099.3	4.4298	0	4.4298

Superheated Water Tables

T Temp. °C	v m³/kg	u kJ/kg	h kJ/kg	s kJ/(kg·K)	v m³/kg	u kJ/kg	h kJ/kg	s kJ/(kg·K)
	p = 0.01 MPa (45.81°C)				p = 0.05 MPa (81.33°C)			
Sat.	14.674	2437.9	2584.7	8.1502	3.240	2483.9	2645.9	7.5939
50	14.869	2443.9	2592.6	8.1749				
100	17.196	2515.5	2687.5	8.4479	3.418	2511.6	2682.5	7.6947
150	19.512	2587.9	2783.0	8.6882	3.889	2585.6	2780.1	7.9401
200	**21.825**	**2661.3**	**2879.5**	**8.9038**	**4.356**	**2659.9**	**2877.7**	**8.1580**
250	24.136	2736.0	2977.3	9.1002	4.820	2735.0	2976.0	8.3556
300	26.445	2812.1	3076.5	9.2813	5.284	2811.3	3075.5	8.5373
400	31.063	2968.9	3279.6	9.6077	6.209	2968.5	3278.9	8.8642
500	35.679	3132.3	3489.1	9.8978	7.134	3132.0	3488.7	9.1546
600	40.295	3302.5	3705.4	10.1608	8.057	3302.2	3705.1	9.4178
700	44.911	3479.6	3928.7	10.4028	8.981	3479.4	3928.5	9.6599
800	49.526	3663.8	4159.0	10.6281	9.904	3663.6	4158.9	9.8852
900	54.141	3855.0	4396.4	10.8396	10.828	3854.9	4396.3	10.0967
1000	58.757	4053.0	4640.6	11.0393	11.751	4052.9	4640.5	10.2964
1100	**63.372**	**4257.5**	**4891.2**	**11.2287**	**12.674**	**4257.4**	**4891.1**	**10.4859**
1200	67.987	4467.9	5147.8	11.4091	13.597	4467.8	5147.7	10.6662
1300	72.602	4683.7	5409.7	11.5811	14.521	4683.6	5409.6	10.8382
	p = 0.10 MPa (99.63°C)				p = 0.20 MPa (120.23°C)			
Sat.	1.6940	2506.1	2675.5	7.3594	0.8857	2529.5	2706.7	7.1272
100	1.6958	2506.7	2676.2	7.3614				
150	1.9364	2582.8	2776.4	7.6134	0.9596	2576.9	2768.8	7.2795
200	2.172	2658.1	2875.3	7.8343	1.0803	2654.4	2870.5	7.5066
250	**2.406**	**2733.7**	**2974.3**	**8.0333**	**1.1988**	**2731.2**	**2971.0**	**7.7086**
300	2.639	2810.4	3074.3	8.2158	1.3162	2808.6	3071.8	7.8926
400	3.103	2967.9	3278.2	8.5435	1.5493	2966.7	3276.6	8.2218
500	3.565	3131.6	3488.1	8.8342	1.7814	3130.8	3487.1	8.5133
600	4.028	3301.9	3704.4	9.0976	2.013	3301.4	3704.0	8.7770
700	**4.490**	**3479.2**	**3928.2**	**9.3398**	**2.244**	**3478.8**	**3927.6**	**9.0194**
800	4.952	3663.5	4158.6	9.5652	2.475	3663.1	4158.2	9.2449
900	5.414	3854.8	4396.1	9.7767	2.705	3854.5	4395.8	9.4566
1000	5.875	4052.8	4640.3	9.9764	2.937	4052.5	4640.0	9.6563
1100	6.337	4257.3	4891.0	10.1659	3.168	4257.0	4890.7	9.8458
1200	**6.799**	**4467.7**	**5147.6**	**10.3463**	**3.399**	**4467.5**	**5147.5**	**10.0262**
1300	7.260	4683.5	5409.5	10.5183	3.630	4683.2	5409.3	10.1982
	p = 0.40 MPa (143.63°C)				p = 0.60 MPa (158.85°C)			
Sat.	0.4625	2553.6	2738.6	6.8959	0.3157	2567.4	2756.8	6.7600
150	0.4708	2564.5	2752.8	6.9299				
200	0.5342	2646.8	2860.5	7.1706	0.3520	2638.9	2850.1	6.9665
250	0.5951	2726.1	2964.2	7.3789	0.3938	2720.9	2957.2	7.1816
300	**0.6548**	**2804.8**	**3066.8**	**7.5662**	**0.4344**	**2801.0**	**3061.6**	**7.3724**
350	0.7137	2884.6	3170.1	7.7324	0.4742	2881.2	3165.7	7.5464
400	0.7726	2964.4	3273.4	7.8985	0.5137	2962.1	3270.3	7.7079
500	0.8893	3129.2	3484.9	8.1913	0.5920	3127.6	3482.8	8.0021
600	1.0055	3300.2	3702.4	8.4558	0.6697	3299.1	3700.9	8.2674
700	**1.1215**	**3477.9**	**3926.5**	**8.6987**	**0.7472**	**3477.0**	**3925.3**	**8.5107**
800	1.2372	3662.4	4157.3	8.9244	0.8245	3661.8	4156.5	8.7367
900	1.3529	3853.9	4395.1	9.1362	0.9017	3853.4	4394.4	8.9486
1000	1.4685	4052.0	4639.4	9.3360	0.9788	4051.5	4638.8	9.1485
1100	1.5840	4256.5	4890.2	9.5256	1.0559	4256.1	4889.6	9.3381
1200	**1.6996**	**4467.0**	**5146.8**	**9.7060**	**1.1330**	**4466.5**	**5146.3**	**9.5185**
1300	1.8151	4682.8	5408.8	9.8780	1.2101	4682.3	5408.3	9.6906
	p = 0.80 MPa (170.43°C)				p = 1.00 MPa (179.91°C)			
Sat.	0.2404	2576.8	2769.1	6.6628	0.194 44	2583.6	2778.1	6.5865
200	0.2608	2630.6	2839.3	6.8158	0.2060	2621.9	2827.9	6.6940
250	0.2931	2715.5	2950.0	7.0384	0.2327	2709.9	2942.6	6.9247
300	0.3241	2797.2	3056.5	7.2328	0.2579	2793.2	3051.2	7.1229
350	**0.3544**	**2878.2**	**3161.7**	**7.4089**	**0.2825**	**2875.2**	**3157.7**	**7.3011**
400	0.3843	2959.7	3267.1	7.5716	0.3066	2957.3	3263.9	7.4651
500	0.4433	3126.0	3480.6	7.8673	0.3541	3124.4	3478.5	7.7622
600	0.5018	3297.9	3699.4	8.1333	0.4011	3296.8	3697.9	8.0290
700	0.5601	3476.2	3924.2	8.3770	0.4478	3475.3	3923.1	8.2731
800	**0.6181**	**3661.1**	**4155.6**	**8.6033**	**0.4943**	**3660.4**	**4154.7**	**8.4996**
900	0.6761	3852.8	4393.7	8.8153	0.5407	3852.2	4392.9	8.7118
1000	0.7340	4051.0	4638.2	9.0153	0.5871	4050.5	4637.6	8.9119
1100	0.7919	4255.6	4889.1	9.2050	0.6335	4255.1	4888.6	9.1017
1200	0.8497	4466.1	5145.9	9.3855	0.6798	4465.6	5145.4	9.2822
1300	**0.9076**	**4681.8**	**5407.9**	**9.5575**	**0.7261**	**4681.3**	**5407.4**	**9.4543**

P-*h* DIAGRAM FOR REFRIGERANT HFC-134a
(metric units)
(Reproduced by permission of the DuPont Company)

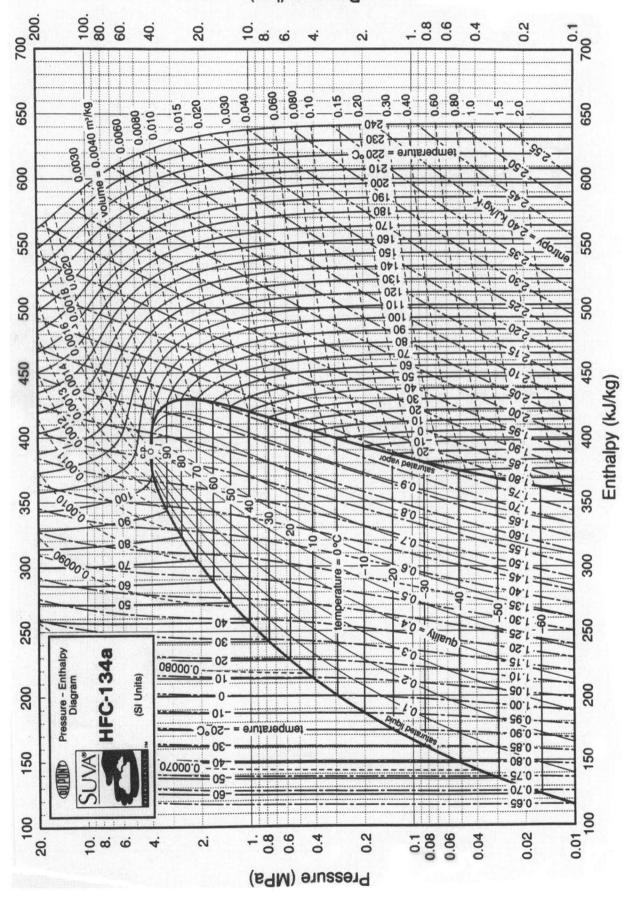

ASHRAE PSYCHROMETRIC CHART NO. 1
(metric units)
(Reproduced by permission of ASHRAE)

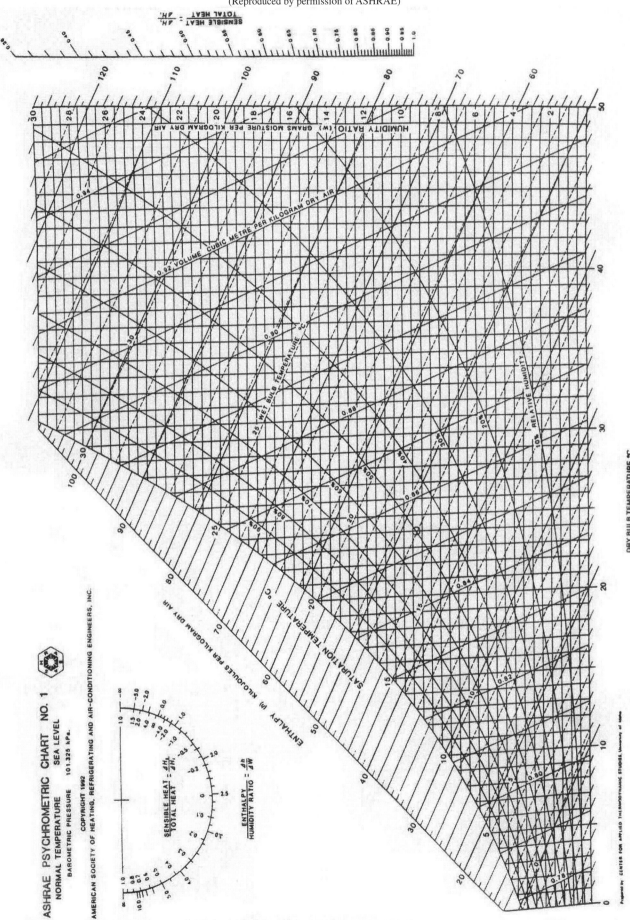

THERMAL AND PHYSICAL PROPERTY TABLES

(at room temperature)

GASES								
Substance	Mol wt	c_p		c_v		k	R	
		kJ/(kg·K)	Btu/(lbm-°R)	kJ/(kg·K)	Btu/(lbm-°R)		kJ/(kg·K)	
Gases								
Air	29	1.00	0.240	0.718	0.171	1.40	0.2870	
Argon	40	0.520	0.125	0.312	0.0756	1.67	0.2081	
Butane	58	1.72	0.415	1.57	0.381	1.09	0.1430	
Carbon dioxide	44	0.846	0.203	0.657	0.158	1.29	0.1889	
Carbon monoxide	28	1.04	0.249	0.744	0.178	1.40	0.2968	
Ethane	30	1.77	0.427	1.49	0.361	1.18	0.2765	
Helium	4	5.19	1.25	3.12	0.753	1.67	2.0769	
Hydrogen	2	14.3	3.43	10.2	2.44	1.40	4.1240	
Methane	16	2.25	0.532	1.74	0.403	1.30	0.5182	
Neon	20	1.03	0.246	0.618	0.148	1.67	0.4119	
Nitrogen	28	1.04	0.248	0.743	0.177	1.40	0.2968	
Octane vapor	114	1.71	0.409	1.64	0.392	1.04	0.0729	
Oxygen	32	0.918	0.219	0.658	0.157	1.40	0.2598	
Propane	44	1.68	0.407	1.49	0.362	1.12	0.1885	
Steam	18	1.87	0.445	1.41	0.335	1.33	0.4615	

SELECTED LIQUIDS AND SOLIDS				
Substance	c_p		Density	
	kJ/(kg·K)	Btu/(lbm-°R)	kg/m³	lbm/ft³
Liquids				
Ammonia	4.80	1.146	602	38
Mercury	0.139	0.033	13,560	847
Water	4.18	1.000	997	62.4
Solids				
Aluminum	0.900	0.215	2,700	170
Copper	0.386	0.092	8,900	555
Ice (0°C; 32°F)	2.11	0.502	917	57.2
Iron	0.450	0.107	7,840	490
Lead	0.128	0.030	11,310	705

HEAT TRANSFER

There are three modes of heat transfer: conduction, convection, and radiation.

BASIC HEAT TRANSFER RATE EQUATIONS

Conduction
Fourier's Law of Conduction

$$\dot{Q} = -kA\frac{dT}{dx}, \text{ where}$$

$\dot{Q}$ = rate of heat transfer (W)
k = the thermal conductivity [W/(m•K)]
A = the surface area perpendicular to direction of heat transfer (m^2)

Convection
Newton's Law of Cooling

$$\dot{Q} = hA(T_w - T_\infty), \text{ where}$$

h = the convection heat transfer coefficient of the fluid [W/(m^2•K)]
A = the convection surface area (m^2)
T_w = the wall surface temperature (K)
T_∞ = the bulk fluid temperature (K)

Radiation
The radiation emitted by a body is given by

$$\dot{Q} = \varepsilon\sigma AT^4, \text{ where}$$

ε = the emissivity of the body
σ = the Stefan-Boltzmann constant
  = 5.67 × 10^{-8} W/(m^2•K^4)
A = the body surface area (m^2)
T = the absolute temperature (K)

CONDUCTION

Conduction Through a Plane Wall

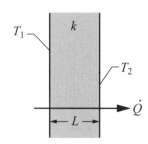

$$\dot{Q} = \frac{-kA(T_2 - T_1)}{L}, \text{ where}$$

A = wall surface area normal to heat flow (m^2)
L = wall thickness (m)
T_1 = temperature of one surface of the wall (K)
T_2 = temperature of the other surface of the wall (K)

Conduction Through a Cylindrical Wall

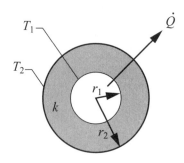

Cylinder (Length = L)

$$\dot{Q} = \frac{2\pi kL(T_1 - T_2)}{\ln\left(\frac{r_2}{r_1}\right)}$$

Critical Insulation Radius

$$r_{cr} = \frac{k_{insulation}}{h_\infty}$$

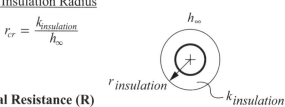

Thermal Resistance (R)

$$\dot{Q} = \frac{\Delta T}{R_{total}}$$

Resistances in series are added: $R_{total} = \Sigma R$, where

Plane Wall Conduction Resistance (K/W): $R = \frac{L}{kA}$, where
L = wall thickness

Cylindrical Wall Conduction Resistance (K/W): $R = \frac{\ln\left(\frac{r_2}{r_1}\right)}{2\pi kL}$, where
L = cylinder length

Convection Resistance (K/W) : $R = \frac{1}{hA}$

Composite Plane Wall

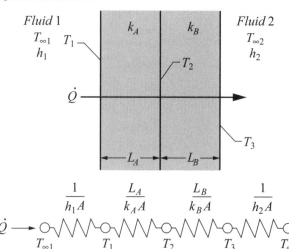

To evaluate Surface or Intermediate Temperatures:

$$\dot{Q} = \frac{T_1 - T_2}{R_A} = \frac{T_2 - T_3}{R_B}$$

Steady Conduction with Internal Energy Generation

The equation for one-dimensional steady conduction is

$$\frac{d^2 T}{dx^2} + \frac{\dot{Q}_{gen}}{k} = 0, \text{ where}$$

$\dot{Q}_{gen}$ = the heat generation rate per unit volume (W/m^3)

For a Plane Wall

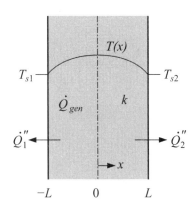

$$T(x) = \frac{\dot{Q}_{gen} L^2}{2k}\left(1 - \frac{x^2}{L^2}\right) + \left(\frac{T_{s2} - T_{s1}}{2}\right)\left(\frac{x}{L}\right) + \left(\frac{T_{s1} - T_{s2}}{2}\right)$$

$$\dot{Q}_1^{''} + \dot{Q}_2^{''} = 2\dot{Q}_{gen} L, \text{ where}$$

$\dot{Q}^{''}$ = the rate of heat transfer per area (heat flux) (W/m^2)

$$\dot{Q}_1^{''} = k\left(\frac{dT}{dx}\right)_{-L} \text{ and } \dot{Q}_2^{''} = k\left(\frac{dT}{dx}\right)_{L}$$

For a Long Circular Cylinder

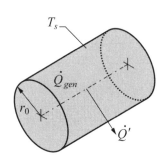

$$\frac{1}{r}\frac{d}{dr}\left(r\frac{dT}{dr}\right) + \frac{\dot{Q}_{gen}}{k} = 0$$

$$T(r) = \frac{\dot{Q}_{gen} r_0^2}{4k}\left(1 - \frac{r^2}{r_0^2}\right) + T_s$$

$$\dot{Q}' = \pi r_0^2 \dot{Q}_{gen}, \text{ where}$$

$\dot{Q}'$ = the heat transfer rate from the cylinder per unit length of the cylinder (W/m)

Transient Conduction Using the Lumped Capacitance Method

The lumped capacitance method is valid if

$$\text{Biot number, Bi} = \frac{hV}{kA_s} \ll 1, \text{ where}$$

h = the convection heat transfer coefficient of the fluid [W/(m^2•K)]
V = the volume of the body (m^3)
k = thermal conductivity of the body [W/(m•K)]
A_s = the surface area of the body (m^2)

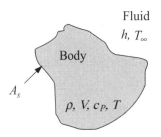

Constant Fluid Temperature

If the temperature may be considered uniform within the body at any time, the heat transfer rate at the body surface is given by

$$\dot{Q} = hA_s(T - T_\infty) = -\rho V(c_P)\left(\frac{dT}{dt}\right), \text{ where}$$

T = the body temperature (K)
T_∞ = the fluid temperature (K)
ρ = the density of the body (kg/m^3)
c_P = the heat capacity of the body [J/(kg•K)]
t = time (s)

The temperature variation of the body with time is

$$T - T_\infty = (T_i - T_\infty)e^{-\beta t}, \text{ where}$$

$$\beta = \frac{hA_s}{\rho V c_P} \qquad \text{where } \beta = \frac{1}{\tau} \text{ and}$$
$$\tau = \text{time constant } (s)$$

The total heat transferred (Q_{total}) up to time t is

$$Q_{total} = \rho V c_P(T_i - T), \text{ where}$$

T_i = initial body temperature (K)

Variable Fluid Temperature

If the ambient fluid temperature varies periodically according to the equation

$$T_\infty = T_{\infty, mean} + \frac{1}{2}\left(T_{\infty, max} - T_{\infty, min}\right)\cos(\omega t)$$

The temperature of the body, after initial transients have died away, is

$$T = \frac{\beta\left[\frac{1}{2}\left(T_{\infty, max} - T_{\infty, min}\right)\right]}{\sqrt{\omega^2 + \beta^2}}\cos\left[\omega t - \tan^{-1}\left(\frac{\omega}{\beta}\right)\right] + T_{\infty, mean}$$

Fins

For a straight fin with uniform cross section (assuming negligible heat transfer from tip),

$$\dot{Q} = \sqrt{hPkA_c}\left(T_b - T_\infty\right)\tanh(mL_c), \text{ where}$$

h = the convection heat transfer coefficient of the fluid [W/(m²•K)]
P = perimeter of exposed fin cross section (m)
k = fin thermal conductivity [W/(m•K)]
A_c = fin cross-sectional area (m²)
T_b = temperature at base of fin (K)
T_∞ = fluid temperature (K)

$$m = \sqrt{\frac{hP}{kA_c}}$$

$L_c = L + \dfrac{A_c}{P}$, corrected length of fin (m)

Rectangular Fin

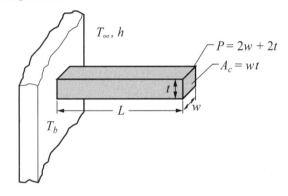

Pin Fin

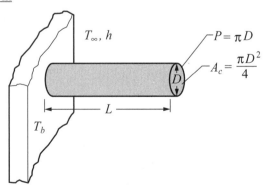

CONVECTION

Terms

D = diameter (m)
$\bar{h}$ = average convection heat transfer coefficient of the fluid [W/(m²•K)]
L = length (m)
$\overline{Nu}$ = average Nusselt number
Pr = Prandtl number = $\dfrac{c_P \mu}{k}$
u_m = mean velocity of fluid (m/s)
u_∞ = free stream velocity of fluid (m/s)
μ = dynamic viscosity of fluid [kg/(s•m)]
ρ = density of fluid (kg/m³)

External Flow

In all cases, evaluate fluid properties at average temperature between that of the body and that of the flowing fluid.

Flat Plate of Length L in Parallel Flow

$$Re_L = \frac{\rho u_\infty L}{\mu}$$

$$\overline{Nu}_L = \frac{\bar{h}L}{k} = 0.6640\, Re_L^{1/2}\, Pr^{1/3} \quad \left(Re_L < 10^5\right)$$

$$\overline{Nu}_L = \frac{\bar{h}L}{k} = 0.0366\, Re_L^{0.8}\, Pr^{1/3} \quad \left(Re_L > 10^5\right)$$

Cylinder of Diameter D in Cross Flow

$$Re_D = \frac{\rho u_\infty D}{\mu}$$

$$\overline{Nu}_D = \frac{\bar{h}D}{k} = C\, Re_D^n\, Pr^{1/3}, \text{ where}$$

Re_D	C	n
1 – 4	0.989	0.330
4 – 40	0.911	0.385
40 – 4,000	0.683	0.466
4,000 – 40,000	0.193	0.618
40,000 – 250,000	0.0266	0.805

Flow Over a Sphere of Diameter, D

$$\overline{Nu}_D = \frac{\bar{h}D}{k} = 2.0 + 0.60\, Re_D^{1/2} Pr^{1/3},$$

$$\left(1 < Re_D < 70,000; 0.6 < Pr < 400\right)$$

Internal Flow

$$Re_D = \frac{\rho u_m D}{\mu}$$

Laminar Flow in Circular Tubes

For laminar flow ($Re_D < 2300$), fully developed conditions

$Nu_D = 4.36$ (uniform heat flux)

$Nu_D = 3.66$ (constant surface temperature)

For laminar flow ($Re_D < 2300$), combined entry length with constant surface temperature

$$Nu_D = 1.86 \left(\frac{Re_D Pr}{\frac{L}{D}} \right)^{1/3} \left(\frac{\mu_b}{\mu_s} \right)^{0.14}, \text{ where}$$

L = length of tube (m)

D = tube diameter (m)

μ_b = dynamic viscosity of fluid [kg/(s•m)] at bulk temperature of fluid, T_b

μ_s = dynamic viscosity of fluid [kg/(s•m)] at inside surface temperature of the tube, T_s

Turbulent Flow in Circular Tubes

For turbulent flow ($Re_D > 10^4$, $Pr > 0.7$) for either uniform surface temperature or uniform heat flux condition, Sieder-Tate equation offers good approximation:

$$Nu_D = 0.027 \, Re_D^{0.8} Pr^{1/3} \left(\frac{\mu_b}{\mu_s} \right)^{0.14}$$

Non-Circular Ducts

In place of the diameter, D, use the equivalent (hydraulic) diameter (D_H) defined as

$$D_H = \frac{4 \times \text{cross−sectional area}}{\text{wetted perimeter}}$$

Circular Annulus ($D_o > D_i$)

In place of the diameter, D, use the equivalent (hydraulic) diameter (D_H) defined as

$$D_H = D_o - D_i$$

Liquid Metals ($0.003 < Pr < 0.05$)

$$Nu_D = 6.3 + 0.0167 \, Re_D^{0.85} Pr^{0.93} \text{ (uniform heat flux)}$$

$$Nu_D = 7.0 + 0.025 \, Re_D^{0.8} Pr^{0.8} \text{ (constant wall temperature)}$$

Condensation of a Pure Vapor

On a Vertical Surface

$$\overline{Nu_L} = \frac{\overline{h}L}{k} = 0.943 \left[\frac{\rho_l^2 g h_{fg} L^3}{\mu_l k_l (T_{sat} - T_s)} \right]^{0.25}, \text{ where}$$

ρ_l = density of liquid phase of fluid (kg/m^3)

g = gravitational acceleration (9.81 m/s^2)

h_{fg} = latent heat of vaporization [J/kg]

L = length of surface [m]

μ_l = dynamic viscosity of liquid phase of fluid [kg/(s•m)]

k_l = thermal conductivity of liquid phase of fluid [W/(m•K)]

T_{sat} = saturation temperature of fluid [K]

T_s = temperature of vertical surface [K]

Note: Evaluate all liquid properties at the average temperature between the saturated temperature, T_{sat}, and the surface temperature, T_s.

Outside Horizontal Tubes

$$\overline{Nu_D} = \frac{\overline{h}D}{k} = 0.729 \left[\frac{\rho_l^2 g h_{fg} D^3}{\mu_l k_l (T_{sat} - T_s)} \right]^{0.25}, \text{ where}$$

D = tube outside diameter (m)

Note: Evaluate all liquid properties at the average temperature between the saturated temperature, T_{sat}, and the surface temperature, T_s.

Natural (Free) Convection

Vertical Flat Plate in Large Body of Stationary Fluid

Equation also can apply to vertical cylinder of sufficiently large diameter in large body of stationary fluid.

$$\overline{h} = C \left(\frac{k}{L} \right) Ra_L^n, \text{ where}$$

L = the length of the plate (cylinder) in the vertical direction

Ra_L = Rayleigh Number = $\dfrac{g\beta(T_s - T_\infty)L^3}{v^2} Pr$

T_s = surface temperature (K)

T_∞ = fluid temperature (K)

β = coefficient of thermal expansion (1/K)

(For an ideal gas: $\beta = \dfrac{2}{T_s + T_\infty}$ with T in absolute temperature)

v = kinematic viscosity (m^2/s)

Range of Ra_L	C	n
$10^4 - 10^9$	0.59	1/4
$10^9 - 10^{13}$	0.10	1/3

Long Horizontal Cylinder in Large Body of Stationary Fluid

$$\overline{h} = C \left(\frac{k}{D} \right) Ra_D^n, \text{ where}$$

$$Ra_D = \frac{g\beta(T_s - T_\infty)D^3}{v^2} Pr$$

Ra_D	C	n
$10^{-3} - 10^2$	1.02	0.148
$10^2 - 10^4$	0.850	0.188
$10^4 - 10^7$	0.480	0.250
$10^7 - 10^{12}$	0.125	0.333

Heat Exchangers

The rate of heat transfer in a heat exchanger is

$$\dot{Q} = UAF\Delta T_{lm}, \text{ where}$$

A = any convenient reference area (m^2)

F = heat exchanger configuration correction factor ($F = 1$ if temperature change of one fluid is negligible)

U = overall heat transfer coefficient based on area A and the log mean temperature difference [W/(m^2•K)]

ΔT_{lm} = log mean temperature difference (K)

Heat Exchangers (cont.)

Overall Heat Transfer Coefficient for Concentric Tube and Shell-and-Tube Heat Exchangers

$$\frac{1}{UA} = \frac{1}{h_i A_i} + \frac{R_{fi}}{A_i} + \frac{\ln\left(\frac{D_o}{D_i}\right)}{2\pi kL} + \frac{R_{fo}}{A_o} + \frac{1}{h_o A_o}, \text{ where}$$

A_i = inside area of tubes (m²)

A_o = outside area of tubes (m²)

D_i = inside diameter of tubes (m)

D_o = outside diameter of tubes (m)

h_i = convection heat transfer coefficient for inside of tubes [W/(m²•K)]

h_o = convection heat transfer coefficient for outside of tubes [W/(m²•K)]

k = thermal conductivity of tube material [W/(m•K)]

R_{fi} = fouling factor for inside of tube [(m²•K)/W]

R_{fo} = fouling factor for outside of tube [(m²•K)/W]

Log Mean Temperature Difference (LMTD)

For *counterflow* in tubular heat exchangers

$$\Delta T_{lm} = \frac{(T_{Ho} - T_{Ci}) - (T_{Hi} - T_{Co})}{\ln\left(\frac{T_{Ho} - T_{Ci}}{T_{Hi} - T_{Co}}\right)}$$

For *parallel flow* in tubular heat exchangers

$$\Delta T_{lm} = \frac{(T_{Ho} - T_{Co}) - (T_{Hi} - T_{Ci})}{\ln\left(\frac{T_{Ho} - T_{Co}}{T_{Hi} - T_{Ci}}\right)}, \text{ where}$$

ΔT_{lm} = log mean temperature difference (K)

T_{Hi} = inlet temperature of the hot fluid (K)

T_{Ho} = outlet temperature of the hot fluid (K)

T_{Ci} = inlet temperature of the cold fluid (K)

T_{Co} = outlet temperature of the cold fluid (K)

Heat Exchanger Effectiveness, ε

$$\varepsilon = \frac{\dot{Q}}{\dot{Q}_{max}} = \frac{\text{actual heat transfer rate}}{\text{maximum possible heat transfer rate}}$$

$$\varepsilon = \frac{C_H(T_{Hi} - T_{Ho})}{C_{min}(T_{Hi} - T_{Ci})} \quad \text{or} \quad \varepsilon = \frac{C_C(T_{Co} - T_{Ci})}{C_{min}(T_{Hi} - T_{Ci})}$$

where

$C = \dot{m}c_P$ = heat capacity rate (W/K)

C_{min} = smaller of C_C or C_H

Number of Transfer Units (*NTU*)

$$NTU = \frac{UA}{C_{min}}$$

Effectiveness-NTU Relations

$$C_r = \frac{C_{min}}{C_{max}} = \text{heat capacity ratio}$$

For *parallel flow concentric tube* heat exchanger

$$\varepsilon = \frac{1 - \exp[-NTU(1 + C_r)]}{1 + C_r}$$

$$NTU = -\frac{\ln[1 - \varepsilon(1 + C_r)]}{1 + C_r}$$

For *counterflow concentric tube* heat exchanger

$$\varepsilon = \frac{1 - \exp[-NTU(1 - C_r)]}{1 - C_r\exp[-NTU(1 - C_r)]} \qquad (C_r < 1)$$

$$\varepsilon = \frac{NTU}{1 + NTU} \qquad (C_r = 1)$$

$$NTU = \frac{1}{C_r - 1}\ln\left(\frac{\varepsilon - 1}{\varepsilon C_r - 1}\right) \qquad (C_r < 1)$$

$$NTU = \frac{\varepsilon}{1 - \varepsilon} \qquad (C_r = 1)$$

RADIATION

Types of Bodies

Any Body

For any body, $\alpha + \rho + \tau = 1$, where

α = absorptivity (ratio of energy absorbed to incident energy)

ρ = reflectivity (ratio of energy reflected to incident energy)

τ = transmissivity (ratio of energy transmitted to incident energy)

Opaque Body

For an opaque body: $\alpha + \rho = 1$

Gray Body

A gray body is one for which

$\alpha = \varepsilon, (0 < \alpha < 1; 0 < \varepsilon < 1)$, where

ε = the emissivity of the body

For a gray body: $\varepsilon + \rho = 1$

Real bodies are frequently approximated as gray bodies.

Black body

A black body is defined as one which absorbs all energy incident upon it. It also emits radiation at the maximum rate for a body of a particular size at a particular temperature. For such a body

$\alpha = \varepsilon = 1$

Shape Factor (View Factor, Configuration Factor) Relations

Reciprocity Relations

$$A_iF_{ij} = A_jF_{ji}, \text{ where}$$

A_i = surface area (m^2) of surface i

F_{ij} = shape factor (view factor, configuration factor); fraction of the radiation leaving surface i that is intercepted by surface j; $0 \leq F_{ij} \leq 1$

Summation Rule for N Surfaces

$$\sum_{j=1}^{N} F_{ij} = 1$$

Net Energy Exchange by Radiation between Two Bodies

Body Small Compared to its Surroundings

$$\dot{Q}_{12} = \varepsilon\sigma A\left(T_1^4 - T_2^4\right), \text{ where}$$

$\dot{Q}_{12}$ = the net heat transfer rate from the body (W)

ε = the emissivity of the body

σ = the Stefan-Boltzmann constant
$[\sigma = 5.67 \times 10^{-8} \text{ W/(m}^2\cdot\text{K}^4)]$

A = the body surface area (m^2)

T_1 = the absolute temperature [K] of the body surface

T_2 = the absolute temperature [K] of the surroundings

Net Energy Exchange by Radiation between Two Black Bodies

The net energy exchange by radiation between two black bodies that see each other is given by

$$\dot{Q}_{12} = A_1F_{12}\sigma\left(T_1^4 - T_2^4\right)$$

Net Energy Exchange by Radiation between Two Diffuse-Gray Surfaces that Form an Enclosure

Generalized Cases

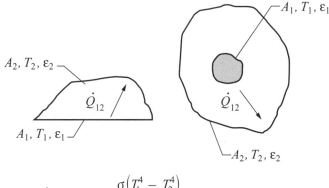

$$\dot{Q}_{12} = \frac{\sigma\left(T_1^4 - T_2^4\right)}{\dfrac{1 - \varepsilon_1}{\varepsilon_1 A_1} + \dfrac{1}{A_1F_{12}} + \dfrac{1 - \varepsilon_2}{\varepsilon_2 A_2}}$$

One-Dimensional Geometry with Thin Low-Emissivity Shield Inserted between Two Parallel Plates

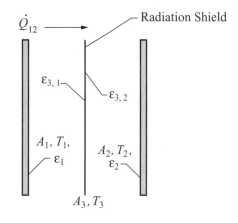

$$\dot{Q}_{12} = \frac{\sigma\left(T_1^4 - T_2^4\right)}{\dfrac{1 - \varepsilon_1}{\varepsilon_1 A_1} + \dfrac{1}{A_1F_{13}} + \dfrac{1 - \varepsilon_{3,1}}{\varepsilon_{3,1} A_3} + \dfrac{1 - \varepsilon_{3,2}}{\varepsilon_{3,2} A_3} + \dfrac{1}{A_3F_{32}} + \dfrac{1 - \varepsilon_2}{\varepsilon_2 A_2}}$$

Reradiating Surface

Reradiating Surfaces are considered to be insulated or adiabatic $\left(\dot{Q}_R = 0\right)$.

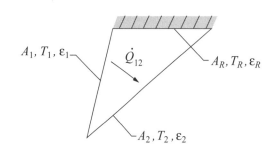

$$\dot{Q}_{12} = \frac{\sigma\left(T_1^4 - T_2^4\right)}{\dfrac{1 - \varepsilon_1}{\varepsilon_1 A_1} + \dfrac{1}{A_1F_{12} + \left[\left(\dfrac{1}{A_1F_{1R}}\right) + \left(\dfrac{1}{A_2F_{2R}}\right)\right]^{-1}} + \dfrac{1 - \varepsilon_2}{\varepsilon_2 A_2}}$$

TRANSPORT PHENOMENA

MOMENTUM, HEAT, AND MASS TRANSFER ANALOGY

For the equations which apply to **turbulent flow in circular tubes**, the following definitions apply:

Nu = Nusselt Number $\left[\dfrac{hD}{k}\right]$

Pr = Prandtl Number $(c_p\mu/k)$,

Re = Reynolds Number $(DV\rho/\mu)$,

Sc = Schmidt Number $[\mu/(\rho D_m)]$,

Sh = Sherwood Number $(k_m D/D_m)$,

St = Stanton Number $[h/(c_p G)]$,

c_m = concentration (mol/m^3),

c_p = heat capacity of fluid $[\text{J}/(\text{kg}\cdot\text{K})]$,

D = tube inside diameter (m),

D_m = diffusion coefficient (m^2/s),

$(dc_m/dy)_w$ = concentration gradient at the wall (mol/m^4),

$(dT/dy)_w$ = temperature gradient at the wall (K/m),

$(dv/dy)_w$ = velocity gradient at the wall (s^{-1}),

f = Moody friction factor,

G = mass velocity $[\text{kg}/(\text{m}^2\cdot\text{s})]$,

h = heat-transfer coefficient at the wall $[\text{W}/(\text{m}^2\cdot\text{K})]$,

k = thermal conductivity of fluid $[\text{W}/(\text{m}\cdot\text{K})]$,

k_m = mass-transfer coefficient (m/s),

L = length over which pressure drop occurs (m),

$(N/A)_w$ = inward mass-transfer flux at the wall $[\text{mol}/(\text{m}^2\cdot\text{s})]$,

$\left(\dot{Q}/A\right)_w$ = inward heat-transfer flux at the wall (W/m^2),

y = distance measured from inner wall toward centerline (m),

Δc_m = concentration difference between wall and bulk fluid (mol/m^3),

ΔT = temperature difference between wall and bulk fluid (K),

μ = absolute dynamic viscosity $(\text{N}\cdot\text{s/m}^2)$, and

τ_w = shear stress (momentum flux) at the tube wall (N/m^2).

Definitions already introduced also apply.

Rate of transfer as a function of gradients at the wall

Momentum Transfer:

$$\tau_w = -\mu\left(\frac{dv}{dy}\right)_w = -\frac{f\rho V^2}{8} = \left(\frac{D}{4}\right)\left(-\frac{\Delta p}{L}\right)_f$$

Heat Transfer:

$$\left(\frac{\dot{Q}}{A}\right)_w = -k\left(\frac{dT}{dy}\right)_w$$

Mass Transfer in Dilute Solutions:

$$\left(\frac{N}{A}\right)_w = -D_m\left(\frac{dc_m}{dy}\right)_w$$

Rate of transfer in terms of coefficients

Momentum Transfer:

$$\tau_w = \frac{f\rho V^2}{8}$$

Heat Transfer:

$$\left(\frac{\dot{Q}}{A}\right)_w = h\Delta T$$

Mass Transfer:

$$\left(\frac{N}{A}\right)_w = k_m\Delta c_m$$

Use of friction factor (f) to predict heat-transfer and masstransfer coefficients (turbulent flow)

Heat Transfer:

$$j_H = \left(\frac{\text{Nu}}{\text{RePr}}\right)\text{Pr}^{2/3} = \frac{f}{8}$$

Mass Transfer:

$$j_M = \left(\frac{\text{Sh}}{\text{ReSc}}\right)\text{Sc}^{2/3} = \frac{f}{8}$$

BIOLOGY

For more information on Biology see the **ENVIRONMENTAL ENGINEERING** section.

CELLULAR BIOLOGY

♦

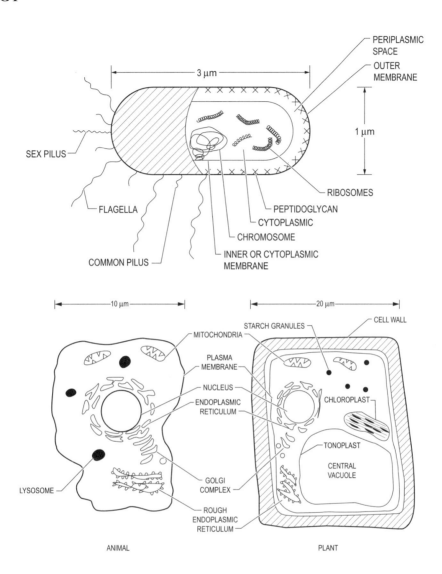

ANIMAL

PLANT

• **Primary Subdivisions of Biological Organisms**

Group	Cell structure	Properties	Constituent groups
Eucaryotes	Eucaryotic	Multicellular; extensive differentiation of cells and tissues Unicellular, coenocytic or mycelial; little or no tissue differentiation	Plants (seed plants, ferns, mosses) Animals (vertebrates, invertebrates) Protists (algae, fungi, protozoa)
Eubacteria	Procaryotic	Cell chemistry similar to eucaryotes	Most bacteria
Archaebacteria	Procaryotic	Distinctive cell chemistry	Methanogens, halophiles, thermoacidophiles

♦ Shuler, Michael L., & Fikret Kargi, *Bioprocess Engineering Basic Concepts*, Prentice Hall PTR, New Jersey, 1992.

• Stanier, Roger; Adelberg, Edward A; Wheelis, Mark L; Decastells; Painter, Page R; Ingraham, John L; *The Microbial World*, 5th ed., 1986. Reprinted by permission of Pearson Education, Inc., Upper Saddle River, NJ.

Biochemical Catabolic Pathways

Catabolism is the breakdown of nutrients to obtain energy and precursors for biosynthesis. Carbohydrates are the most important class of carbonaceous nutrients for fermentations, although some microbial species can also utilize amino acids, hydrocarbons and other compounds. As an illustration of microbial diversity, almost any carbohydrate or related compound can be fermented by some microbe.

Embden-Meyerhof-Parnas (EMP)

♦

The Embden-Meyerhof-Parnas (EMP) pathway. Notice that each six-carbon glucose substrate molecule yields two three-carbon intermediates, each of which passes through the reaction sequence on the right-hand side.

♦ Bailey, J.E. and D.F. Olis. *Biochemical Engineering Fundamentals,* 2nd ed., McGraw-Hill, New York, 1986.

Overall Stoichiometry

$$C_6H_{12}O_6 + P_i + 2ADP + 2NAD^+ \rightarrow 2C_3H_4O_3 + 2ATP + 2\left(NADH + H^+\right)$$

where

ADP is Adenosine diphosphate

ATP is Adenosine triphosphate. ADP and ATP are the primary energy carriers in biosynthesis. Each mole of ATP carries –7.3 kcal free energy in the process of being reduced to ADP.

NAD is Nicotinamide adenine dinucleotide (+ denotes the oxidized form and NADH denotes the reduced form). The role of this group is to serve as an electron carrier.

$C_3H_4O_3$ is pyruvate, a compound of central importance in biosynthesis reactions.

The EMP pathway, among several other biochemical pathways, provides carbon skeletons for cellular biosynthesis and provides energy sources via substrate level phosphorylation. The EMP pathway is perhaps the most common carbohydrate catabolic pathway. Other catabolic pathways include the pentose phosphate cycle, yielding 1.67 ATP for each mole of glucose and the Entner-Doudoroff (ED) pathway, yielding 1 ATP per mole of glucose.

Pathways are called upon by the elaborate control mechanisms within the organism depending on needed synthesis materials and available nutrients.

Biochemical Respiratory Pathways

Respiration is an energy producing process in which organic or reduced inorganic compounds are oxidized by inorganic compounds. If oxygen is the oxidizer, the processes yields carbon dioxide and water and is denoted as aerobic. Otherwise the process is denoted as facultative or anaerobic. Respiratory processes also serve as producers of precursors for biosynthesis. A summary of reductants and oxidants in bacterial respirations are shown in the table below.

TABLE 5.4 REDUCTANTS AND OXIDANTS IN BACTERIAL RESPIRATIONS [†]

REDUCTANT	OXIDANT	PRODUCTS	ORGANISM
H_2	O_2	H_2O	HYDROGEN BACTERIA
H_2	SO_4^{2-}	$H_2O + S^{2-}$	*Desulfovibrio*
ORGANIC COMPOUNDS	O_2	$CO_2 + H_2O$	MANY BACTERIA, ALL PLANTS AND ANIMALS
NH_3	O_2	$NO_2^- + H_2O$	NITRIFYING BACTERIA
NO_2^-	O_2	$NO_3^- + H_2O$	NITRIFYING BACTERIA
ORGANIC COMPOUNDS	NO_3^-	$N_2 + CO_2$	DENITRIFYING BACTERIA
Fe^{2+}	O_2	Fe^{3+}	*Ferrobacillus* (iron bacteria)
S^{2-}	O_2	$SO_4^{2-} + H_2O$	*Thiobacillus* (sulfur bacteria)

[†] From W.R. Sistrom, *Microbial Life*, 2d ed., table 4-2, p. 53, Holt, Rinehart, and Winston, Inc. New York, 1969.

The most important of the respiratory cycles is the tricarboxylic acid (denoted as TCA, citric acid cycle or Krebs cycle). The TCA cycle is shown in the figure below.

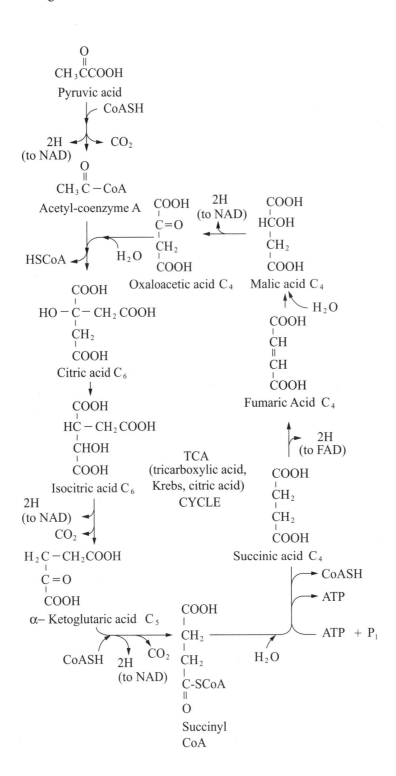

The tricarboxylic acid cycle

Bailey, J.E. and D.F. Olis, *Biochemical Engineering Fundamentals,* 2nd ed., McGraw-Hill, New York, 1986.

Important associated respiratory reactions is the oxidative phosphorylation in which ATP is regenerated.

Adding the EMP reactions with the TCA and oxidative phosphorylation reactions (not shown), results in a stoichiometry giving a net upper bound on ATP yield from glucose in a respiring cell with glucose as the primary carbon source and oxygen as the electron donor.

$$C_6H_{12}O_6 + 38ADP + 38P_i + 6O_2 \rightarrow 6CO_2 + 38ATP + 44H_2O$$

The free energy is approximately 38 moles (-7.3 kcal/mole) or -277 kcal/mole glucose.

Anaerobic reactions, where the carbon source and/or electron donor may be other inorganic or organic substrates, provide less free energy and thus result in lower product yields and lower biomass production.

The energy capture efficiency of the EMP-TCA sequence compared to the inorganic combustion reaction of glucose (net free energy of -686 kcal/mole glucose with the negative sign indicating energy evolved) is about 40%.

Photosynthesis

Photosynthesis is a most important process form synthesizing glucose from carbon dioxide. It also produces oxygen. The most important photosynthesis reaction is summarized as follows.

$$6CO_2 + 6H_2O + light \rightarrow C_6H_{12}O_6 + 6O_2$$

The light is required to be in the 400- to 700-nm range (visible light). Chlorophyll is the primary photosynthesis compound and it is found in organisms ranging from tree and plant leaves to single celled algae. The reaction above requires an energy equivalent of roughly 1,968 kcal per mole of glucose produced. The inorganic synthesis of glucose requires $+686$ kcal/mole glucose produced. Hence the efficiency of photosynthesis is roughly (686 kcal/mole glucose inorganic)/(1,968 kcal/mole glucose photosynthesis)

$$\frac{686 \text{ kcal/mole}}{1968 \text{ kcal/mole}} \text{ or } 35\%.$$

Other bacterial photosynthesis reactions may occur where the carbon source is other organic or inorganic substrates.

♦ **Organismal Growth in Batch Culture**

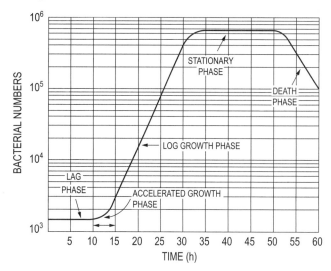

Exponential (log) growth with constant specific growth rate, μ

$$\mu = \left(\frac{1}{x}\right)\left(\frac{dx}{dt}\right), \text{ where}$$

x = the cell/organism number or cell/organism concentration
t = time (hr)
μ = the specific growth rate (time^{-1}) while in the exponential growth phase.

Logistic Growth–Batch Growth including initial into stationary phase

$$\frac{dx}{dt} = kx\left(1 - \frac{x}{x_\infty}\right)$$

$$x = \frac{x_0}{x_\infty}\left(1 - e^{kt}\right), \text{ where}$$

where,
k = logistic growth constant (h^{-1}),
x_0 = initial concentration (g/l)
x_∞ = carrying capacity (g/l).

♦ Davis, M.L., *Principles of Environmental Engineering*, McGraw-Hill, New York, 2004.

Characteristics of Selected Microbial Cells

Organism genus or type	Type	Metabolism[1]	Gram reaction[2]	Morphological characteristics[3]
Escherichia	Bacteria	Chemoorganotroph-facultative	Negative	Rod–may or may not be motile, variable extracellular material
Enterobacter	Bacteria	Chemoorganotroph-facultative	Negative	Rod–motile; significant extracellular material
Bacillus	Bacteria	Chemoorganotroph-aerobic	Positive	Rod–usually motile; spore; can be significant extracellular material
Lactobacillus	Bacteria	Chemoorganotroph-facultative	Variable	Rod–chains–usually nonmotile; little extracellular material
Staphylococcus	Bacteria	Chemoorganotroph-facultative	Positive	Cocci–nonmotile; moderate extracellular material
Nitrobacter	Bacteria	Chemoautotroph-aerobic; can use nitrite as electron donor	Negative	Short rod–usually nonmotile; little extracellular material
Rhizobium	Bacteria	Chemoorganotroph-aerobic; nitrogen fixing	Negative	Rods–motile; copious extracellular slime
Pseudomonas	Bacteria	Chemoorganotroph-aerobic and some chemolithotroph facultative (using NO_3 as electron acceptor)	Negative	Rods–motile; little extracellular slime
Thiobacillus	Bacteria	Chemoautotroph-facultative	Negative	Rods–motile; little extracellular slime
Clostridium	Bacteria	Chemoorganotroph-anaerobic	Positive	Rods–usually motile; spore; some extracellular slime
Methanobacterium	Bacteria	Chemoautotroph-anaerobic	Unknown	Rods or cocci–motility unknown; some extracellular slime
Chromatium	Bacteria	Photoautotroph-anaerobic	N/A	Rods–motile; some extracellular material
Spirogyra	Alga	Photoautotroph-aerobic	N/A	Rod/filaments; little extracellular material
Aspergillus	Mold	Chemoorganotroph-aerobic and facultative	--	Filamentous fanlike or cylindrical conidia and various spores
Candida	Yeast	Chemoorganotroph-aerobic and facultative	--	Usually oval but can form elongated cells, mycelia, and various spores
Saccharomyces	Yeast	Chemoorganotroph-facultative	--	Spherical or ellipsoidal; reproduced by budding; can form various spores

[1] Aerobic – requires or can use oxygen as an electron receptor.

Facultative – can vary the electron receptor from oxygen to organic materials.

Anaerobic – organic or inorganics other than oxygen serve as electron acceptor.

Chemoorganotrophs – derive energy and carbon from organic materials.

Chemoautotrophs – derive energy from organic carbons and carbon from carbon dioxide. Some species can also derive energy from inorganic sources.

Photolithotrophs – derive energy from light and carbon from CO_2. May be aerobic or anaerobic.

[2] Gram negative indicates a complex cell wall with a lipopolysaccharide outer layer. Gram positive indicates a less complicated cell wall with a peptide-based outer layer.

[3] Extracellular material production usually increases with reduced oxygen levels (e.g., facultative). Carbon source also affects production; extracellular material may be polysaccharides and/or proteins; statements above are to be understood as general in nature.

Adapted from Pelczar, M.J., R.D. Reid, and E.C.S. Chan, *Microbiology: Concepts and Applications,* McGraw-Hill, New York, 1993.

Transfer Across Membrane Barriers

Mechanisms

Passive diffusion – affected by lipid solubility (high solubility increases transport), molecular size (decreased with molecular size), and ionization (decreased with ionization).

Passive diffusion is influenced by:

1. Partition coefficient (indicates lipid solubility; high lipid solubility characterizes materials that easily penetrate skin and other membranes).

2. Molecular size is important in that small molecules tend to transport much easier than do large molecules.

3. Degree of ionization is important because, in most cases, only unionized forms of materials transport easily through membranes. Ionization is described by the following relationships:

Acids

$$pK_a - pH = \log_{10}\left[\frac{\text{nonionized form}}{\text{ionized form}}\right] = \log_{10}\frac{HA}{A}$$

Base

$$pK_a - pH = \log_{10}\left[\frac{\text{ionized form}}{\text{nonionized form}}\right] = \log_{10}\frac{HB^+}{B}$$

Facilitated diffusion – requires participation of a protein carrier molecule. This mode of transport is highly compound dependent.

Active diffusion – requires protein carrier and energy and is similarly affected by ionization and is highly compound dependent.

Other – includes the specialized mechanisms occurring in lungs, liver, and spleen.

BIOPROCESSING

Stoichiometry of Selected Biological Systems

Aerobic Production of Biomass and a Single Extracellular Product

$$C_{Ncs}H_{m*Ncs}O_{n*Ncs} + aO_2 + bNH_3 \rightarrow cC_{Ncc}H_{\alpha*Ncc}O_{\beta*Ncc}N_{\delta*Ncc} + dC_{Ncp}H_{x*Ncp}O_{y*Ncp}N_{z*Ncp} + eH_2O + fCO_2$$

$$\quad\quad \text{Substrate} \quad\quad\quad\quad\quad\quad\quad\quad \text{Biomass} \quad\quad\quad\quad\quad\quad \text{Bioproduct}$$

where

Ncs = the number of carbons in the substrate equivalent molecule

Ncc = the number of carbons in the biomass equivalent molecule

Ncp = the number of carbons in the bioproduct equivalent molecule

Coefficients m, n, α, β, x, y, z as shown above are multipliers useful in energy balances. The coefficient times the respective number of carbons is the equivalent number of constituent atoms in the representative molecule as shown in the following table.

Degrees of Reduction (available electrons per unit of carbon)

$\gamma_s = 4 + m - 2n$

$\gamma_b = 4 + \alpha - 2\beta - 3\delta$

$\gamma_p = 4 + x - 2y - 3z$

Subscripts refer to substrate (s), biomass (b), or product (p).

A high degree of reduction denotes a low degree of oxidation.

Carbon balance

$$cNcc + dNcp + f = Ncs$$

Nitrogen balance

$$c\delta Ncc + dzNcp = b$$

Electron balance

$$c\gamma_b \, Ncb + d\gamma_p \, Ncp = \gamma_s \, Ncs - 4a$$

Energy balance

$$Q_o c\gamma_b \, Ncc + Q_o d\gamma_p \, Ncp = Q_o\gamma_s \, Ncs - Q_o 4a,$$

Q_o = heat evolved per unit of equivalent of available electrons

≈ 26.95 kcal/mole O_2 consumed

Respiratory quotient (RQ) is the CO_2 produced per unit of O_2

$$RQ = \frac{f}{a}$$

Yield coefficient = c (grams of cells per gram substrate, $Y_{X/S}$)
or = d (grams of product per gram substrate, $Y_{X/XP}$)

Satisfying the carbon, nitrogen, and electron balances plus knowledge of the respiratory coefficient and a yield coefficient is sufficient to solve for a, b, c, d, and f coefficients.

Composition Data for Biomass and Selected Organic Compounds

♦

COMPOUND	MOLECULAR FORMULA	DEGREE OF REDUCTION, γ	WEIGHT, m
BIOMASS	$CH_{1.64}N_{0.16}O_{0.52}$ $P_{0.0054}S_{0.005}$[a]	4.17 (NH_3) 4.65 (N_3) 5.45 (HNO_3)	24.5
METHANE	CH_4	8	16.0
n-ALKANE	C_4H_{32}	6.13	14.1
METHANOL	CH_4O	6.0	32.0
ETHANOL	C_2H_6O	6.0	23.0
GLYCEROL	$C_2H_6O_3$	4.67	30.7
MANNITOL	$C_6H_{14}O_6$	4.33	30.3
ACETIC ACID	$C_2H_4O_2$	4.0	30.0
LACTIC ACID	$C_3H_6O_3$	4.0	30.0
GLUCOSE	$C_6H_{12}O_6$	4.0	30.0
FORMALDEHYDE	CH_2O	4.0	30.0
GLUCONIC ACID	$C_6H_{12}O_7$	3.67	32.7
SUCCINIC ACID	$C_4H_6O_4$	3.50	29.5
CITRIC ACID	$C_6H_8O_7$	3.0	32.0
MALIC ACID	$C_4H_6O_5$	3.0	33.5
FORMIC ACID	CH_2O_2	2.0	46.0
OXALIC ACID	$C_2H_2O_4$	1.0	45.0

The weights in column 4 of the above table are molecular weights expressed as a molecular weight per unit carbon; frequently denoted as reduced molecular weight. For example, glucose has a molecular weight of 180 g/mole and 6 carbons for a reduced molecular weight of 180/6 = 30.

Except for biomass, complete formulas for compounds are given in column 2 of the above table. Biomass may be represented by:

$$C_{4.4}H_{7.216}N_{0.704}O_{2.288}P_{0.0237}S_{0.022}$$

Aerobic conversion coefficients for an experimental bioconversion of a glucose substrate to oxalic acid with indicated nitrogen sources and conversion ratios with biomass and one product produced.

Theoretical yield coefficients[a]	Nitrogen Source		
	NH_3 (shown in the above equation)	HNO_3	N_2
a	2.909	3.569	3.929
b	0.48	0.288	0.144
c	0.6818	0.409	0.409
d	0.00167	0.00167	0.00167
e	4.258	4.66	4.522
f	2.997	4.197	4.917
Substrate to cells conversion ratio	0.5	0.3	0.3
Substrate to bioproduct conversion ratio	0.1	0.1	0.1
Respiration Quotient (RQ)	1.03	1.17	1.06

[a] The stoichiometric coefficients represent theoretical balance conditions. Product yields in practice are frequently lower and may be adjusted by the application of yield coefficients.

Energy balance

$$Q_o c \gamma_b Ncc = Q_o \gamma_s Ncs - Q_o 4a$$

$$Q_o \approx 26.95 \text{ kcal/mole } O_2 \text{ consumed}$$

Aerobic Biodegradation of Glucose with No Product, Ammonia Nitrogen Source, Cell Production Only, $RQ = 1.1$

$$\underset{\text{Substrate}}{C_6H_{12}O_6} + aO_2 + bNH_3 \rightarrow \underset{\text{Cells}}{cCH_{1.8}O_{0.5}N_{0.2}} + dCO_2 + eH_2O$$

For the above conditions, one finds that:

    a = 1.94
    b = 0.77
    c = 3.88
    d = 2.13
    e = 3.68

The c coefficient represents a theoretical maximum yield coefficient, which may be reduced by a yield factor.

Anaerobic Biodegradation of Organic Wastes, Incomplete Stabilization

$$C_aH_bO_cN_d \rightarrow nC_wH_xO_yN_z + mCH_4 + sCO_2 + rH_2O + (d - nx)NH_3$$

$$s = a - nw - m$$

$$r = c - ny - 2s$$

Knowledge of product composition, yield coefficient (n) and a methane/CO_2 ratio is needed.

Anaerobic Biodegradation of Organic Wastes, Complete Stabilization

$$C_aH_bO_cN_d + rH_2O \rightarrow mCH_4 + sCO_2 + dNH_3$$

$$r = \frac{4a - b - 2c + 3d}{4} \qquad m = \frac{4a + b - 2c - 3d}{8}$$

$$s = \frac{4a - b + 2c + 3d}{8}$$

♦ B. Atkinson and F. Mavitona, *Biochemical Engineering and Biotechnology Handbook*, Macmillan, Inc., 1983. Used with permission of Nature Publishing Group (www.nature.com).

CHEMISTRY

Avogadro's Number: The number of elementary particles in a mol of a substance.

$$1 \text{ mol} = 1 \text{ gram mole}$$
$$1 \text{ mol} = 6.02 \times 10^{23} \text{ particles}$$

A *mol* is defined as an amount of a substance that contains as many particles as 12 grams of ^{12}C (carbon 12). The elementary particles may be atoms, molecules, ions, or electrons.

ACIDS, BASES, and pH (aqueous solutions)

$$pH = \log_{10}\left(\frac{1}{[H^+]}\right), \text{ where}$$

$[H^+]$ = molar concentration of hydrogen ion, in gram moles per liter

Acids have pH < 7.

Bases have pH > 7.

ELECTROCHEMISTRY

Cathode – The electrode at which reduction occurs.

Anode – The electrode at which oxidation occurs.

Oxidation – The loss of electrons.

Reduction – The gaining of electrons.

Oxidizing Agent – A species that causes others to become oxidized.

Reducing Agent – A species that causes others to be reduced.

Cation – Positive ion

Anion – Negative ion

DEFINITIONS

Molarity of Solutions – The number of gram moles of a substance dissolved in a liter of solution.

Molality of Solutions – The number of gram moles of a substance per 1,000 grams of solvent.

Normality of Solutions – The product of the molarity of a solution and the number of valence changes taking place in a reaction.

Equivalent Mass – The number of parts by mass of an element or compound which will combine with or replace directly or indirectly 1.008 parts by mass of hydrogen, 8.000 parts of oxygen, or the equivalent mass of any other element or compound. For all elements, the atomic mass is the product of the equivalent mass and the valence.

Molar Volume of an Ideal Gas [at 0°C (32°F) and 1 atm (14.7 psia)]; 22.4 L/(g mole) [359 ft^3/(lb mole)].

Mole Fraction of a Substance – The ratio of the number of moles of a substance to the total moles present in a mixture of substances. Mixture may be a solid, a liquid solution, or a gas.

Equilibrium Constant of a Chemical Reaction

$$aA + bB \rightleftharpoons cC + dD$$

$$K_{eq} = \frac{[C]^c [D]^d}{[A]^a [B]^b}$$

Le Chatelier's Principle for Chemical Equilibrium – When a stress (such as a change in concentration, pressure, or temperature) is applied to a system in equilibrium, the equilibrium shifts in such a way that tends to relieve the stress.

Heats of Reaction, Solution, Formation, and Combustion – Chemical processes generally involve the absorption or evolution of heat. In an endothermic process, heat is absorbed (enthalpy change is positive). In an exothermic process, heat is evolved (enthalpy change is negative).

Solubility Product of a slightly soluble substance *AB*:

$$A_mB_n \rightarrow mA^{n+} + nB^{m-}$$

Solubility Product Constant = $K_{SP} = [A^+]^m [B^-]^n$

Metallic Elements – In general, metallic elements are distinguished from nonmetallic elements by their luster, malleability, conductivity, and usual ability to form positive ions.

Nonmetallic Elements – In general, nonmetallic elements are not malleable, have low electrical conductivity, and rarely form positive ions.

Faraday's Law – In the process of electrolytic changes, equal quantities of electricity charge or discharge equivalent quantities of ions at each electrode. One gram equivalent weight of matter is chemically altered at each electrode for 96,485 coulombs, or one Faraday, of electricity passed through the electrolyte.

A *catalyst* is a substance that alters the rate of a chemical reaction and may be recovered unaltered in nature and amount at the end of the reaction. The catalyst does not affect the position of equilibrium of a reversible reaction.

The *atomic number* is the number of protons in the atomic nucleus. The atomic number is the essential feature which distinguishes one element from another and determines the position of the element in the periodic table.

Boiling Point Elevation – The presence of a nonvolatile solute in a solvent raises the boiling point of the resulting solution compared to the pure solvent; i.e., to achieve a given vapor pressure, the temperature of the solution must be higher than that of the pure substance.

Freezing Point Depression – The presence of a solute lowers the freezing point of the resulting solution compared to that of the pure solvent.

PERIODIC TABLE OF ELEMENTS

Atomic Number			
Symbol			
Atomic Weight			

1	2	3	4	5	6	7	8	9	10	11	12	13	14	15	16	17	18
1 **H** 1.0079																	2 **He** 4.0026
3 **Li** 6.941	4 **Be** 9.0122											5 **B** 10.811	6 **C** 12.011	7 **N** 14.007	8 **O** 15.999	9 **F** 18.998	10 **Ne** 20.179
11 **Na** 22.990	12 **Mg** 24.305											13 **Al** 26.981	14 **Si** 28.086	15 **P** 30.974	16 **S** 32.066	17 **Cl** 35.453	18 **Ar** 39.948
19 **K** 39.098	20 **Ca** 40.078	21 **Sc** 44.956	22 **Ti** 47.88	23 **V** 50.941	24 **Cr** 51.996	25 **Mn** 54.938	26 **Fe** 55.847	27 **Co** 58.933	28 **Ni** 58.69	29 **Cu** 63.546	30 **Zn** 65.39	31 **Ga** 69.723	32 **Ge** 72.61	33 **As** 74.921	34 **Se** 78.96	35 **Br** 79.904	36 **Kr** 83.80
37 **Rb** 85.468	38 **Sr** 87.62	39 **Y** 88.906	40 **Zr** 91.224	41 **Nb** 92.906	42 **Mo** 95.94	43 **Tc** (98)	44 **Ru** 101.07	45 **Rh** 102.91	46 **Pd** 106.42	47 **Ag** 107.87	48 **Cd** 112.41	49 **In** 114.82	50 **Sn** 118.71	51 **Sb** 121.75	52 **Te** 127.60	53 **I** 126.90	54 **Xe** 131.29
55 **Cs** 132.91	56 **Ba** 137.33	57* **La** 138.91	72 **Hf** 178.49	73 **Ta** 180.95	74 **W** 183.85	75 **Re** 186.21	76 **Os** 190.2	77 **Ir** 192.22	78 **Pt** 195.08	79 **Au** 196.97	80 **Hg** 200.59	81 **Tl** 204.38	82 **Pb** 207.2	83 **Bi** 208.98	84 **Po** (209)	85 **At** (210)	86 **Rn** (222)
87 **Fr** (223)	88 **Ra** 226.02	89** **Ac** 227.03	104 **Rf** (261)	105 **Ha** (262)													

*Lanthanide Series

58 **Ce** 140.12	59 **Pr** 140.91	60 **Nd** 144.24	61 **Pm** (145)	62 **Sm** 150.36	63 **Eu** 151.96	64 **Gd** 157.25	65 **Tb** 158.92	66 **Dy** 162.50	67 **Ho** 164.93	68 **Er** 167.26	69 **Tm** 168.93	70 **Yb** 173.04	71 **Lu** 174.97

**Actinide Series

90 **Th** 232.04	91 **Pa** 231.04	92 **U** 238.03	93 **Np** 237.05	94 **Pu** (244)	95 **Am** (243)	96 **Cm** (247)	97 **Bk** (247)	98 **Cf** (251)	99 **Es** (252)	100 **Fm** (257)	101 **Md** (258)	102 **No** (259)	103 **Lr** (260)

IMPORTANT FAMILIES OF ORGANIC COMPOUNDS

		FAMILY										
	Alkane	**Alkene**	**Alkyne**	**Arene**	**Haloalkane**	**Alcohol**	**Ether**	**Amine**	**Aldehyde**	**Ketone**	**Carboxylic Acid**	**Ester**
Specific Example	CH_3CH_3	$H_2C=CH_2$	$HC\equiv CH$	(benzene ring)	CH_3CH_2Cl	CH_3CH_2OH	CH_3OCH_3	CH_3NH_2	$CH_3\overset{O}{\overset{\|}{C}}H$	$CH_3\overset{O}{\overset{\|}{C}}CH_3$	$CH_3\overset{O}{\overset{\|}{C}}OH$	$CH_3\overset{O}{\overset{\|}{C}}OCH_3$
IUPAC Name	Ethane	Ethene or Ethylene	Ethyne or Acetylene	Benzene	Chloroethane	Ethanol	Methoxy-methane	Methan-amine	Ethanal	Acetone	Ethanoic Acid	Methyl ethanoate
Common Name	Ethane	Ethylene	Acetylene	Benzene	Ethyl chloride	Ethyl alcohol	Dimethyl ether	Methyl-amine	Acetal-dehyde	Dimethyl ketone	Acetic Acid	Methyl acetate
General Formula	RH	$RCH=CH_2$ $RCH=CHR$ $R_2C=CHR$ $R_2C=CR_2$	$RC\equiv CH$ $RC\equiv CR$	ArH	RX	ROH	ROR	RNH_2 R_2NH R_3N	$R\overset{O}{\overset{\|}{C}}H$	$R_1\overset{O}{\overset{\|}{C}}R_2$	$R\overset{O}{\overset{\|}{C}}OH$	$R\overset{O}{\overset{\|}{C}}OR$
Functional Group	C–H and C–C bonds	$\diagup C=C \diagdown$	$-C\equiv C-$	Aromatic Ring	$-\overset{\|}{\underset{\|}{C}}-X$	$-\overset{\|}{\underset{\|}{C}}-OH$	$-\overset{\|}{\underset{\|}{C}}-O-\overset{\|}{\underset{\|}{C}}-$	$-\overset{\|}{\underset{\|}{C}}-\overset{\|}{N}-$	$-\overset{O}{\overset{\|}{C}}-H$	$-\overset{O}{\overset{\|}{C}}-$	$-\overset{O}{\overset{\|}{C}}-OH$	$-\overset{O}{\overset{\|}{C}}-O-C-$

Standard Oxidation Potentials for Corrosion Reactions*	
Corrosion Reaction	**Potential, E_o, Volts vs. Normal Hydrogen Electrode**
$Au \rightarrow Au^{3+} + 3e^-$	−1.498
$2H_2O \rightarrow O_2 + 4H^+ + 4e^-$	−1.229
$Pt \rightarrow Pt^{2+} + 2e^-$	−1.200
$Pd \rightarrow Pd^{2+} + 2e^-$	−0.987
$Ag \rightarrow Ag^+ + e^-$	−0.799
$2Hg \rightarrow Hg_2^{2+} + 2e^-$	−0.788
$Fe^{2+} \rightarrow Fe^{3+} + e^-$	−0.771
$4(OH)^- \rightarrow O_2 + 2H_2O + 4e^-$	−0.401
$Cu \rightarrow Cu^{2+} + 2e^-$	−0.337
$Sn^{2+} \rightarrow Sn^{4+} + 2e^-$	−0.150
$H_2 \rightarrow 2H^+ + 2e^-$	0.000
$Pb \rightarrow Pb^{2+} + 2e^-$	+0.126
$Sn \rightarrow Sn^{2+} + 2e^-$	+0.136
$Ni \rightarrow Ni^{2+} + 2e^-$	+0.250
$Co \rightarrow Co^{2+} + 2e^-$	+0.277
$Cd \rightarrow Cd^{2+} + 2e^-$	+0.403
$Fe \rightarrow Fe^{2+} + 2e^-$	+0.440
$Cr \rightarrow Cr^{3+} + 3e^-$	+0.744
$Zn \rightarrow Zn^{2+} + 2e^-$	+0.763
$Al \rightarrow Al^{3+} + 3e^-$	+1.662
$Mg \rightarrow Mg^{2+} + 2e^-$	+2.363
$Na \rightarrow Na^+ + e^-$	+2.714
$K \rightarrow K^+ + e^-$	+2.925

* Measured at 25°C. Reactions are written as anode half-cells.
Arrows are reversed for cathode half-cells.

Flinn, Richard A. and Trojan, Paul K., *Engineering Materials and Their Applications*, 4th ed., Houghton Mifflin Company, 1990.

NOTE: In some chemistry texts, the reactions and the signs of the values (in this table) are reversed; for example, the half-cell potential of zinc is given as −0.763 volt for the reaction $Zn^{2+} + 2e^- \rightarrow Zn$. When the potential E_o is positive, the reaction proceeds spontaneously as written.

MATERIALS SCIENCE/STRUCTURE OF MATTER

ATOMIC BONDING

Primary Bonds

      Ionic (e.g., salts, metal oxides)

      Covalent (e.g., within polymer molecules)

      Metallic (e.g., metals)

CORROSION

A table listing the standard electromotive potentials of metals is shown on the previous page.

For corrosion to occur, there must be an anode and a cathode in electrical contact in the presence of an electrolyte.

Anode Reaction (Oxidation) of a Typical Metal, M

$M^{o} \rightarrow M^{n+} + ne^{-}$

Possible Cathode Reactions (Reduction)

$$\frac{1}{2} O_2 + 2 e^- + H_2O \rightarrow 2 OH^-$$
$$\frac{1}{2} O_2 + 2 e^- + 2 H_3O^+ \rightarrow 3 H_2O$$
$$2 e^- + 2 H_3O^+ \rightarrow 2 H_2O + H_2$$

When dissimilar metals are in contact, the more electropositive one becomes the anode in a corrosion cell. Different regions of carbon steel can also result in a corrosion reaction: e.g., cold-worked regions are anodic to noncoldworked; different oxygen concentrations can cause oxygen-deficient regions to become cathodic to oxygen-rich regions; grain boundary regions are anodic to bulk grain; in multiphase alloys, various phases may not have the same galvanic potential.

DIFFUSION

Diffusion Coefficient

    $D = D_0 \, e^{-Q/(RT)}$, where

D = diffusion coefficient,

D_0 = proportionality constant,

Q = activation energy,

R = gas constant [8.314 J/(mol•K)], and

T = absolute temperature.

THERMAL AND MECHANICAL PROCESSING

Cold working (plastically deforming) a metal increases strength and lowers ductility.

Raising temperature causes (1) recovery (stress relief), (2) recrystallization, and (3) grain growth. *Hot working* allows these processes to occur simultaneously with deformation.

Quenching is rapid cooling from elevated temperature, preventing the formation of equilibrium phases.

In steels, quenching austenite [FCC (γ) iron] can result in martensite instead of equilibrium phases—ferrite [BCC (α) iron] and cementite (iron carbide).

TESTING METHODS

Standard Tensile Test

Using the standard tensile test, one can determine elastic modulus, yield strength, ultimate tensile strength, and ductility (% elongation). (See Mechanics of Materials section.)

Endurance Test

Endurance tests (fatigue tests to find endurance limit) apply a cyclical loading of constant maximum amplitude. The plot (usually semi-log or log-log) of the maximum stress (σ) and the number (N) of cycles to failure is known as an *S-N* plot. The figure below is typical of steel but may not be true for other metals; i.e., aluminum alloys, etc.

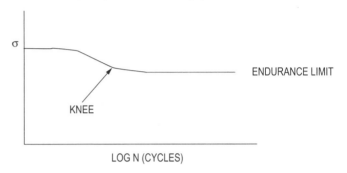

The *endurance stress* (*endurance limit* or *fatigue limit*) is the maximum stress which can be repeated indefinitely without causing failure. The *fatigue life* is the number of cycles required to cause failure for a given stress level.

Impact Test

The *Charpy Impact Test* is used to find energy required to fracture and to identify ductile to brittle transition.

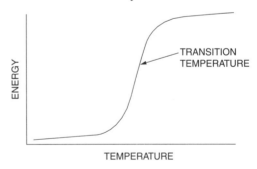

Impact tests determine the amount of energy required to cause failure in standardized test samples. The tests are repeated over a range of temperatures to determine the *ductile to brittle transition temperature*.

Creep

Creep occurs under load at elevated temperatures. The general equation describing creep is:

$$\frac{d\varepsilon}{dt} = A\sigma^n e^{-Q/(RT)}$$

where:

ε = strain,

t = time,

A = pre-exponential constant,

σ = applied stress,

n = stress sensitivity.

For polymers below, the glass transition temperature, T_g, n is typically between 2 and 4, and Q is $\geq$ 100 kJ/mol. Above T_g, n is typically between 6 and 10, and Q is ~ 30 kJ/mol.

For metals and ceramics, n is typically between 3 and 10, and Q is between 80 and 200 kJ/mol.

STRESS CONCENTRATION IN BRITTLE MATERIALS

When a crack is present in a material loaded in tension, the stress is intensified in the vicinity of the crack tip. This phenomenon can cause significant loss in overall ability of a member to support a tensile load.

$$K_I = y\sigma\sqrt{\pi a}$$

K_I = the stress intensity in tension, MPa•m$^{1/2}$,

y = is a geometric parameter,

$\quad y = 1$ for interior crack

$\quad y = 1.1$ for exterior crack

σ = is the nominal applied stress, and

a = is crack length as shown in the two diagrams below.

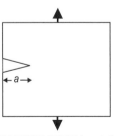

EXTERIOR CRACK ($y = 1.1$)

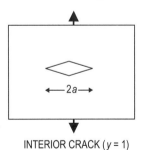

INTERIOR CRACK ($y = 1$)

The critical value of stress intensity at which catastrophic crack propagation occurs, K_{Ic}, is a material property.

Representative Values of Fracture Toughness

Material	K_{Ic} (MPa•m$^{1/2}$)	K_{Ic} (ksi•in$^{1/2}$)
A1 2014-T651	24.2	22
A1 2024-T3	44	40
52100 Steel	14.3	13
4340 Steel	46	42
Alumina	4.5	4.1
Silicon Carbide	3.5	3.2

HARDENABILITY OF STEELS

Hardenability is the "ease" with which hardness may be attained. *Hardness* is a measure of resistance to plastic deformation.

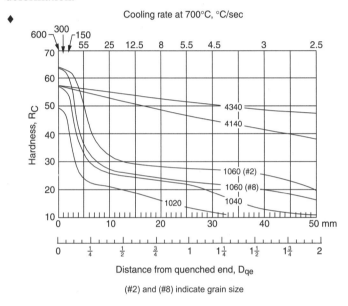

(#2) and (#8) indicate grain size

JOMINY HARDENABILITY CURVES FOR SIX STEELS

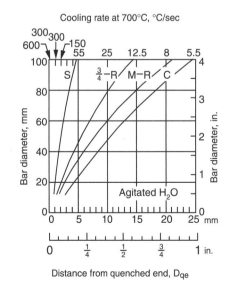

COOLING RATES FOR BARS QUENCHED IN AGITATED WATER

◆ Van Vlack, L., *Elements of Materials Science & Engineering*, Addison-Wesley, Boston, 1989.

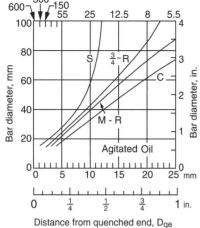

Cooling rate at 700°C, °C/sec

COOLING RATES FOR BARS QUENCHED IN AGITATED OIL

RELATIONSHIP BETWEEN HARDNESS AND TENSILE STRENGTH

For steels, there is a general relationship between Brinell hardness and tensile strength as follows:

$$TS(psi) \simeq 500 \, BHN$$

$$TS(MPa) \simeq 3.5 \, BHN$$

ASTM GRAIN SIZE

$$S_V = 2P_L$$

$$N_{(0.0645 \, mm^2)} = 2^{(n-1)}$$

$$\frac{N_{actual}}{Actual \ Area} = \frac{N}{(0.0645 \ mm^2)}, \ where$$

S_V = grain-boundary surface per unit volume,

P_L = number of points of intersection per unit length between the line and the boundaries,

N = number of grains observed in a area of $0.0645 \, mm^2$, and

n = grain size (nearest integer > 1).

COMPOSITE MATERIALS

$$\rho_c = \Sigma f_i \rho_i$$

$$C_c = \Sigma f_i c_i$$

$$\left[\Sigma \frac{f_i}{E_i} \right]^{-1} \leq E_c \leq \Sigma f_i E_i$$

$$\sigma_c = \Sigma f_i \sigma_i$$

ρ_c = density of composite,

C_c = heat capacity of composite per unit volume,

E_c = Young's modulus of composite,

f_i = volume fraction of individual material,

c_i = heat capacity of individual material per unit volume, and

E_i = Young's modulus of individual material

σ_c = strength parallel to fiber direction.

Also, for axially oriented, long, fiber-reinforced composites, the strains of the two components are equal.

$$(\Delta L/L)_1 = (\Delta L/L)_2$$

ΔL = change in length of the composite,

L = original length of the composite.

HALF-LIFE

$$N = N_o e^{-0.693t/\tau}, \ where$$

N_o = original number of atoms,

N = final number of atoms,

t = time, and

τ = half-life.

Material	Density ρ Mg/m^3	Young's Modulus E GPa	E/ρ $N \cdot m/g$
Aluminum	2.7	70	26,000
Steel	7.8	205	26,000
Magnesium	1.7	45	26,000
Glass	2.5	70	28,000
Polystyrene	1.05	2	2,700
Polyvinyl Chloride	1.3	< 4	< 3,500
Alumina fiber	3.9	400	100,000
Aramide fiber	1.3	125	100,000
Boron fiber	2.3	400	170,000
Beryllium fiber	1.9	300	160,000
BeO fiber	3.0	400	130,000
Carbon fiber	2.3	700	300,000
Silicon Carbide fiber	3.2	400	120,000

CONCRETE

Portland Cement Concrete

Concrete is a mixture of portland cement, fine aggregate, coarse aggregate, air, and water. It is a temporarily plastic material, which can be cast or molded, but is later converted to a solid mass by chemical reaction.

Water-cement (W/C) ratio is the primary factor affecting the strength of concrete. The figure below shows how W/C, expressed as a ratio by weight, affects the compressive strength for both air-entrained and non-air-entrained concrete. Strength decreases with an increase in W/C in both cases.

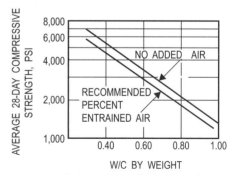

Concrete strength decreases with increases in water-cement ratio for concrete with and without entrained air.

(From *Concrete Manual*, 8th ed., U.S. Bureau of Reclamation, 1975.)

♦ Van Vlack, L., *Elements of Materials Science & Engineering*, Addison-Wesley, Boston, 1989.

Water Content affects workability. However, an increase in water without a corresponding increase in cement reduces the concrete strength. Air entrainment is the preferred method of increasing workability.

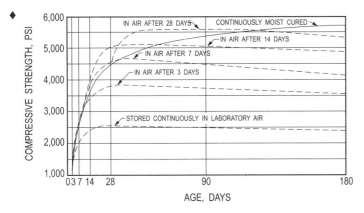

Concrete compressive strength varies with moist-curing conditions. Mixes tested had a water-cement ratio of 0.50, a slump of 3.5 in., cement content of 556 lb/yd³, sand content of 36%, and air content of 4%.

POLYMERS

Classification of Polymers

Polymers are materials consisting of high molecular weight carbon-based chains, often thousands of atoms long. Two broad classifications of polymers are thermoplastics or thermosets. Thermoplastic materials can be heated to high temperature and then reformed. Thermosets, such as vulcanized rubber or epoxy resins, are cured by chemical or thermal processes which cross link the polymer chains, preventing any further re-formation.

Amorphous Materials and Glasses

Silica and some carbon-based polymers can form either crystalline or amorphous solids, depending on their composition, structure, and processing conditions. These two forms exhibit different physical properties. Volume expansion with increasing temperature is shown schematically in the following graph, in which T_m is the melting temperature, and T_g is the glass transition temperature. Below the glass transition temperature, amorphous materials behave like brittle solids. For most common polymers, the glass transition occurs between –40°C and 250°C.

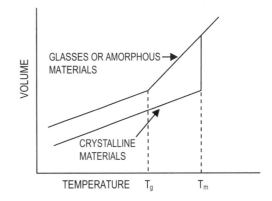

Thermo-Mechanical Properties of Polymers

The curve for the elastic modulus, E, or strength of polymers, σ, behaves according to the following pattern:

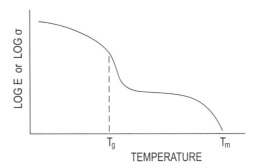

Polymer Additives

Chemicals and compounds are added to polymers to improve properties for commercial use. These substances, such as plasticizers, improve formability during processing, while others increase strength or durability.

Examples of common additives are:

Plasticizers: vegetable oils, low molecular weight polymers or monomers

Fillers: talc, chopped glass fibers

Flame retardants: halogenated paraffins, zinc borate, chlorinated phosphates

Ultraviolet or visible light resistance: carbon black

Oxidation resistance: phenols, aldehydes

♦ Merritt, Frederick S., *Standard Handbook for Civil Engineers*, 3rd ed., McGraw-Hill, 1983.

BINARY PHASE DIAGRAMS

Allows determination of (1) what phases are present at equilibrium at any temperature and average composition, (2) the compositions of those phases, and (3) the fractions of those phases.

Eutectic reaction (liquid → two solid phases)

Eutectoid reaction (solid → two solid phases)

Peritectic reaction (liquid + solid → solid)

Pertectoid reaction (two solid phases → solid)

Lever Rule

The following phase diagram and equations illustrate how the weight of each phase in a two-phase system can be determined:

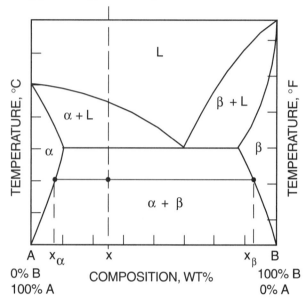

(In diagram, L = liquid.) If x = the average composition at temperature T, then

$$\text{wt}\% \, \alpha = \frac{x_\beta - x}{x_\beta - x_\alpha} \times 100$$

$$\text{wt}\% \, \beta = \frac{x - x_\alpha}{x_\beta - x_\alpha} \times 100$$

Iron-Iron Carbide Phase Diagram

♦

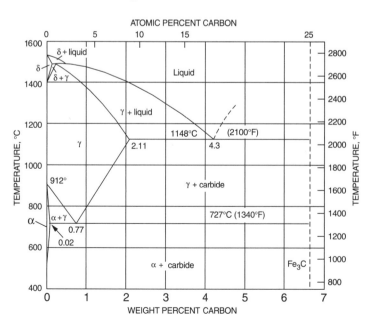

♦ Van Vlack, L., *Elements of Materials Science & Engineering*, Addison-Wesley, Boston, 1989.

COMPUTER SPEADSHEETS

A *spreadsheet* is a collection of items arranged in a tabular format and organized in rows and columns. Typically, rows are assigned numbers (1, 2, 3, …) and columns are assigned letters (A, B, C, …) as illustrated below.

	A	B	C	D	E	F
1						
2						
3					▨	
4						
5						
6						

A *cell* is a unique element identified by an *address* consisting of the column letter and the row number. For example, the address of the shaded cell shown above is E3. A cell may contain a number, formula, or label.

By default, when a cell containing a formula is copied to another cell, the column and row references will automatically be changed (this is called *relative addressing*). The following example demonstrates relative addressing.

$$C3 = B4 + D5$$

If C3 is copied to	The result is
D3	D3 = C4 + E5
C4	C4 = B5 + D6
B4	B4 = A5 + C6
E5	E5 = D6 + F7

If a row or column is referenced using *absolute addressing* (typically indicated with the $ symbol), the row or column reference will not be changed when a cell containing a formula is copied to another cell. The following example illustrates absolute addressing.

$$C3 = \$B4 + D\$5 + \$A\$1$$

If C3 is copied to	Result is
D3	D3 = $B4 + E$5 + A1
C4	C4 = $B5 + D$5 + A1
B4	B4 = $B5 + C$5 + A1
E5	E5 = $B6 + F$5 + A1

MEASUREMENT AND CONTROLS

MEASUREMENT

Definitions:

Transducer – a device used to convert a physical parameter such as temperature, pressure, flow, light intensity, etc. into an electrical signal (also called a *sensor*).

Transducer sensitivity – the ratio of change in electrical signal magnitude to the change in magnitude of the physical parameter being measured.

Resistance Temperature Detector (RTD) – a device used to relate change in resistance to change in temperature. Typically made from platinum, the controlling equation for an RTD is given by:

$$R_T = R_0\big[1 + \alpha(T - T_0)\big], \text{ where}$$

R_T is the resistance of the RTD at temperature T (measured in °C)

R_0 is the resistance of the RTD at the reference temperature T_0 (usually 0° C)

α is the temperature coefficient of the RTD

Strain Gauge – a device whose electrical resistance varies in proportion to the amount of strain in the device.

Gauge factor (*GF*) – the ratio of fractional change in electrical resistance to the fractional change in length (strain):

$$GF = \frac{\Delta R/R}{\Delta L/L} = \frac{\Delta R/R}{\varepsilon}, \text{ where}$$

R is the nominal resistance of the strain gauge at nominal length L.

ΔR is the change in resistance due the change in length ΔL.

ε is the normal strain sensed by the gauge.

The gauge factor for metallic strain gauges is typically around 2.

Wheatstone Bridge – an electrical circuit used to measure changes in resistance.

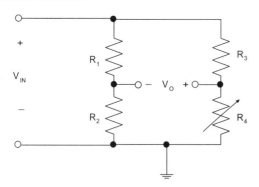

WHEATSTONE BRIDGE

If $R_1 R_4 = R_2 R_3$ then $V_0 = 0$ V and the bridge is said to be *balanced*. If $R_1 = R_2 = R_3 = R$ and $R_4 = R + \Delta R$ where

$$\Delta R \ll R, \text{ then } V_0 \approx \frac{\Delta R}{4R} \cdot V_{IN}.$$

SAMPLING

When a continuous-time or analog signal is sampled using a discrete-time method, certain basic concepts should be considered. The sampling rate or frequency is given by

$$f_s = \frac{1}{\Delta t}$$

Shannon's sampling theorem states that in order to accurately reconstruct the analog signal from the discrete sample points, the sample rate must be larger than twice the highest frequency contained in the measured signal. Denoting this frequency, which is called the Nyquist frequency, as f_N, the sampling theorem requires that

$$f_s > 2f_N$$

When the above condition is not met, the higher frequencies in the measured signal will not be accurately represented and will appear as lower frequencies in the sampled data. These are known as alias frequencies.

Analog-to-Digital Conversion

When converting an analog signal to digital form, the resolution of the conversion is an important factor. For a measured analog signal over the nominal range $[V_L, V_H]$, where V_L is the low end of the voltage range and V_H is the nominal high end of the voltage range, the voltage resolution is given by

$$\varepsilon_V = \frac{V_H - V_L}{2^n}$$

where n is the number of conversion bits of the A/D converter with typical values of 4, 8, 10, 12, or 16. This number is a key design parameter. After converting an analog signal, the A/D converter produces an integer number of n bits. Call this number N. Note that the range of N is $[0, 2^n - 1]$. When calculating the discrete voltage, V, using the reading, N, from the A/D converter the following equation is used.

$$V = \varepsilon_V N + V_L$$

Note that with this strategy, the highest measurable voltage is one voltage resolution less than V_H, or $V_H - \varepsilon_V$.

Signal Conditioning

Signal conditioning of the measured analog signal is often required to prevent alias frequencies and to reduce measurement errors. For information on these signal conditioning circuits, also known as filters, see the **ELECTRICAL AND COMPUTER ENGINEERING** section.

MEASUREMENT UNCERTAINTY

Suppose that a calculated result R depends on measurements whose values are $x_1 \pm w_1$, $x_2 \pm w_2$, $x_3 \pm w_3$, etc., where $R = f(x_1, x_2, x_3, \dots x_n)$, x_i is the measured value, and w_i is the uncertainty in that value. The uncertainty in R, w_R, can be estimated using the Kline-McClintock equation:

$$w_R = \sqrt{\left(w_1 \frac{\partial f}{\partial x_1}\right)^2 + \left(w_2 \frac{\partial f}{\partial x_2}\right)^2 + \dots + \left(w_n \frac{\partial f}{\partial x_n}\right)^2}$$

CONTROL SYSTEMS

The linear time-invariant transfer function model represented by the block diagram

can be expressed as the ratio of two polynomials in the form

$$\frac{Y(s)}{X(s)} = G(s) = \frac{N(s)}{D(s)} = K \frac{\prod_{m=1}^{M}(s - z_m)}{\prod_{n=1}^{N}(s - p_n)}$$

where the M zeros, z_m, and the N poles, p_n, are the roots of the numerator polynomial, $N(s)$, and the denominator polynomial, $D(s)$, respectively.

One classical negative feedback control system model block diagram is

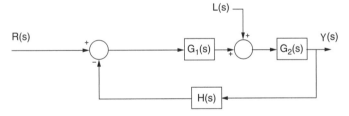

where $G_1(s)$ is a controller or compensator, $G_2(s)$ represents a plant model, and $H(s)$ represents the measurement dynamics. $Y(s)$ represents the controlled variable, $R(s)$ represents the reference input, and $L(s)$ represents a load disturbance. $Y(s)$ is related to $R(s)$ and $L(s)$ by

$$Y(s) = \frac{G_1(s) G_2(s)}{1 + G_1(s) G_2(s) H(s)} R(s) + \frac{G_2(s)}{1 + G_1(s) G_2(s) H(s)} L(s)$$

$G_1(s)\,G_2(s)\,H(s)$ is the open-loop transfer function. The closed-loop characteristic equation is

$$1 + G_1(s)\,G_2(s)\,H(s) = 0$$

System performance studies normally include

1. Steady-state analysis using constant inputs based on the Final Value Theorem. If all poles of a $G(s)$ function have negative real parts, then

$$\text{DC Gain} = \lim_{s \to 0} G(s)$$

Note that $G(s)$ could refer to either an open-loop or a closed-loop transfer function.

For the unity feedback control system model

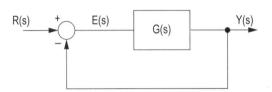

with the open-loop transfer function defined by

$$G(s) = \frac{K_B}{s^T} \times \frac{\prod_{m=1}^{M}\left(1 + s/\omega_m\right)}{\prod_{n=1}^{N}\left(1 + s/\omega_n\right)}$$

The following steady-state error analysis table can be constructed where T denotes the type of system, i.e., type 0, type 1, etc.

Steady-State Error e_{ss}			
Input \ Type	$T = 0$	$T = 1$	$T = 2$
Unit Step	$1/(K_B + 1)$	0	0
Ramp	∞	$1/K_B$	0
Acceleration	∞	∞	$1/K_B$

2. Frequency response evaluations to determine dynamic performance and stability. For example, relative stability can be quantified in terms of

a. Gain margin (GM), which is the additional gain required to produce instability in the unity gain feedback control system. If at $\omega = \omega_{180}$,

$$\angle G(j\omega_{180}) = -180°; \text{ then}$$
$$\text{GM} = -20\log_{10}\left(\left|G(j\omega_{180})\right|\right)$$

b. Phase margin (PM), which is the additional phase required to produce instability. Thus,
$$\text{PM} = 180° + \angle G(j\omega_{0dB})$$

where ω_{0dB} is the ω that satisfies $\left|G(j\omega)\right| = 1$.

3. Transient responses are obtained by using Laplace transforms or computer solutions with numerical integration.

Common Compensator/Controller forms are

PID Controller $G_C(s) = K\left(1 + \frac{1}{T_I s} + T_D s\right)$

Lag or Lead Compensator $G_C(s) = K\left(\frac{1 + sT_1}{1 + sT_2}\right)$ depending on the ratio of T_1/T_2.

Routh Test
For the characteristic equation

$$a_n s^n + a_{n-1} s^{n-1} + a_{n-2} s^{n-2} + \ldots + a_0 = 0$$

the coefficients are arranged into the first two rows of an array. Additional rows are computed. The array and coefficient computations are defined by:

$$
\begin{array}{cccccc}
a_n & a_{n-2} & a_{n-4} & \cdots & \cdots & \cdots \\
a_{n-1} & a_{n-3} & a_{n-5} & \cdots & \cdots & \cdots \\
b_1 & b_2 & b_3 & \cdots & \cdots & \cdots \\
c_1 & c_2 & c_3 & \cdots & \cdots & \cdots
\end{array}
$$

where

$$b_1 = \frac{a_{n-1} a_{n-2} - a_n a_{n-3}}{a_{n-1}} \qquad c_1 = \frac{a_{n-3} b_1 - a_{n-1} b_2}{b_1}$$

$$b_2 = \frac{a_{n-1} a_{n-4} - a_n a_{n-5}}{a_{n-1}} \qquad c_2 = \frac{a_{n-5} b_1 - a_{n-1} b_3}{b_1}$$

The necessary and sufficient conditions for all the roots of the equation to have negative real parts is that all the elements in the first column be of the same sign and nonzero.

First-Order Control System Models
The transfer function model for a first-order system is

$$\frac{Y(s)}{R(s)} = \frac{K}{\tau s + 1},$$ where

K = steady-state gain,

τ = time constant

The step response of a first-order system to a step input of magnitude M is

$$y(t) = y_0 e^{-t/\tau} + KM\left(1 - e^{-t/\tau}\right)$$

In the chemical process industry, y_0 is typically taken to be zero, and $y(t)$ is referred to as a deviation variable.
For systems with time delay (dead time or transport lag) θ, the transfer function is

$$\frac{Y(s)}{R(s)} = \frac{Ke^{-\theta s}}{\tau s + 1}$$

The step response for $t \geq \theta$ to a step of magnitude M is

$$y(t) = \left[y_0 e^{-(t-\theta)/\tau} + KM\left(1 - e^{-(t-\theta)/\tau}\right)\right] u(t - \theta),$$ where $u(t)$ is the unit step function.

Second-Order Control System Models
One standard second-order control system model is

$$\frac{Y(s)}{R(s)} = \frac{K\omega_n^2}{s^2 + 2\zeta\omega_n s + \omega_n^2},$$ where

K = steady-state gain,

ζ = the damping ratio,

ω_n = the undamped natural ($\zeta = 0$) frequency,

$\omega_d = \omega_n \sqrt{1 - \zeta^2}$, the damped natural frequency,

and

$\omega_r = \omega_n \sqrt{1 - 2\zeta^2}$, the damped resonant frequency.

If the damping ratio ζ is less than unity, the system is said to be underdamped; if ζ is equal to unity, it is said to be critically damped; and if ζ is greater than unity, the system is said to be overdamped.

For a unit step input to a normalized underdamped second-order control system, the time required to reach a peak value t_p and the value of that peak M_p are given by

$$t_p = \pi/(\omega_n)\sqrt{1 - \zeta^2}$$
$$M_p = 1 + e^{-\pi\zeta/\sqrt{1 - \zeta^2}}$$

The percent overshoot (% OS) of the response is given by

$$\% \, OS = 100 e^{-\pi\zeta/\sqrt{1 - \zeta^2}}$$

For an underdamped second-order system, the logarithmic decrement is

$$\delta = \frac{1}{m} \ln\left(\frac{x_k}{x_{k+m}}\right) = \frac{2\pi\zeta}{\sqrt{1 - \zeta^2}}$$

where x_k and x_{k+m} are the amplitudes of oscillation at cycles k and $k + m$, respectively. The period of oscillation τ is related to ω_d by

$$\omega_d \tau = 2\pi$$

The time required for the output of a second-order system to settle to within 2% of its final value is defined to be

$$T_s = \frac{4}{\zeta\omega_n}$$

An alternative form commonly employed in the chemical process industry is

$$\frac{Y(s)}{R(s)} = \frac{K}{\tau^2 s^2 + 2\zeta\tau s + 1}, \text{ where}$$

K = steady-state gain,

ζ = the damping ratio,

τ = the inverse natural frequency.

Root Locus
The root locus is the locus of points in the complex s-plane satisfying

$$1 + K\frac{(s - z_1)(s - z_2)...(s - z_m)}{(s - p_1)(s - p_2)...(s - p_n)} = 0 \quad m \leq n$$

as K is varied. The p_i and z_j are the open-loop poles and zeros, respectively. When K is increased from zero, the locus has the following properties.

1. Locus branches exist on the real axis to the left of an odd number of open-loop poles and/or zeros.

2. The locus originates at the open-loop poles $p_1, ..., p_n$ and terminates at the zeros $z_1, ..., z_m$. If $m < n$ then $(n - m)$ branches terminate at infinity at asymptote angles

$$\alpha = \frac{(2k + 1)180°}{n - m} \quad k = 0, \pm 1, \pm 2, \pm 3, ...$$

with the real axis.

3. The intersection of the real axis with the asymptotes is called the asymptote centroid and is given by

$$\sigma_A = \frac{\sum\limits_{i=1}^{n} \text{Re}(p_i) - \sum\limits_{i=1}^{m} \text{Re}(z_i)}{n - m}$$

4. If the locus crosses the imaginary (ω) axis, the values of K and ω are given by letting $s = j\omega$ in the defining equation.

State-Variable Control System Models
One common state-variable model for dynamic systems has the form

$$\dot{\mathbf{x}}(t) = \mathbf{A}\mathbf{x}(t) + \mathbf{B}\mathbf{u}(t) \quad \text{(state equation)}$$
$$\mathbf{y}(t) = \mathbf{C}\mathbf{x}(t) + \mathbf{D}\mathbf{u}(t) \quad \text{(output equation)}$$

where

$\mathbf{x}(t)$ = N by 1 state vector (N state variables),

$\mathbf{u}(t)$ = R by 1 input vector (R inputs),

$\mathbf{y}(t)$ = M by 1 output vector (M outputs),

$\mathbf{A}$ = system matrix,

$\mathbf{B}$ = input distribution matrix,

$\mathbf{C}$ = output matrix, and

$\mathbf{D}$ = feed-through matrix.

The orders of the matrices are defined via variable definitions.

State-variable models automatically handle multiple inputs and multiple outputs. Furthermore, state-variable models can be formulated for open-loop system components or the complete closed-loop system.

The Laplace transform of the time-invariant state equation is

$$s\mathbf{X}(s) - \mathbf{x}(0) = \mathbf{A}\mathbf{X}(s) + \mathbf{B}\mathbf{U}(s)$$

from which

$$\mathbf{X}(s) = \Phi(s)\,\mathbf{x}(0) + \Phi(s)\,\mathbf{B}\mathbf{U}(s)$$

where the Laplace transform of the state transition matrix is

$$\Phi(s) = [s\mathbf{I} - \mathbf{A}]^{-1}.$$

The state-transition matrix

$$\Phi(t) = L^{-1}\{\Phi(s)\}$$

(also defined as $e^{\mathbf{A}t}$) can be used to write

$$\mathbf{x}(t) = \Phi(t)\,\mathbf{x}(0) + \int_0^t \Phi(t - \tau)\,\mathbf{B}\mathbf{u}(\tau)\,d\tau$$

The output can be obtained with the output equation; e.g., the Laplace transform output is

$$\mathbf{Y}(s) = \{\mathbf{C}\Phi(s)\,\mathbf{B} + \mathbf{D}\}\mathbf{U}(s) + \mathbf{C}\Phi(s)\,\mathbf{x}(0)$$

The latter term represents the output(s) due to initial conditions, whereas the former term represents the output(s) due to the $\mathbf{U}(s)$ inputs and gives rise to transfer function definitions.

ENGINEERING ECONOMICS

Factor Name	Converts	Symbol	Formula
Single Payment Compound Amount	to F given P	$(F/P, i\%, n)$	$(1+i)^n$
Single Payment Present Worth	to P given F	$(P/F, i\%, n)$	$(1+i)^{-n}$
Uniform Series Sinking Fund	to A given F	$(A/F, i\%, n)$	$\dfrac{i}{(1+i)^n - 1}$
Capital Recovery	to A given P	$(A/P, i\%, n)$	$\dfrac{i(1+i)^n}{(1+i)^n - 1}$
Uniform Series Compound Amount	to F given A	$(F/A, i\%, n)$	$\dfrac{(1+i)^n - 1}{i}$
Uniform Series Present Worth	to P given A	$(P/A, i\%, n)$	$\dfrac{(1+i)^n - 1}{i(1+i)^n}$
Uniform Gradient Present Worth	to P given G	$(P/G, i\%, n)$	$\dfrac{(1+i)^n - 1}{i^2(1+i)^n} - \dfrac{n}{i(1+i)^n}$
Uniform Gradient † Future Worth	to F given G	$(F/G, i\%, n)$	$\dfrac{(1+i)^n - 1}{i^2} - \dfrac{n}{i}$
Uniform Gradient Uniform Series	to A given G	$(A/G, i\%, n)$	$\dfrac{1}{i} - \dfrac{n}{(1+i)^n - 1}$

NOMENCLATURE AND DEFINITIONS

A .... Uniform amount per interest period

B .... Benefit

BV .... Book value

C .... Cost

d .... Combined interest rate per interest period

D_j .... Depreciation in year j

F .... Future worth, value, or amount

f .... General inflation rate per interest period

G .... Uniform gradient amount per interest period

i .... Interest rate per interest period

i_e .... Annual effective interest rate

m .... Number of compounding periods per year

n .... Number of compounding periods; or the expected life of an asset

P .... Present worth, value, or amount

r .... Nominal annual interest rate

S_n .... Expected salvage value in year n

Subscripts

j .... at time j

n .... at time n

† .... $F/G = (F/A - n)/i = (F/A) \times (A/G)$

NON-ANNUAL COMPOUNDING

$$i_e = \left(1 + \frac{r}{m}\right)^m - 1$$

BREAK-EVEN ANALYSIS

By altering the value of any one of the variables in a situation, holding all of the other values constant, it is possible to find a value for that variable that makes the two alternatives equally economical. This value is the break-even point.

Break-even analysis is used to describe the percentage of capacity of operation for a manufacturing plant at which income will just cover expenses.

The payback period is the period of time required for the profit or other benefits of an investment to equal the cost of the investment.

INFLATION

To account for inflation, the dollars are deflated by the general inflation rate per interest period f, and then they are shifted over the time scale using the interest rate per interest period i. Use a combined interest rate per interest period d for computing present worth values P and Net P.

The formula for d is $d = i + f + (i \times f)$

DEPRECIATION

Straight Line

$$D_j = \frac{C - S_n}{n}$$

Accelerated Cost Recovery System (ACRS)

$$D_j = (\text{factor}) \, C$$

A table of modified factors is provided below.

Sum of the Years Digits

$$D_j = \frac{n + 1 - j}{\sum\limits_{j=1}^{n} j}(C - S_n)$$

BOOK VALUE

BV = initial cost $- \Sigma \, D_j$

TAXATION

Income taxes are paid at a specific rate on taxable income. Taxable income is total income less depreciation and ordinary expenses. Expenses do not include capital items, which should be depreciated.

CAPITALIZED COSTS

Capitalized costs are present worth values using an assumed perpetual period of time.

$$\text{Capitalized Costs} = P = \frac{A}{i}$$

BONDS

Bond Value equals the present worth of the payments the purchaser (or holder of the bond) receives during the life of the bond at some interest rate i.

Bond Yield equals the computed interest rate of the bond value when compared with the bond cost.

RATE-OF-RETURN

The minimum acceptable rate-of-return (MARR) is that interest rate that one is willing to accept, or the rate one desires to earn on investments. The rate-of-return on an investment is the interest rate that makes the benefits and costs equal.

BENEFIT-COST ANALYSIS

In a benefit-cost analysis, the benefits B of a project should exceed the estimated costs C.

$$B - C \geq 0, \text{ or } B/C \geq 1$$

MODIFIED ACRS FACTORS				
	Recovery Period (Years)			
	3	5	7	10
Year	Recovery Rate (Percent)			
1	33.3	20.0	14.3	10.0
2	44.5	32.0	24.5	18.0
3	14.8	19.2	17.5	14.4
4	7.4	11.5	12.5	11.5
5		11.5	8.9	**9.2**
6		5.8	8.9	7.4
7			8.9	6.6
8			4.5	6.6
9				6.5
10				6.5
11				3.3

Factor Table - $i = 0.50\%$

n	P/F	P/A	P/G	F/P	F/A	A/P	A/F	A/G
1	0.9950	0.9950	0.0000	1.0050	1.0000	1.0050	1.0000	0.0000
2	0.9901	1.9851	0.9901	1.0100	2.0050	0.5038	0.4988	0.4988
3	0.9851	2.9702	2.9604	1.0151	3.0150	0.3367	0.3317	0.9967
4	0.9802	3.9505	5.9011	1.0202	4.0301	0.2531	0.2481	1.4938
5	0.9754	4.9259	9.8026	1.0253	5.0503	0.2030	0.1980	1.9900
6	0.9705	5.8964	14.6552	1.0304	6.0755	0.1696	0.1646	2.4855
7	0.9657	6.8621	20.4493	1.0355	7.1059	0.1457	0.1407	2.9801
8	0.9609	7.8230	27.1755	1.0407	8.1414	0.1278	0.1228	3.4738
9	0.9561	8.7791	34.8244	1.0459	9.1821	0.1139	0.1089	3.9668
10	0.9513	9.7304	43.3865	1.0511	10.2280	0.1028	0.0978	4.4589
11	0.9466	10.6770	52.8526	1.0564	11.2792	0.0937	0.0887	4.9501
12	0.9419	11.6189	63.2136	1.0617	12.3356	0.0861	0.0811	5.4406
13	0.9372	12.5562	74.4602	1.0670	13.3972	0.0796	0.0746	5.9302
14	0.9326	13.4887	86.5835	1.0723	14.4642	0.0741	0.0691	6.4190
15	0.9279	14.4166	99.5743	1.0777	15.5365	0.0694	0.0644	6.9069
16	0.9233	15.3399	113.4238	1.0831	16.6142	0.0652	0.0602	7.3940
17	0.9187	16.2586	128.1231	1.0885	17.6973	0.0615	0.0565	7.8803
18	0.9141	17.1728	143.6634	1.0939	18.7858	0.0582	0.0532	8.3658
19	0.9096	18.0824	160.0360	1.0994	19.8797	0.0553	0.0503	8.8504
20	0.9051	18.9874	177.2322	1.1049	20.9791	0.0527	0.0477	9.3342
21	0.9006	19.8880	195.2434	1.1104	22.0840	0.0503	0.0453	9.8172
22	0.8961	20.7841	214.0611	1.1160	23.1944	0.0481	0.0431	10.2993
23	0.8916	21.6757	233.6768	1.1216	24.3104	0.0461	0.0411	10.7806
24	0.8872	22.5629	254.0820	1.1272	25.4320	0.0443	0.0393	11.2611
25	0.8828	23.4456	275.2686	1.1328	26.5591	0.0427	0.0377	11.7407
30	0.8610	27.7941	392.6324	1.1614	32.2800	0.0360	0.0310	14.1265
40	0.8191	36.1722	681.3347	1.2208	44.1588	0.0276	0.0226	18.8359
50	0.7793	44.1428	1,035.6966	1.2832	56.6452	0.0227	0.0177	23.4624
60	0.7414	51.7256	1,448.6458	1.3489	69.7700	0.0193	0.0143	28.0064
100	0.6073	78.5426	3,562.7934	1.6467	129.3337	0.0127	0.0077	45.3613

Factor Table - $i = 1.00\%$

n	P/F	P/A	P/G	F/P	F/A	A/P	A/F	A/G
1	0.9901	0.9901	0.0000	1.0100	1.0000	1.0100	1.0000	0.0000
2	0.9803	1.9704	0.9803	1.0201	2.0100	0.5075	0.4975	0.4975
3	0.9706	2.9410	2.9215	1.0303	3.0301	0.3400	0.3300	0.9934
4	0.9610	3.9020	5.8044	1.0406	4.0604	0.2563	0.2463	1.4876
5	0.9515	4.8534	9.6103	1.0510	5.1010	0.2060	0.1960	1.9801
6	0.9420	5.7955	14.3205	1.0615	6.1520	0.1725	0.1625	2.4710
7	0.9327	6.7282	19.9168	1.0721	7.2135	0.1486	0.1386	2.9602
8	0.9235	7.6517	26.3812	1.0829	8.2857	0.1307	0.1207	3.4478
9	0.9143	8.5650	33.6959	1.0937	9.3685	0.1167	0.1067	3.9337
10	0.9053	9.4713	41.8435	1.1046	10.4622	0.1056	0.0956	4.4179
11	0.8963	10.3676	50.8067	1.1157	11.5668	0.0965	0.0865	4.9005
12	0.8874	11.2551	60.5687	1.1268	12.6825	0.0888	0.0788	5.3815
13	0.8787	12.1337	71.1126	1.1381	13.8093	0.0824	0.0724	5.8607
14	0.8700	13.0037	82.4221	1.1495	14.9474	0.0769	0.0669	6.3384
15	0.8613	13.8651	94.4810	1.1610	16.0969	0.0721	0.0621	6.8143
16	0.8528	14.7179	107.2734	1.1726	17.2579	0.0679	0.0579	7.2886
17	0.8444	15.5623	120.7834	1.1843	18.4304	0.0643	0.0543	7.7613
18	0.8360	16.3983	134.9957	1.1961	19.6147	0.0610	0.0510	8.2323
19	0.8277	17.2260	149.8950	1.2081	20.8109	0.0581	0.0481	8.7017
20	0.8195	18.0456	165.4664	1.2202	22.0190	0.0554	0.0454	9.1694
21	0.8114	18.8570	181.6950	1.2324	23.2392	0.0530	0.0430	9.6354
22	0.8034	19.6604	198.5663	1.2447	24.4716	0.0509	0.0409	10.0998
23	0.7954	20.4558	216.0660	1.2572	25.7163	0.0489	0.0389	10.5626
24	0.7876	21.2434	234.1800	1.2697	26.9735	0.0471	0.0371	11.0237
25	0.7798	22.0232	252.8945	1.2824	28.2432	0.0454	0.0354	11.4831
30	0.7419	25.8077	355.0021	1.3478	34.7849	0.0387	0.0277	13.7557
40	0.6717	32.8347	596.8561	1.4889	48.8864	0.0305	0.0205	18.1776
50	0.6080	39.1961	879.4176	1.6446	64.4632	0.0255	0.0155	22.4363
60	0.5504	44.9550	1,192.8061	1.8167	81.6697	0.0222	0.0122	26.5333
100	0.3697	63.0289	2,605.7758	2.7048	170.4814	0.0159	0.0059	41.3426

Factor Table - i = 1.50%

n	P/F	P/A	P/G	F/P	F/A	A/P	A/F	A/G
1	0.9852	0.9852	0.0000	1.0150	1.0000	1.0150	1.0000	0.0000
2	0.9707	1.9559	0.9707	1.0302	2.0150	0.5113	0.4963	0.4963
3	0.9563	2.9122	2.8833	1.0457	3.0452	0.3434	0.3284	0.9901
4	0.9422	3.8544	5.7098	1.0614	4.0909	0.2594	0.2444	1.4814
5	0.9283	4.7826	9.4229	1.0773	5.1523	0.2091	0.1941	1.9702
6	0.9145	5.6972	13.9956	1.0934	6.2296	0.1755	0.1605	2.4566
7	0.9010	6.5982	19.4018	1.1098	7.3230	0.1516	0.1366	2.9405
8	0.8877	7.4859	26.6157	1.1265	8.4328	0.1336	0.1186	3.4219
9	0.8746	8.3605	32.6125	1.1434	9.5593	0.1196	0.1046	3.9008
10	0.8617	9.2222	40.3675	1.1605	10.7027	0.1084	0.0934	4.3772
11	0.8489	10.0711	48.8568	1.1779	11.8633	0.0993	0.0843	4.8512
12	0.8364	10.9075	58.0571	1.1956	13.0412	0.0917	0.0767	5.3227
13	0.8240	11.7315	67.9454	1.2136	14.2368	0.0852	0.0702	5.7917
14	0.8118	12.5434	78.4994	1.2318	15.4504	0.0797	0.0647	6.2582
15	0.7999	13.3432	89.6974	1.2502	16.6821	0.0749	0.0599	6.7223
16	0.7880	14.1313	101.5178	1.2690	17.9324	0.0708	0.0558	7.1839
17	0.7764	14.9076	113.9400	1.2880	19.2014	0.0671	0.0521	7.6431
18	0.7649	15.6726	126.9435	1.3073	20.4894	0.0638	0.0488	8.0997
19	0.7536	16.4262	140.5084	1.3270	21.7967	0.0609	0.0459	8.5539
20	0.7425	17.1686	154.6154	1.3469	23.1237	0.0582	0.0432	9.0057
21	0.7315	17.9001	169.2453	1.3671	24.4705	0.0559	0.0409	9.4550
22	0.7207	18.6208	184.3798	1.3876	25.8376	0.0537	0.0387	9.9018
23	0.7100	19.3309	200.0006	1.4084	27.2251	0.0517	0.0367	10.3462
24	0.6995	20.0304	216.0901	1.4295	28.6335	0.0499	0.0349	10.7881
25	0.6892	20.7196	232.6310	1.4509	30.0630	0.0483	0.0333	11.2276
30	0.6398	24.0158	321.5310	1.5631	37.5387	0.0416	0.0266	13.3883
40	0.5513	29.9158	524.3568	1.8140	54.2679	0.0334	0.0184	17.5277
50	0.4750	34.9997	749.9636	2.1052	73.6828	0.0286	0.0136	21.4277
60	0.4093	39.3803	988.1674	2.4432	96.2147	0.0254	0.0104	25.0930
100	0.2256	51.6247	1,937.4506	4.4320	228.8030	0.0194	0.0044	37.5295

Factor Table - i = 2.00%

n	P/F	P/A	P/G	F/P	F/A	A/P	A/F	A/G
1	0.9804	0.9804	0.0000	1.0200	1.0000	1.0200	1.0000	0.0000
2	0.9612	1.9416	0.9612	1.0404	2.0200	0.5150	0.4950	0.4950
3	0.9423	2.8839	2.8458	1.0612	3.0604	0.3468	0.3268	0.9868
4	0.9238	3.8077	5.6173	1.0824	4.1216	0.2626	0.2426	1.4752
5	0.9057	4.7135	9.2403	1.1041	5.2040	0.2122	0.1922	1.9604
6	0.8880	5.6014	13.6801	1.1262	6.3081	0.1785	0.1585	2.4423
7	0.8706	6.4720	18.9035	1.1487	7.4343	0.1545	0.1345	2.9208
8	0.8535	7.3255	24.8779	1.1717	8.5830	0.1365	0.1165	3.3961
9	0.8368	8.1622	31.5720	1.1951	9.7546	0.1225	0.1025	3.8681
10	0.8203	8.9826	38.9551	1.2190	10.9497	0.1113	0.0913	4.3367
11	0.8043	9.7868	46.9977	1.2434	12.1687	0.1022	0.0822	4.8021
12	0.7885	10.5753	55.6712	1.2682	13.4121	0.0946	0.0746	5.2642
13	0.7730	11.3484	64.9475	1.2936	14.6803	0.0881	0.0681	5.7231
14	0.7579	12.1062	74.7999	1.3195	15.9739	0.0826	0.0626	6.1786
15	0.7430	12.8493	85.2021	1.3459	17.2934	0.0778	0.0578	6.6309
16	0.7284	13.5777	96.1288	1.3728	18.6393	0.0737	0.0537	7.0799
17	0.7142	14.2919	107.5554	1.4002	20.0121	0.0700	0.0500	7.5256
18	0.7002	14.9920	119.4581	1.4282	21.4123	0.0667	0.0467	7.9681
19	0.6864	15.6785	131.8139	1.4568	22.8406	0.0638	0.0438	8.4073
20	0.6730	16.3514	144.6003	1.4859	24.2974	0.0612	0.0412	8.8433
21	0.6598	17.0112	157.7959	1.5157	25.7833	0.0588	0.0388	9.2760
22	0.6468	17.6580	171.3795	1.5460	27.2990	0.0566	0.0366	9.7055
23	0.6342	18.2922	185.3309	1.5769	28.8450	0.0547	0.0347	10.1317
24	0.6217	18.9139	199.6305	1.6084	30.4219	0.0529	0.0329	10.5547
25	0.6095	19.5235	214.2592	1.6406	32.0303	0.0512	0.0312	10.9745
30	0.5521	22.3965	291.7164	1.8114	40.5681	0.0446	0.0246	13.0251
40	0.4529	27.3555	461.9931	2.2080	60.4020	0.0366	0.0166	16.8885
50	0.3715	31.4236	642.3606	2.6916	84.5794	0.0318	0.0118	20.4420
60	0.3048	34.7609	823.6975	3.2810	114.0515	0.0288	0.0088	23.6961
100	0.1380	43.0984	1,464.7527	7.2446	312.2323	0.0232	0.0032	33.9863

Factor Table - *i* = 4.00%

n	P/F	P/A	P/G	F/P	F/A	A/P	A/F	A/G
1	0.9615	0.9615	0.0000	1.0400	1.0000	1.0400	1.0000	0.0000
2	0.9246	1.8861	0.9246	1.0816	2.0400	0.5302	0.4902	0.4902
3	0.8890	2.7751	2.7025	1.1249	3.1216	0.3603	0.3203	0.9739
4	0.8548	3.6299	5.2670	1.1699	4.2465	0.2755	0.2355	1.4510
5	0.8219	4.4518	8.5547	1.2167	5.4163	0.2246	0.1846	1.9216
6	0.7903	5.2421	12.5062	1.2653	6.6330	0.1908	0.1508	2.3857
7	0.7599	6.0021	17.0657	1.3159	7.8983	0.1666	0.1266	2.8433
8	0.7307	6.7327	22.1806	1.3686	9.2142	0.1485	0.1085	3.2944
9	0.7026	7.4353	27.8013	1.4233	10.5828	0.1345	0.0945	3.7391
10	0.6756	8.1109	33.8814	1.4802	12.0061	0.1233	0.0833	4.1773
11	0.6496	8.7605	40.3772	1.5395	13.4864	0.1141	0.0741	4.6090
12	0.6246	9.3851	47.2477	1.6010	15.0258	0.1066	0.0666	5.0343
13	0.6006	9.9856	54.4546	1.6651	16.6268	0.1001	0.0601	5.4533
14	0.5775	10.5631	61.9618	1.7317	18.2919	0.0947	0.0547	5.8659
15	0.5553	11.1184	69.7355	1.8009	20.0236	0.0899	0.0499	6.2721
16	0.5339	11.6523	77.7441	1.8730	21.8245	0.0858	0.0458	6.6720
17	0.5134	12.1657	85.9581	1.9479	23.6975	0.0822	0.0422	7.0656
18	0.4936	12.6593	94.3498	2.0258	25.6454	0.0790	0.0390	7.4530
19	0.4746	13.1339	102.8933	2.1068	27.6712	0.0761	0.0361	7.8342
20	0.4564	13.5903	111.5647	2.1911	29.7781	0.0736	0.0336	8.2091
21	0.4388	14.0292	120.3414	2.2788	31.9692	0.0713	0.0313	8.5779
22	0.4220	14.4511	129.2024	2.3699	34.2480	0.0692	0.0292	8.9407
23	0.4057	14.8568	138.1284	2.4647	36.6179	0.0673	0.0273	9.2973
24	0.3901	15.2470	147.1012	2.5633	39.0826	0.0656	0.0256	9.6479
25	0.3751	15.6221	156.1040	2.6658	41.6459	0.0640	0.0240	9.9925
30	0.3083	17.2920	201.0618	3.2434	56.0849	0.0578	0.0178	11.6274
40	0.2083	19.7928	286.5303	4.8010	95.0255	0.0505	0.0105	14.4765
50	0.1407	21.4822	361.1638	7.1067	152.6671	0.0466	0.0066	16.8122
60	0.0951	22.6235	422.9966	10.5196	237.9907	0.0442	0.0042	18.6972
100	0.0198	24.5050	563.1249	50.5049	1,237.6237	0.0408	0.0008	22.9800

Factor Table - *i* = 6.00%

n	P/F	P/A	P/G	F/P	F/A	A/P	A/F	A/G
1	0.9434	0.9434	0.0000	1.0600	1.0000	1.0600	1.0000	0.0000
2	0.8900	1.8334	0.8900	1.1236	2.0600	0.5454	0.4854	0.4854
3	0.8396	2.6730	2.5692	1.1910	3.1836	0.3741	0.3141	0.9612
4	0.7921	3.4651	4.9455	1.2625	4.3746	0.2886	0.2286	1.4272
5	0.7473	4.2124	7.9345	1.3382	5.6371	0.2374	0.1774	1.8836
6	0.7050	4.9173	11.4594	1.4185	6.9753	0.2034	0.1434	2.3304
7	0.6651	5.5824	15.4497	1.5036	8.3938	0.1791	0.1191	2.7676
8	0.6274	6.2098	19.8416	1.5938	9.8975	0.1610	0.1010	3.1952
9	0.5919	6.8017	24.5768	1.6895	11.4913	0.1470	0.0870	3.6133
10	0.5584	7.3601	29.6023	1.7908	13.1808	0.1359	0.0759	4.0220
11	0.5268	7.8869	34.8702	1.8983	14.9716	0.1268	0.0668	4.4213
12	0.4970	8.3838	40.3369	2.0122	16.8699	0.1193	0.0593	4.8113
13	0.4688	8.8527	45.9629	2.1329	18.8821	0.1130	0.0530	5.1920
14	0.4423	9.2950	51.7128	2.2609	21.0151	0.1076	0.0476	5.5635
15	0.4173	9.7122	57.5546	2.3966	23.2760	0.1030	0.0430	5.9260
16	0.3936	10.1059	63.4592	2.5404	25.6725	0.0990	0.0390	6.2794
17	0.3714	10.4773	69.4011	2.6928	28.2129	0.0954	0.0354	6.6240
18	0.3505	10.8276	75.3569	2.8543	30.9057	0.0924	0.0324	6.9597
19	0.3305	11.1581	81.3062	3.0256	33.7600	0.0896	0.0296	7.2867
20	0.3118	11.4699	87.2304	3.2071	36.7856	0.0872	0.0272	7.6051
21	0.2942	11.7641	93.1136	3.3996	39.9927	0.0850	0.0250	7.9151
22	0.2775	12.0416	98.9412	3.6035	43.3923	0.0830	0.0230	8.2166
23	0.2618	12.3034	104.7007	3.8197	46.9958	0.0813	0.0213	8.5099
24	0.2470	12.5504	110.3812	4.0489	50.8156	0.0797	0.0197	8.7951
25	0.2330	12.7834	115.9732	4.2919	54.8645	0.0782	0.0182	9.0722
30	0.1741	13.7648	142.3588	5.7435	79.0582	0.0726	0.0126	10.3422
40	0.0972	15.0463	185.9568	10.2857	154.7620	0.0665	0.0065	12.3590
50	0.0543	15.7619	217.4574	18.4202	290.3359	0.0634	0.0034	13.7964
60	0.0303	16.1614	239.0428	32.9877	533.1282	0.0619	0.0019	14.7909
100	0.0029	16.6175	272.0471	339.3021	5,638.3681	0.0602	0.0002	16.3711

Factor Table - $i = 8.00\%$

n	P/F	P/A	P/G	F/P	F/A	A/P	A/F	A/G
1	0.9259	0.9259	0.0000	1.0800	1.0000	1.0800	1.0000	0.0000
2	0.8573	1.7833	0.8573	1.1664	2.0800	0.5608	0.4808	0.4808
3	0.7938	2.5771	2.4450	1.2597	3.2464	0.3880	0.3080	0.9487
4	0.7350	3.3121	4.6501	1.3605	4.5061	0.3019	0.2219	1.4040
5	0.6806	3.9927	7.3724	1.4693	5.8666	0.2505	0.1705	1.8465
6	0.6302	4.6229	10.5233	1.5869	7.3359	0.2163	0.1363	2.2763
7	0.5835	5.2064	14.0242	1.7138	8.9228	0.1921	0.1121	2.6937
8	0.5403	5.7466	17.8061	1.8509	10.6366	0.1740	0.0940	3.0985
9	0.5002	6.2469	21.8081	1.9990	12.4876	0.1601	0.0801	3.4910
10	0.4632	6.7101	25.9768	2.1589	14.4866	0.1490	0.0690	3.8713
11	0.4289	7.1390	30.2657	2.3316	16.6455	0.1401	0.0601	4.2395
12	0.3971	7.5361	34.6339	2.5182	18.9771	0.1327	0.0527	4.5957
13	0.3677	7.9038	39.0463	2.7196	21.4953	0.1265	0.0465	4.9402
14	0.3405	8.2442	43.4723	2.9372	24.2149	0.1213	0.0413	5.2731
15	0.3152	8.5595	47.8857	3.1722	27.1521	0.1168	0.0368	5.5945
16	0.2919	8.8514	52.2640	3.4259	30.3243	0.1130	0.0330	5.9046
17	0.2703	9.1216	56.5883	3.7000	33.7502	0.1096	0.0296	6.2037
18	0.2502	9.3719	60.8426	3.9960	37.4502	0.1067	0.0267	6.4920
19	0.2317	9.6036	65.0134	4.3157	41.4463	0.1041	0.0241	6.7697
20	0.2145	9.8181	69.0898	4.6610	45.7620	0.1019	0.0219	7.0369
21	0.1987	10.0168	73.0629	5.0338	50.4229	0.0998	0.0198	7.2940
22	0.1839	10.2007	76.9257	5.4365	55.4568	0.0980	0.0180	7.5412
23	0.1703	10.3711	80.6726	5.8715	60.8933	0.0964	0.0164	7.7786
24	0.1577	10.5288	84.2997	6.3412	66.7648	0.0950	0.0150	8.0066
25	0.1460	10.6748	87.8041	6.8485	73.1059	0.0937	0.0137	8.2254
30	0.0994	11.2578	103.4558	10.0627	113.2832	0.0888	0.0088	9.1897
40	0.0460	11.9246	126.0422	21.7245	259.0565	0.0839	0.0039	10.5699
50	0.0213	12.2335	139.5928	46.9016	573.7702	0.0817	0.0017	11.4107
60	0.0099	12.3766	147.3000	101.2571	1,253.2133	0.0808	0.0008	11.9015
100	0.0005	12.4943	155.6107	2,199.7613	27,484.5157	0.0800		12.4545

Factor Table - $i = 10.00\%$

n	P/F	P/A	P/G	F/P	F/A	A/P	A/F	A/G
1	0.9091	0.9091	0.0000	1.1000	1.0000	1.1000	1.0000	0.0000
2	0.8264	1.7355	0.8264	1.2100	2.1000	0.5762	0.4762	0.4762
3	0.7513	2.4869	2.3291	1.3310	3.3100	0.4021	0.3021	0.9366
4	0.6830	3.1699	4.3781	1.4641	4.6410	0.3155	0.2155	1.3812
5	0.6209	3.7908	6.8618	1.6105	6.1051	0.2638	0.1638	1.8101
6	0.5645	4.3553	9.6842	1.7716	7.7156	0.2296	0.1296	2.2236
7	0.5132	4.8684	12.7631	1.9487	9.4872	0.2054	0.1054	2.6216
8	0.4665	5.3349	16.0287	2.1436	11.4359	0.1874	0.0874	3.0045
9	0.4241	5.7590	19.4215	2.3579	13.5735	0.1736	0.0736	3.3724
10	0.3855	6.1446	22.8913	2.5937	15.9374	0.1627	0.0627	3.7255
11	0.3505	6.4951	26.3962	2.8531	18.5312	0.1540	0.0540	4.0641
12	0.3186	6.8137	29.9012	3.1384	21.3843	0.1468	0.0468	4.3884
13	0.2897	7.1034	33.3772	3.4523	24.5227	0.1408	0.0408	4.6988
14	0.2633	7.3667	36.8005	3.7975	27.9750	0.1357	0.0357	4.9955
15	0.2394	7.6061	40.1520	4.1772	31.7725	0.1315	0.0315	5.2789
16	0.2176	7.8237	43.4164	4.5950	35.9497	0.1278	0.0278	5.5493
17	0.1978	8.0216	46.5819	5.0545	40.5447	0.1247	0.0247	5.8071
18	0.1799	8.2014	49.6395	5.5599	45.5992	0.1219	0.0219	6.0526
19	0.1635	8.3649	52.5827	6.1159	51.1591	0.1195	0.0195	6.2861
20	0.1486	8.5136	55.4069	6.7275	57.2750	0.1175	0.0175	6.5081
21	0.1351	8.6487	58.1095	7.4002	64.0025	0.1156	0.0156	6.7189
22	0.1228	8.7715	60.6893	8.1403	71.4027	0.1140	0.0140	6.9189
23	0.1117	8.8832	63.1462	8.9543	79.5430	0.1126	0.0126	7.1085
24	0.1015	8.9847	65.4813	9.8497	88.4973	0.1113	0.0113	7.2881
25	0.0923	9.0770	67.6964	10.8347	98.3471	0.1102	0.0102	7.4580
30	0.0573	9.4269	77.0766	17.4494	164.4940	0.1061	0.0061	8.1762
40	0.0221	9.7791	88.9525	45.2593	442.5926	0.1023	0.0023	9.0962
50	0.0085	9.9148	94.8889	117.3909	1,163.9085	0.1009	0.0009	9.5704
60	0.0033	9.9672	97.7010	304.4816	3,034.8164	0.1003	0.0003	9.8023
100	0.0001	9.9993	99.9202	13,780.6123	137,796.1234	0.1000		9.9927

Factor Table - $i = 12.00\%$

n	P/F	P/A	P/G	F/P	F/A	A/P	A/F	A/G
1	0.8929	0.8929	0.0000	1.1200	1.0000	1.1200	1.0000	0.0000
2	0.7972	1.6901	0.7972	1.2544	2.1200	0.5917	0.4717	0.4717
3	0.7118	2.4018	2.2208	1.4049	3.3744	0.4163	0.2963	0.9246
4	0.6355	3.0373	4.1273	1.5735	4.7793	0.3292	0.2092	1.3589
5	0.5674	3.6048	6.3970	1.7623	6.3528	0.2774	0.1574	1.7746
6	0.5066	4.1114	8.9302	1.9738	8.1152	0.2432	0.1232	2.1720
7	0.4523	4.5638	11.6443	2.2107	10.0890	0.2191	0.0991	2.5515
8	0.4039	4.9676	14.4714	2.4760	12.2997	0.2013	0.0813	2.9131
9	0.3606	5.3282	17.3563	2.7731	14.7757	0.1877	0.0677	3.2574
10	0.3220	5.6502	20.2541	3.1058	17.5487	0.1770	0.0570	3.5847
11	0.2875	5.9377	23.1288	3.4785	20.6546	0.1684	0.0484	3.8953
12	0.2567	6.1944	25.9523	3.8960	24.1331	0.1614	0.0414	4.1897
13	0.2292	6.4235	28.7024	4.3635	28.0291	0.1557	0.0357	4.4683
14	0.2046	6.6282	31.3624	4.8871	32.3926	0.1509	0.0309	4.7317
15	0.1827	6.8109	33.9202	5.4736	37.2797	0.1468	0.0268	4.9803
16	0.1631	6.9740	36.3670	6.1304	42.7533	0.1434	0.0234	5.2147
17	0.1456	7.1196	38.6973	6.8660	48.8837	0.1405	0.0205	5.4353
18	0.1300	7.2497	40.9080	7.6900	55.7497	0.1379	0.0179	5.6427
19	0.1161	7.3658	42.9979	8.6128	63.4397	0.1358	0.0158	5.8375
20	0.1037	7.4694	44.9676	9.6463	72.0524	0.1339	0.0139	6.0202
21	0.0926	7.5620	46.8188	10.8038	81.6987	0.1322	0.0122	6.1913
22	0.0826	7.6446	48.5543	12.1003	92.5026	0.1308	0.0108	6.3514
23	0.0738	7.7184	50.1776	13.5523	104.6029	0.1296	0.0096	6.5010
24	0.0659	7.7843	51.6929	15.1786	118.1552	0.1285	0.0085	6.6406
25	0.0588	7.8431	53.1046	17.0001	133.3339	0.1275	0.0075	6.7708
30	0.0334	8.0552	58.7821	29.9599	241.3327	0.1241	0.0041	7.2974
40	0.0107	8.2438	65.1159	93.0510	767.0914	0.1213	0.0013	7.8988
50	0.0035	8.3045	67.7624	289.0022	2,400.0182	0.1204	0.0004	8.1597
60	0.0011	8.3240	68.8100	897.5969	7,471.6411	0.1201	0.0001	8.2664
100		8.3332	69.4336	83,522.2657	696,010.5477	0.1200		8.3321

Factor Table - $i = 18.00\%$

n	P/F	P/A	P/G	F/P	F/A	A/P	A/F	A/G
1	0.8475	0.8475	0.0000	1.1800	1.0000	1.1800	1.0000	0.0000
2	0.7182	1.5656	0.7182	1.3924	2.1800	0.6387	0.4587	0.4587
3	0.6086	2.1743	1.9354	1.6430	3.5724	0.4599	0.2799	0.8902
4	0.5158	2.6901	3.4828	1.9388	5.2154	0.3717	0.1917	1.2947
5	0.4371	3.1272	5.2312	2.2878	7.1542	0.3198	0.1398	1.6728
6	0.3704	3.4976	7.0834	2.6996	9.4423	0.2859	0.1059	2.0252
7	0.3139	3.8115	8.9670	3.1855	12.1415	0.2624	0.0824	2.3526
8	0.2660	4.0776	10.8292	3.7589	15.3270	0.2452	0.0652	2.6558
9	0.2255	4.3030	12.6329	4.4355	19.0859	0.2324	0.0524	2.9358
10	0.1911	4.4941	14.3525	5.2338	23.5213	0.2225	0.0425	3.1936
11	0.1619	4.6560	15.9716	6.1759	28.7551	0.2148	0.0348	3.4303
12	0.1372	4.7932	17.4811	7.2876	34.9311	0.2086	0.0286	3.6470
13	0.1163	4.9095	18.8765	8.5994	42.2187	0.2037	0.0237	3.8449
14	0.0985	5.0081	20.1576	10.1472	50.8180	0.1997	0.0197	4.0250
15	0.0835	5.0916	21.3269	11.9737	60.9653	0.1964	0.0164	4.1887
16	0.0708	5.1624	22.3885	14.1290	72.9390	0.1937	0.0137	4.3369
17	0.0600	5.2223	23.3482	16.6722	87.0680	0.1915	0.0115	4.4708
18	0.0508	5.2732	24.2123	19.6731	103.7403	0.1896	0.0096	4.5916
19	0.0431	5.3162	24.9877	23.2144	123.4135	0.1881	0.0081	4.7003
20	0.0365	5.3527	25.6813	27.3930	146.6280	0.1868	0.0068	4.7978
21	0.0309	5.3837	26.3000	32.3238	174.0210	0.1857	0.0057	4.8851
22	0.0262	5.4099	26.8506	38.1421	206.3448	0.1848	0.0048	4.9632
23	0.0222	5.4321	27.3394	45.0076	244.4868	0.1841	0.0041	5.0329
24	0.0188	5.4509	27.7725	53.1090	289.4944	0.1835	0.0035	5.0950
25	0.0159	5.4669	28.1555	62.6686	342.6035	0.1829	0.0029	5.1502
30	0.0070	5.5168	29.4864	143.3706	790.9480	0.1813	0.0013	5.3448
40	0.0013	5.5482	30.5269	750.3783	4,163.2130	0.1802	0.0002	5.5022
50	0.0003	5.5541	30.7856	3,927.3569	21,813.0937	0.1800		5.5428
60	0.0001	5.5553	30.8465	20,555.1400	114,189.6665	0.1800		5.5526
100		5.5556	30.8642	15,424,131.91	85,689,616.17	0.1800		5.5555

ETHICS

Engineering is considered to be a "profession" rather than an "occupation" because of several important characteristics shared with other recognized learned professions, law, medicine, and theology: special knowledge, special privileges, and special responsibilities. Professions are based on a large knowledge base requiring extensive training. Professional skills are important to the well-being of society. Professions are self-regulating, in that they control the training and evaluation processes that admit new persons to the field. Professionals have autonomy in the workplace; they are expected to utilize their independent judgment in carrying out their professional responsibilities. Finally, professions are regulated by ethical standards.[1]

The expertise possessed by engineers is vitally important to public welfare. In order to serve the public effectively, engineers must maintain a high level of technical competence. However, a high level of technical expertise without adherence to ethical guidelines is as much a threat to public welfare as is professional incompetence. Therefore, engineers must also be guided by ethical principles.

The ethical principles governing the engineering profession are embodied in codes of ethics. Such codes have been adopted by state boards of registration, professional engineering societies, and even by some private industries. An example of one such code is the NCEES Rules of Professional Conduct, found in Section 240 of the *Model Rules* and presented here. As part of his/her responsibility to the public, an engineer is responsible for knowing and abiding by the code. Additional rules of conduct are also included in the *Model Rules*.

The three major sections of the *Model Rules* address (1) Licensee's Obligation to Society, (2) Licensee's Obligation to Employers and Clients, and (3) Licensee's Obligation to Other Licensees. The principles amplified in these sections are important guides to appropriate behavior of professional engineers.

Application of the code in many situations is not controversial. However, there may be situations in which applying the code may raise more difficult issues. In particular, there may be circumstances in which terminology in the code is not clearly defined, or in which two sections of the code may be in conflict. For example, what constitutes "valuable consideration" or "adequate" knowledge may be interpreted differently by qualified professionals. These types of questions are called *conceptual issues*, in which definitions of terms may be in dispute. In other situations, *factual issues* may also affect ethical dilemmas. Many decisions regarding engineering design may be based upon interpretation of disputed or incomplete information. In addition, *tradeoffs* revolving around competing issues of risk vs. benefit, or safety vs. economics may require judgments that are not fully addressed simply by application of the code.

No code can give immediate and mechanical answers to all ethical and professional problems that an engineer may face. Creative problem solving is often called for in ethics, just as it is in other areas of engineering.

Model Rules, Section 240.15, Rules of Professional Conduct

A. LICENSEE'S OBLIGATION TO SOCIETY

1. Licensees, in the performance of their services for clients, employers, and customers, shall be cognizant that their first and foremost responsibility is to the public welfare.

2. Licensees shall approve and seal only those design documents and surveys that conform to accepted engineering and surveying standards and safeguard the life, health, property, and welfare of the public.

3. Licensees shall notify their employer or client and such other authority as may be appropriate when their professional judgment is overruled under circumstances where the life, health, property, or welfare of the public is endangered.

4. Licensees shall be objective and truthful in professional reports, statements, or testimony. They shall include all relevant and pertinent information in such reports, statements, or testimony.

5. Licensees shall express a professional opinion publicly only when it is founded upon an adequate knowledge of the facts and a competent evaluation of the subject matter.

6. Licensees shall issue no statements, criticisms, or arguments on technical matters which are inspired or paid for by interested parties, unless they explicitly identify the interested parties on whose behalf they are speaking and reveal any interest they have in the matters.

7. Licensees shall not permit the use of their name or firm name by, nor associate in the business ventures with, any person or firm which is engaging in fraudulent or dishonest business or professional practices.

8. Licensees having knowledge of possible violations of any of these Rules of Professional Conduct shall provide the board with the information and assistance necessary to make the final determination of such violation.

---

[1] Harris, C.E., M.S. Pritchard, & M.J. Rabins, *Engineering Ethics: Concepts and Cases*, Wadsworth Publishing Company, pages 27–28, 1995.

B. LICENSEE'S OBLIGATION TO EMPLOYER AND CLIENTS

1. Licensees shall undertake assignments only when qualified by education or experience in the specific technical fields of engineering or surveying involved.

2. Licensees shall not affix their signatures or seals to any plans or documents dealing with subject matter in which they lack competence, nor to any such plan or document not prepared under their direct control and personal supervision.

3. Licensees may accept assignments for coordination of an entire project, provided that each design segment is signed and sealed by the licensee responsible for preparation of that design segment.

4. Licensees shall not reveal facts, data, or information obtained in a professional capacity without the prior consent of the client or employer except as authorized or required by law. Licensees shall not solicit or accept gratuities, directly or indirectly, from contractors, their agents, or other parties in connection with work for employers or clients.

5. Licensees shall make full prior disclosures to their employers or clients of potential conflicts of interest or other circumstances which could influence or appear to influence their judgment or the quality of their service.

6. Licensees shall not accept compensation, financial or otherwise, from more than one party for services pertaining to the same project, unless the circumstances are fully disclosed and agreed to by all interested parties.

7. Licensees shall not solicit or accept a professional contract from a governmental body on which a principal or officer of their organization serves as a member. Conversely, licensees serving as members, advisors, or employees of a government body or department, who are the principals or employees of a private concern, shall not participate in decisions with respect to professional services offered or provided by said concern to the governmental body which they serve.

C. LICENSEE'S OBLIGATION TO OTHER LICENSEES

1. Licensees shall not falsify or permit misrepresentation of their, or their associates', academic or professional qualifications. They shall not misrepresent or exaggerate their degree of responsibility in prior assignments nor the complexity of said assignments. Presentations incident to the solicitation of employment or business shall not misrepresent pertinent facts concerning employers, employees, associates, joint ventures, or past accomplishments.

2. Licensees shall not offer, give, solicit, or receive, either directly or indirectly, any commission, or gift, or other valuable consideration in order to secure work, and shall not make any political contribution with the intent to influence the award of a contract by public authority.

3. Licensees shall not attempt to injure, maliciously or falsely, directly or indirectly, the professional reputation, prospects, practice, or employment of other licensees, nor indiscriminately criticize other licensees' work.

CHEMICAL ENGINEERING

For additional information concerning heat transfer and fluid mechanics, refer to the **HEAT TRANSFER, THERMODYNAMICS, MECHANICAL ENGINEERING** or **FLUID MECHANICS** sections.

For additional information concerning chemical process control, refer to the **MEASUREMENT AND CONTROLS** section.

For additonal information concerning statistical data analysis, refer to the following:

Confidence Intervals
See the subsection in the **ENGINEERING PROBABILITY AND STATISTICS** section of this handbook.

Statistical Quality Control
See the subsection in the **INDUSTRIAL ENGINEERING** section of this handbook.

Linear Regression
See the subsection in the **ENGINEERING PROBABILITY AND STATISTICS** section of this handbook.

One-Way Analysis of Variance (ANOVA)
See the subsection in the **INDUSTRIAL ENGINEERING** section of this handbook.

Microbial Kinetics
See the subsection in the **ENVIRONMENTAL ENGINEERING** section of this handbook.

SELECTED RULES OF NOMENCLATURE IN ORGANIC CHEMISTRY

Alcohols
Three systems of nomenclature are in general use. In the first, the alkyl group attached to the hydroxyl group is named and the separate word *alcohol* is added. In the second system, the higher alcohols are considered as derivatives of the first member of the series, which is called *carbinol*. The third method is the modified Geneva system in which (1) the longest carbon chain containing the hydroxyl group determines the surname, (2) the ending *e* of the corresponding saturated hydrocarbon is replaced by *ol*, (3) the carbon chain is numbered from the end that gives the hydroxyl group the smaller number, and (4) the side chains are named and their positions indicated by the proper number. Alcohols in general are divided into three classes. In *primary* alcohols the hydroxyl group is united to a primary carbon atom, that is, a carbon atom united directly to only one other carbon atom. *Secondary* alcohols have the hydroxyl group united to a secondary carbon atom, that is, one united to two other carbon atoms. *Tertiary* alcohols have the hydroxyl group united to a tertiary carbon atom, that is, one united to three other carbon atoms.

Ethers
Ethers are generally designated by naming the alkyl groups and adding the word *ether*. The group RO is known as an *alkoxyl group*. Ethers may also be named as alkoxy derivatives of hydrocarbons.

Carboxylic Acids
The name of each linear carboxylic acid is unique to the number of carbon atoms it contains. 1: (one carbon atom) Formic. 2: Acetic. 3: Propionic. 4: Butyric. 5: Valeric. 6: Caproic. 7: Enanthic. 8: Caprylic. 9: Pelargonic. 10: Capric.

Aldehydes
The common names of aldehydes are derived from the acids that would be formed on oxidation, that is, the acids having the same number of carbon atoms. In general the *ic acid* is dropped and *aldehyde* added.

Ketones
The common names of ketones are derived from the acid which on pyrolysis would yield the ketone. A second method, especially useful for naming mixed ketones, simply names the alkyl groups and adds the word *ketone*. The name is written as three separate words.

Unsaturated Acyclic Hydrocarbons
The simplest compounds in this class of hydrocarbon chemicals are olefins or alkenes with a single carbon-carbon double bond, having the general formula of C_nH_{2n}. The simplest example in this category is ethylene, C_2H_4. Dienes are acyclic hydrocarbons with two carbon-carbon double bonds, having the general formula of C_nH_{2n-2}; butadiene (C_4H_6) is an example of such. Similarly, trienes have three carbon-carbon double bonds with the general formula of C_nH_{2n-4}; hexatriene (C_6H_8)is such an example. The simplest alkynes have a single carbon-carbon triple bond with the general formula of C_nH_{2n-2}. This series of compounds begins with acetylene or C_2H_2.

Common Names and Molecular Formulas of Some Industrial (Inorganic and Organic) Chemicals

Common Name	Chemical Name	Molecular Formula
Muriatic acid	Hydrochloric acid	HCl
Cumene	Isopropyl benzene	$C_6H_5CH(CH_3)_2$
Styrene	Vinyl benzene	$C_6H_5CH{=}CH_2$
—	Hypochlorite ion	OCl^{-1}
—	Chlorite ion	ClO_2^{-1}
—	Chlorate ion	ClO_3^{-1}
—	Perchlorate ion	ClO_4^{-1}
Gypsum	Calcium sulfate	$CaSO_4$
Limestone	Calcium carbonate	$CaCO_3$
Dolomite	Magnesium carbonate	$MgCO_3$
Bauxite	Aluminum oxide	Al_2O_3
Anatase	Titanium dioxide	TiO_2
Rutile	Titanium dioxide	TiO_2
—	Vinyl chloride	$CH_2{=}CHCl$
—	Ethylene oxide	C_2H_4O
Pyrite	Ferrous sulfide	FeS
Epsom salt	Magnesium sulfate	$MgSO_4$
Hydroquinone	p-Dihydroxy benzene	$C_6H_4(OH)_2$
Soda ash	Sodium carbonate	Na_2CO_3
Salt	Sodium chloride	$NaCl$
Potash	Potassium carbonate	K_2CO_3
Baking soda	Sodium bicarbonate	$NaHCO_3$
Lye	Sodium hydroxide	$NaOH$
Caustic soda	Sodium hydroxide	$NaOH$
—	Vinyl alcohol	$CH_2{=}CHOH$
Carbolic acid	Phenol	C_6H_5OH
Aniline	Aminobenzene	$C_6H_5NH_2$
—	Urea	$(NH_2)_2CO$
Toluene	Methyl benzene	$C_6H_5CH_3$
Xylene	Dimethyl benzene	$C_6H_4(CH_3)_2$
—	Silane	SiH_4
—	Ozone	O_3
Neopentane	2,2-Dimethylpropane	$CH_3C(CH_3)_2CH_3$
Magnetite	Ferrous/ferric oxide	Fe_3O_4
Quicksilver	Mercury	Hg
Heavy water	Deuterium oxide	$(H^2)_2O$
—	Borane	BH_3
Eyewash	Boric acid (solution)	H_3BO_3
—	Deuterium	H^2
—	Tritium	H^3
Laughing gas	Nitrous oxide	N_2O
—	Phosgene	$COCl_2$
Wolfram	Tungsten	W
—	Permanganate ion	MnO_4^{-1}
—	Dichromate ion	$Cr_2O_7^{-2}$
—	Hydronium ion	H_3O^{+1}
Brine	Sodium chloride (solution)	$NaCl$
Battery acid	Sulfuric acid	H_2SO_4

CHEMICAL THERMODYNAMICS

Vapor-Liquid Equilibrium

For a multi-component mixture at equilibrium

$$\hat{f}_i^V = \hat{f}_i^L, \text{ where}$$

$\hat{f}_i^V$ = fugacity of component i in the vapor phase, and

$\hat{f}_i^L$ = fugacity of component i in the liquid phase.

Fugacities of component i in a mixture are commonly calculated in the following ways:

For a liquid $\hat{f}_i^L = x_i \gamma_i f_i^L$, where

x_i = mole fraction of component i,

γ_i = activity coefficient of component i, and

f_i^L = fugacity of pure liquid component i.

For a vapor $\hat{f}_i^V = y_i \hat{\Phi}_i P$, where

y_i = mole fraction of component i in the vapor,

$\hat{\Phi}_i$ = fugacity coefficient of component i in the vapor, and

P = system pressure.

The activity coefficient γ_i is a correction for liquid phase non-ideality. Many models have been proposed for γ_i such as the Van Laar model:

$$\ln\gamma_1 = A_{12}\left(1 + \frac{A_{12}x_1}{A_{21}x_2}\right)^{-2}$$

$$\ln\gamma_2 = A_{21}\left(1 + \frac{A_{21}x_2}{A_{12}x_1}\right)^{-2}, \text{ where}$$

γ_1 = activity coefficient of component 1 in a two-component system,

γ_2 = activity coefficient of component 2 in a two-component system, and

A_{12}, A_{21} = constants, typically fitted from experimental data.

The pure component fugacity is calculated as:

$$f_i^L = \Phi_i^{\text{sat}} P_i^{\text{sat}} \exp\left\{v_i^L\left(P - P_i^{\text{sat}}\right)/(RT)\right\}, \text{ where}$$

Φ_i^{sat} = fugacity coefficient of pure saturated i,

P_i^{sat} = saturation pressure of pure i,

v_i^L = specific volume of pure liquid i,

R = Ideal Gas Law Constant, and

T = absolute temperature.

Often at system pressures close to atmospheric:

$$f_i^L \cong P_i^{\text{sat}}$$

The fugacity coefficient $\hat{\Phi}_i$ for component i in the vapor is calculated from an equation of state (e.g., Virial). Sometimes it is approximated by a pure component value from a correlation. Often at pressures close to atmospheric, $\hat{\Phi}_i = 1$. The fugacity coefficient is a correction for vapor phase non-ideality.

For sparingly soluble gases the liquid phase is sometimes represented as:

$$\hat{f}_i^L = x_i k_i$$

where k_i is a constant set by experiment (Henry's constant). Sometimes other concentration units are used besides mole fraction with a corresponding change in k_i.

Reactive Systems

Conversion: moles reacted/moles fed

Extent: For each species in a reaction, the mole balance may be written:
$$\text{moles}_{i,out} = \text{moles}_{i,in} + v_i\xi \text{ where}$$
ξ is the **extent** in moles and v_i is the stoichiometric coefficient of the ith species, the sign of which is negative for reactants and positive for products.

Limiting reactant: reactant that would be consumed first if the reaction proceeded to completion. Other reactants are **excess reactants**.

Selectivity: moles of desired product formed/moles of undesired product formed.

Yield: moles of desired product formed/moles that would have been formed if there were no side reactions and the limiting reactant had reacted completely.

Chemical Reaction Equilibrium

For reaction

$$aA + bB \rightleftharpoons cC + dD$$

$$\Delta G^\circ = -RT \ln K_a$$

$$K_a = \frac{\left(\hat{a}_C^c\right)\left(\hat{a}_D^d\right)}{\left(\hat{a}_A^a\right)\left(\hat{a}_B^b\right)} = \prod_i\left(\hat{a}_i\right)^{v_i}, \text{ where}$$

$\hat{a}_i$ = activity of component $i = \dfrac{\hat{f}_i}{f_i^\circ}$

f_i° = fugacity of pure i in its standard state

v_i = stoichiometric coefficient of component i

ΔG° = standard Gibbs energy change of reaction

K_a = chemical equilibrium constant

For mixtures of ideal gases:

$f_i^° =$ unit pressure, often 1 bar

$\hat{f}_i = y_i P = p_i$

where $p_i =$ partial pressure of component i.

Then $K_a = K_p = \dfrac{\left(p_C^c\right)\left(p_D^d\right)}{\left(p_A^a\right)\left(p_B^b\right)} = P^{c+d-a-b} \dfrac{\left(y_C^c\right)\left(y_D^d\right)}{\left(y_A^a\right)\left(y_B^b\right)}$

For solids $\hat{a}_i = 1$

For liquids $\hat{a}_i = x_i \gamma_i$

The effect of temperature on the equilibrium constant is

$$\frac{d \ln K}{dT} = \frac{\Delta H^°}{RT^2}$$

where $\Delta H^° =$ standard enthalpy change of reaction.

HEATS OF REACTION

For a chemical reaction the associated energy can be defined in terms of heats of formation of the individual species $\Delta \hat{H}_f^°$ at the standard state

$$\left(\Delta \hat{H}_r^°\right) = \underset{\text{products}}{\Sigma}\, \upsilon_i \left(\Delta \hat{H}_f^°\right)_i - \underset{\text{reactants}}{\Sigma}\, \upsilon_i \left(\Delta \hat{H}_f^°\right)_i$$

The standard state is 25°C and 1 atm.

The heat of formation is defined as the enthalpy change associated with the formation of a compound from its atomic species as they normally occur in nature [i.e., $O_2(g)$, $H_2(g)$, C(solid), etc.]

The heat of reaction varies with the temperature as follows:

$$\Delta H_r^°(T) = \Delta H_r^°(T_{\text{ref}}) + \int_{T_{\text{ref}}}^{T} \Delta C_P dT$$

where T_{ref} is some reference temperature (typically 25°C or 298 K), and:

$$\Delta C_P = \underset{\text{products}}{\Sigma}\, v_i C_{P,i} - \underset{\text{reactants}}{\Sigma}\, v_i C_{P,i}$$

and $C_{P,i}$ is the molar heat capacity of component i.

The heat of reaction for a combustion process using oxygen is also known as the heat of combustion. The principal products are $CO_2(g)$ and $H_2O(l)$.

CHEMICAL REACTION ENGINEERING

A chemical reaction may be expressed by the general equation

$a\text{A} + b\text{B} \leftrightarrow c\text{C} + d\text{D}$.

The rate of reaction of any component is defined as the moles of that component formed per unit time per unit volume.

$-r_A = -\dfrac{1}{V}\dfrac{dN_A}{dt}$ 　[negative because A disappears]

$-r_A = \dfrac{-dC_A}{dt}$ 　if V is constant

The rate of reaction is frequently expressed by

$-r_A = k f_r(C_A, C_B, ...)$, where

k = reaction rate constant and

C_I = concentration of component I.

In the conversion of A, the fractional conversion X_A is defined as the moles of A reacted per mole of A fed.

$X_A = (C_{A0} - C_A)/C_{A0}$ 　if V is constant

The Arrhenius equation gives the dependence of k on temperature

$k = Ae^{-E_a/\overline{R}T}$, where

A = pre-exponential or frequency factor,

E_a = activation energy (J/mol, cal/mol),

T = temperature (K), and

$\overline{R}$ = gas law constant = 8.314 J/(mol•K).

For values of rate constant (k_i) at two temperatures (T_i),

$$E_a = \frac{RT_1 T_2}{(T_1 - T_2)} \ln\left(\frac{k_1}{k_2}\right)$$

Reaction Order

If $-r_A = k C_A^x C_B^y$

the reaction is x order with respect to reactant A and y order with respect to reactant B. The overall order is

$n = x + y$

BATCH REACTOR, CONSTANT T AND V
Zero-Order Irreversible Reaction

$$
\begin{aligned}
-r_A &= kC_A^0 = k(1) \\
-dC_A/dt &= k &&\text{or} \\
C_A &= C_{A0} - kt \\
dX_A/dt &= k/C_{A0} &&\text{or} \\
C_{A0}X_A &= kt
\end{aligned}
$$

First-Order Irreversible Reaction

$$
\begin{aligned}
-r_A &= kC_A \\
-dC_A/dt &= kC_A &&\text{or} \\
\ln(C_A/C_{A0}) &= -kt \\
dX_A/dt &= k(1 - X_A) &&\text{or} \\
\ln(1 - X_A) &= -kt
\end{aligned}
$$

Second-Order Irreversible Reaction

$$-r_A = kC_A^2$$
$$-dC_A/dt = kC_A^2 \quad \text{or}$$
$$1/C_A - 1/C_{A0} = kt$$
$$dX_A/dt = kC_{A0}(1 - X_A)^2 \quad \text{or}$$
$$X_A/[C_{A0}(1 - X_A)] = kt$$

First-Order Reversible Reactions

$$A \underset{k_2}{\overset{k_1}{\rightleftharpoons}} R$$

$$-r_A = -\frac{dC_A}{dt} = k_1 C_A - k_2 C_R$$

$$K_c = k_1/k_2 = \hat{C}_R/\hat{C}_A$$

$$M = C_{R_0}/C_{A0}$$

$$\frac{d\hat{X}_A}{dt} = \frac{k_1(M + 1)}{M + \hat{X}_A}(\hat{X}_A - X_A)$$

$$-\ln\left(1 - \frac{X_A}{\hat{X}_A}\right) = -\ln\frac{C_A - \hat{C}_A}{C_{A0} - \hat{C}_A}$$

$$= \frac{(M + 1)}{(M + \hat{X}_A)}k_1 t$$

Reactions of Shifting Order

$$-r_A = \frac{k_1 C_A}{1 + k_2 C_A}$$

$$\ln\left(\frac{C_{A_o}}{C_A}\right) + k_2(C_{A_o} - C_A) = k_1 t$$

$$\frac{\ln(C_{A_o}/C_A)}{C_{A_o} - C_A} = -k_2 + \frac{k_1 t}{C_{A_o} - C_A}$$

This form of the rate equation is used for elementary enzyme-catalyzed reactions and for elementary surfaced-catalyzed reactions.

Batch Reactor, General

For a well-mixed, constant-volume batch reactor

$$-r_A = -dC_A/dt$$

$$t = -C_{A0}\int_0^{X_A} dX_A/(-r_A)$$

If the volume of the reacting mass varies with the conversion (such as a variable-volume batch reactor) according to

$$V = V_{X_A = 0}(1 + \varepsilon_A X_A)$$

(ie., at constant pressure), where

$$\varepsilon_A = \frac{V_{X_A = 1} - V_{X_A = 0}}{V_{X_A = 0}} = \frac{\Delta V}{V_{X_A = 0}}$$

then at any time

$$C_A = C_{A0}\left[\frac{1 - X_A}{1 + \varepsilon_A X_A}\right]$$

and

$$t = -C_{A0}\int_0^{X_A} dX_A/\left[(1 + \varepsilon_A X_A)(-r_A)\right]$$

For a first-order irreversible reaction,

$$kt = -\ln(1 - X_A) = -\ln\left(1 - \frac{\Delta V}{\varepsilon_A V_{XA = 0}}\right)$$

FLOW REACTORS, STEADY STATE

Space-time τ is defined as the reactor volume divided by the inlet volumetric feed rate. Space-velocity SV is the reciprocal of space-time, $SV = 1/\tau$.

Plug-Flow Reactor (PFR)

$$\tau = \frac{C_{A0}V_{PFR}}{F_{A0}} = C_{A0}\int_0^{X_A}\frac{dX_A}{(-r_A)}, \text{ where}$$

F_{A0} = moles of A fed per unit time.

Continuous-Stirred Tank Reactor (CSTR)

For a constant volume, well-mixed CSTR

$$\frac{\tau}{C_{A0}} = \frac{V_{CSTR}}{F_{A0}} = \frac{X_A}{-r_A}, \text{ where}$$

$-r_A$ is evaluated at exit stream conditions.

Continuous-Stirred Tank Reactors in Series

With a first-order reaction $A \rightarrow R$, no change in volume.

$$\tau_{N\text{-reactors}} = N\tau_{individual}$$

$$= \frac{N}{k}\left[\left(\frac{C_{A0}}{C_{AN}}\right)^{1/N} - 1\right], \text{ where}$$

N = number of CSTRs (equal volume) in series, and
C_{AN} = concentration of A leaving the Nth CSTR.

Two Irreversible Reactions in Parallel

$$A \overset{k_D}{\rightarrow} D\text{(desired)}$$

$$A \overset{k_U}{\rightarrow} U\text{(undesired)}$$

$$-r_A = -dc_A/dt = k_D C_A^x + k_U C_A^y$$

$$r_D = dc_D/dt = k_D C_A^x$$

$$r_U = dc_U/dt = k_U C_A^y$$

Y_D = instantaneous fractional yield of D

$$= dC_D/(-dC_A)$$

$\bar{Y}_D$ = overall fractional yield of D

$$= N_{D_f}/(N_{A_0} - N_{A_f})$$

where N_{A_f} and N_{D_f} are measured at the outlet of the flow reactor.

$\bar{S}_{DU}$ = overall selectivity to D

$$= N_{D_f}/N_{U_f}$$

Two First-Order Irreversible Reactions in Series

$$A \xrightarrow{k_D} D \xrightarrow{k_U} U$$

$$r_A = -dC_A/dt = k_D C_A$$

$$r_D = dC_D/dt = k_D C_A - k_U C_D$$

$$r_U = dC_U/dt = k_U C_D$$

Yield and selectivity definitions are identical to those for two irreversible reactions in parallel. The optimal yield of D in a PFR is

$$\frac{C_{D,\max}}{C_{A0}} = \left(\frac{k_D}{k_U}\right)^{k_U/(k_U - k_D)}$$

at time

$$\tau_{\max} = \frac{1}{k_{\log\,mean}} = \frac{\ln(k_U/k_D)}{(k_U - k_D)}$$

The optimal yield of D in a CSTR is

$$\frac{C_{D,\max}}{C_{A0}} = \frac{1}{\left[(k_U/k_D)^{1/2} + 1\right]^2}$$

at time

$$\tau_{\max} = 1\Big/\sqrt{k_D k_U^D}$$

MASS TRANSFER

Diffusion

Molecular Diffusion

Gas: $N_A = \dfrac{p_A}{P}(N_A + N_B) - \dfrac{D_m}{RT}\dfrac{\partial p_A}{\partial z}$

Liquid: $N_A = x_A(N_A + N_B) - CD_m\dfrac{\partial x_A}{\partial z}$

Unidirectional Diffusion of a Gas A Through a Second Stagnant Gas B ($N_b = 0$)

$$N_A = \frac{D_m P}{RT(p_B)_{lm}} \times \frac{(p_{A2} - p_{A1})}{z_2 - z_1}$$

in which $(p_B)_{lm}$ is the log mean of p_{B2} and p_{B1},

N_i = diffusive flux [mole/(time × area)] of component i through area A, in z direction,

D_m = mass diffusivity,

p_I = partial pressure of species I, and

C = concentration (mole/volume)

EQUIMOLAR COUNTER-DIFFUSION (GASES)
($N_B = -N_A$)

$$N_A = D_m/(RT) \times \left[(p_{A1} - p_{A2})/(\Delta z)\right]$$

$$N_A = D_m (C_{A1} - C_{A2})/\Delta z$$

CONVECTION
Two-Film Theory (for Equimolar Counter-Diffusion)

$$\begin{aligned}
N_A &= k'_G(p_{AG} - p_{Ai})\\
&= k'_L(C_{Ai} - C_{AL})\\
&= K'_G(p_{AG} - p_A^*)\\
&= K'_L(C_A^* - C_{AL})
\end{aligned}$$

where p_A^* is the partial pressure in equilibrium with C_{AL}, and C_A^* = concentration in equilibrium with p_{AG}.

$P_A^* = H \bullet C_{AL}$ and $C_A^* = p_{AG}/H$, where H is the Henry's Law Constant.

Overall Coefficients

$$1/K'_G = 1/k'_G + H/k'_L$$
$$1/K'_L = 1/Hk'_G + 1/k'_L$$

Dimensionless Group Equation (Sherwood)

For the turbulent flow inside a tube the Sherwood number

$$\text{Sh} = \left(\frac{k_m D}{D_m}\right) = 0.023\left(\frac{DV\rho}{\mu}\right)^{0.8}\left(\frac{\mu}{\rho D_m}\right)^{1/3}$$

where,

D = inside diameter,

D_m = diffusion coefficient,

V = average velocity in the tube,

ρ = fluid density,

μ = fluid viscosity, and

k_m = mass transfer coefficient.

Distillation

Definitions:

α = relative volatility,

B = molar bottoms-product rate,

D = molar overhead-product rate,

F = molar feed rate,

L = molar liquid downflow rate,

R_D = ratio of reflux to overhead product,

V = molar vapor upflow rate,

W = total moles in still pot,

x = mole fraction of the more volatile component in the liquid phase, and

y = mole fraction of the more volatile component in the vapor phase.

Subscripts:

B = bottoms product,

D = overhead product,

F = feed,

m = any plate in stripping section of column,

$m+1$ = plate below plate m,

n = any plate in rectifying section of column,

$n+1$ = plate below plate n, and

o = original charge in still pot.

Flash (or equilibrium) Distillation

Component material balance:
$$Fz_F = yV + xL$$

Overall material balance:
$$F = V + L$$

Differential (Simple or Rayleigh) Distillation

$$\ln\left(\frac{W}{W_o}\right) = \int_{x_o}^{x} \frac{dx}{y - x}$$

When the relative volatility α is constant,
$$y = \alpha x / [1 + (\alpha - 1)x]$$

can be substituted to give
$$\ln\left(\frac{W}{W_o}\right) = \frac{1}{(\alpha - 1)} \ln\left[\frac{x(1 - x_o)}{x_o(1 - x)}\right] + \ln\left[\frac{1 - x_o}{1 - x}\right]$$

For binary system following Raoult's Law
$$\alpha = (y/x)_a / (y/x)_b = p_a / p_b, \text{ where}$$

p_i = partial pressure of component i.

Continuous Distillation (binary system)

Constant molal overflow is assumed
(trays counted downward)

OVERALL MATERIAL BALANCES
   Total Material:
   $$F = D + B$$

   Component A:
   $$Fz_F = Dx_D + Bx_B$$

OPERATING LINES
Rectifying Section
   Total Material:
   $$V_{n+1} = L_n + D$$

   Component A:
   $$V_{n+1}y_{n+1} = L_n x_n + Dx_D$$
   $$y_{n+1} = [L_n/(L_n + D)]x_n + Dx_D/(L_n + D)$$

Stripping Section
   Total Material:
   $$L_m = V_{m+1} + B$$

   Component A:
   $$L_m x_m = V_{m+1}y_{m+1} + Bx_B$$
   $$y_{m+1} = [L_m/(L_m - B)]x_m - Bx_B/(L_m - B)$$

Reflux Ratio
   Ratio of reflux to overhead product
   $$R_D = L_R/D = (V_R - D)/D$$

Minimum reflux ratio is defined as that value which results in an infinite number of contact stages. For a binary system the equation of the operating line is

$$y = \frac{R_{\min}}{R_{\min} + 1}x + \frac{x_D}{R_{\min} + 1}$$

Feed Condition Line
   slope $= q/(q - 1)$, where

   $$q = \frac{\text{heat to convert one mol of feed to saturated vapor}}{\text{molar heat of vaporization}}$$

Murphree Plate Efficiency

$$E_{ME} = (y_n - y_{n+1})/(y_n^* - y_{n+1}), \text{ where}$$

y_n = concentration of vapor above plate n,

y_{n+1} = concentration of vapor entering from plate below n, and

y_n^* = concentration of vapor in equilibrium with liquid leaving plate n.

A similar expression can be written for the stripping section by replacing n with m.

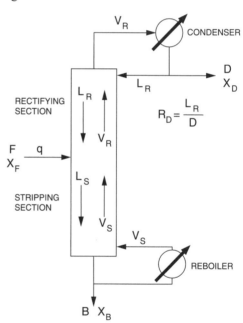

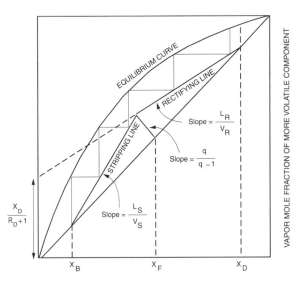

LIQUID MOLE FRACTION OF MORE VOLATILE COMPONENT

Absorption (packed columns)

Continuous Contact Columns

$$Z = NTU_G \cdot HTU_G = NTU_L \cdot HTU_L = N_{EQ} \cdot HETP$$

Z = column height

NTU_G = number of transfer units (gas phase)

NTU_L = number of transfer units (liquid phase)

N_{EQ} = number of equilibrium stages

HTU_G = height of transfer unit (gas phase)

HTU_L = height of transfer unit (liquid phase)

$HETP$ = height equivalent to theoretical plate (stage)

$$HTU_G = \frac{G}{K'_G a} \qquad HTU_L = \frac{L}{K'_L a}$$

G = gas phase mass velocity (mass or moles/flow area • time)

L = liquid phase mass velocity (mass or moles/flow area • time)

K'_G = overall gas phase mass transfer coefficient (mass or moles/mass transfer area • time)

K'_L = overall liquid phase mass transfer coefficient (mass or moles/mass transfer area • time)

a = mass transfer area/volume of column (length^{-1})

$$NTU_G = \int_{y_1}^{y_2} \frac{dy}{(y - y^*)} \qquad NTU_L = \int_{x_1}^{x_2} \frac{dx}{(x^* - x)}$$

y = gas phase solute mole fraction

x = liquid phase solute mole fraction

y^* = $K \cdot x$, where K = equilibrium constant

x^* = y/K, where K = equilibrium constant

y_2, x_2 = mole fractions at the lean end of column

y_1, x_1 = mole fractions at the rich end of column

For dilute solutions (constant G/L and constant K value for entire column):

$$NTU_G = \frac{y_1 - y_2}{(y - y^*)_{LM}}$$

$$(y - y^*)_{LM} = \frac{(y_1 - y_1^*) - (y_2 - y_2^*)}{\ln\left(\frac{y_1 - y_1^*}{y_2 - y_2^*}\right)}$$

For a chemically reacting system—absorbed solute reacts in the liquid phase—the preceeding relation simplifies to:

$$NTU_G = \ln\left(\frac{y_1}{y_2}\right)$$

Other Mass Transfer Operations

Refer to the **ENVIRONMENTAL ENGINEERING— Water Treatment Technologies** section of this book for the following operations:

Reverse Osmosis

Ultrafiltration

Electrodialysis

Adsorption

Solid/Fluid Separations

Refer to the **ENVIRONMENTAL ENGINEERING** section of this book for information on Cyclones, Baghouses, Electrostatic Precipitators, and Particle Settling.

COST ESTIMATION

Cost Indexes

Cost indexes are used to update historical cost data to the present. If a purchase cost is available for an item of equipment in year M, the equivalent current cost would be found by:

$$\text{Current \$} = (\text{Cost in year } M)\left(\frac{\text{Current Index}}{\text{Index in year } M}\right)$$

Component	Range
Direct costs	
Purchased equipment-delivered (including fabricated equipment and process machinery such as pumps and compressors)	100
Purchased-equipment installation	39–47
Instrumentation and controls (installed)	9–18
Piping (installed)	16–66
Electrical (installed)	10–11
Buildings (including services)	18–29
Yard improvements	10–13
Service facilities (installed)	40–70
Land (if purchase is required)	6
Total direct plant cost	264–346
Indirect costs	
Engineering and supervision	32–33
Construction expenses	34–41
Total direct and indirect plant costs	336–420
Contractor's fee (about 5% of direct and indirect plant costs)	17–21
Contingency (about 10% of direct and indirect plant costs)	36–42
Fixed-capital investment	387–483
Working capital (about 15% of total capital investment)	68–86
Total capital investment	455–569

Scaling of Equipment Costs

The cost of Unit A at one capacity related to the cost of a similar Unit B with X times the capacity of Unit A is approximately X^n times the cost of Unit B.

$$\text{Cost of Unit A} = \text{cost of Unit B}\left(\frac{\text{capacity of Unit A}}{\text{capacity of Unit B}}\right)^n$$

TYPICAL EXPONENTS (n) FOR EQUIPMENT COST VS. CAPACITY		
Equipment	**Size range**	**Exponent**
Dryer, drum, single vacuum	10–10^2 ft^2	0.76
Dryer, drum, single atmospheric	10–10^2 ft^2	0.40
Fan, centrifugal	10^3–10^4 ft^3/min	0.44
Fan, centrifugal	2×10^4–7×10^4 ft^3/min	1.17
Heat exchanger, shell and tube, floating head, c.s.	100–400 ft^2	0.60
Heat exchanger, shell and tube, fixed sheet, c.s.	100–400 ft^2	0.44
Motor, squirrel cage, induction, 440 volts, explosion proof	5–20 hp	0.69
Motor, squirrel cage, induction, 440 volts, explosion proof	20–200 hp	0.99
Tray, bubble cup, c.s.	3–10 ft diameter	1.20
Tray, sieve, c.s.	3–10 ft diameter	0.86

CHEMICAL PROCESS SAFETY

Threshold Limit Value (TLV)

TLV is the highest dose (ppm by volume in the atmosphere) the body is able to detoxify without any detectable effects.

Examples are:

Compound	TLV
Ammonia	25
Chlorine	0.5
Ethyl Chloride	1,000
Ethyl Ether	400

Flammability

LFL = lower flammability limit (volume % in air)
UFL = upper flammability limit (volume % in air)

A vapor-air mixture will only ignite and burn over the range of concentrations between LFL and UFL. Examples are:

Compound	LFL	UFL
Ethyl alcohol	3.3	19
Ethyl ether	1.9	36.0
Ethylene	2.7	36.0
Methane	5	15
Propane	2.1	9.5

Concentrations of Vaporized Liquids

Vaporization Rate (Q_m, mass/time) from a Liquid Surface

$$Q_m = [MKA_SP^{sat}/(R_gT_L)]$$

M = molecular weight of volatile substance

K = mass transfer coefficient

A_S = area of liquid surface

P^{sat} = saturation vapor pressure of the pure liquid at T_L

R_g = ideal gas constant

T_L = absolute temperature of the liquid

Mass Flow Rate of Liquid from a Hole in the Wall of a Process Unit

$$Q_m = A_HC_0(2\rho g_cP_g)^{1/2}$$

A_H = area of hole

C_0 = discharge coefficient

ρ = density of the liquid

g_c = gravitational constant

P_g = gauge pressure within the process unit

Concentration (C_{ppm}) of Vaporized Liquid in Ventilated Space

$$C_{ppm} = [Q_mR_gT \times 10^6/(kQ_VPM)]$$

T = absolute ambient temperature

k = nonideal mixing factor

Q_V = ventilation rate

P = absolute ambient pressure

Concentration in the Atmosphere

See "Atmospheric Dispersion Modeling" under
AIR POLLUTION in the **ENVIRONMENTAL ENGINEERING** section.

Sweep-Through Concentration Change in a Vessel

$$Q_Vt = V\ln[(C_1 - C_0)/(C_2 - C_0)]$$

Q_V = volumetric flow rate

t = time

V = vessel volume

C_0 = inlet concentration

C_1 = initial concentration

C_2 = final concentration

CIVIL ENGINEERING

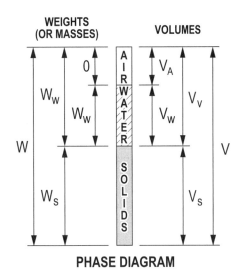

WEIGHTS (OR MASSES)   **VOLUMES**

PHASE DIAGRAM

W_W = weight of water

W_S = weight of solids

W = total weight

V = total volume

V_A = volume of air

V_S = volume of soil solids

V_W = volume of water

V_V = volume of voids = $V_A + V_W$

γ = total unit weight = W/V

γ_W = unit weight of water (62.4 lb/ft^3 or 9.81 kN/m^3)

γ_{sat} = saturated unit weight

     = $(G_s + e)\gamma_W/(1 + e) = \gamma(G_s + e)/(1 + \omega)$

     where γ, e, and ω are in-situ soil properties

γ' = effective (submerged) unit weight = $\gamma_{sat} - \gamma_W$

γ_S = unit weight of soil solids = W_S/V_S

γ_D = dry unit weight = W_S/V

     = $G_S\gamma_W/(1+e) = \gamma/(1+\omega)$

$\gamma_{D\text{-MAX}}$ = maximum dry unit weight

$\gamma_{D\text{-MIN}}$ = minimum dry unit weight

ω = water content (%) = $(W_W/W_S) \times 100$

G_S = specific gravity of soil solids = $(W_S/V_S)/\gamma_W$

e = void ratio = V_V/V_S

e_{MAX} = maximum void ratio

e_{MIN} = minimum void ratio

n = porosity = $V_V/V = e/(1+e)$

S = degree of saturation (%) = $(V_W/V_V) \times 100$

D_r = relative density (%)

     = $[(e_{MAX} - e)/(e_{MAX} - e_{MIN})] \times 100$

     = $[(\gamma_{D\text{-FIELD}} - \gamma_{D\text{-MIN}})/(\gamma_{D\text{-MAX}} - \gamma_{D\text{-MIN}})][\gamma_{D\text{-MAX}}/\gamma_{D\text{-FIELD}}] \times 100$

RC = relative compaction (%) = $(\gamma_{D\text{-FIELD}}/\gamma_{D\text{-MAX}}) \times 100$

PL = plastic limit

LL = liquid limit

PI = plasticity index = $LL - PL$

D_{10} = effective grain size, grain size corresponding to 10% finer on grain size curve

D_{30} = grain size corresponding to 30% finer

D_{60} = grain size corresponding to 60% finer

C_U = coefficient of uniformity = D_{60}/D_{10}

C_G = coefficient of concavity (or curvature)

     = $(D_{30})^2/[(D_{10})(D_{60})]$

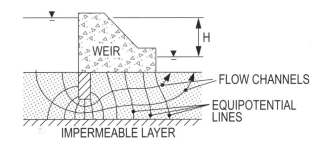

WEIR — FLOW CHANNELS — EQUIPOTENTIAL LINES — IMPERMEABLE LAYER

FLOW NET

k = coefficient of permeability

     = hydraulic conductivity

     = $Q/(i\,A\,t_E)$ from constant head test

     = $2.303[(aL)/(At_E)]\log_{10}(h_1/h_2)$ from falling head test

Q = rate of seepage discharge = $v\,A$

i = hydraulic gradient = dH/dL (parallel to flow)

A = cross-sectional area (perpendicular to flow)

a = area of reservoir tube during falling head test

t_E = elapsed time during falling head test

h_1 = head at time $t = 0$ during falling head test

h_2 = head at time $t = t_E$ during falling head test

v = discharge velocity = $k\,i$

Q = $kH(N_F/N_D)$ for flow nets (Q per unit width)

H = total hydraulic head differential (potential)

L = length of flow through a flow channel

N_F = number of flow channels

N_D = number of equipotential drops

i_C = critical gradient = $(\gamma - \gamma_W)/\gamma_W$

i_E = seepage exit gradient (in vertical direction)

FS_S = factor of safety against seepage liquefaction = i_C/i_E

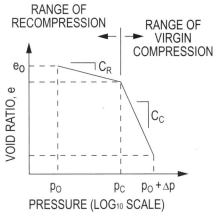

SOIL CONSOLIDATION CURVE

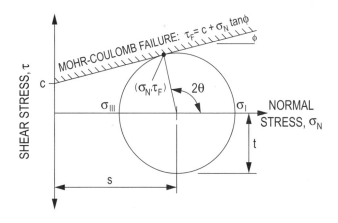

C_C = compression index = slope, $\Delta e/\Delta\log_{10}p$, in virgin compression portion of consolidation curve
= 0.009(LL – 10), by correlation to liquid limit

C_R = recompression index = slope, $\Delta e/\Delta\log 10 p$, in recompression portion of consoliation curve
= $C_C/6$, by correlation to C_C

e_O = initial void ratio (prior to consolidation)

Δe = change in void ratio (by sign convention, positive change in void ratio corresponds to a decrease in total volume of soil).

p_O = initial effective consolidation stress, or effective in situ overburden stress at center of consolidating stratum at beginning of consolidation.

p_C = past maximum consolidation stress

Δp = induced change in consolidation stress at center of consolidating stratum = $I\, q_S$

I = stress influence value at center of consolidating stratum

q_S = applied surface stress causing consolidation

Δe_{TOT} = total change in void ratio due to recompression (if any) and due to virgin compression (if any)

ε_V = average ultimate strain in soil layer
= $\Delta e_{TOT}/(1+e_O)$

S_{ULT} = ultimate consolidation settlement in soil layer
= $\varepsilon_V H_S$

H_S = thickness of soil layer

c_V = coefficient of consolidation

T_V = time factor for vertical consolidation
= $(c_V\, t_C)/H_{DR}{}^2$

t_C = elapsed time from beginning of consolidation load

H_{DR} = length of pore water drainage path

U_{AV} = average degree of consolidation (see Degree of Consolidation Curve later in this section)

S_T = approximate settlement at $t = t_C$
= $U_{AV} S_{ULT}$

σ_N = total normal stress = P/A

P = normal force

A = cross-sectional area over which force acts

τ = shear stress = T/A

T = shearing force

u = pore water pressure = $h_U\gamma_W$

h_U = uplift or pressure head

σ' = effective stress = $\sigma - u$

s = mean normal stress

t = maximum shear stress

σ_I = major principal stress

σ_{III} = minor principal stress

θ = orientation angle between plane of existing normal stress and plane of major principal stress

τ_F = shear stress at failure
= $c + \sigma_N\tan\phi$, by Mohr-Coulomb criterion

c = cohesion

ϕ = angle of internal friction

q_U = unconfined compressive strength = 2c in unconfined compression test

K_F = σ_{III}/σ_I (at failure)

q_{ULT} = net ultimate bearing capacity (strip footing)
= $cN_C + \gamma D_f(N_q - 1) + \frac{1}{2}\gamma B N_\gamma$

N_C = bearing capacity factor for cohesion

N_q = bearing capacity factor for depth

N_γ = bearing capacity factor for unit weight

q_{ALL} = allowable bearing capacity = q_{ULT}/FS_q

FS_q = factor of safety for bearing capacity

B = width of strip footing

D_f = depth of footing below ground surface

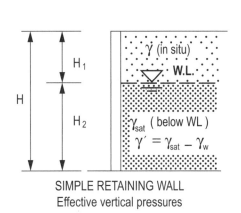

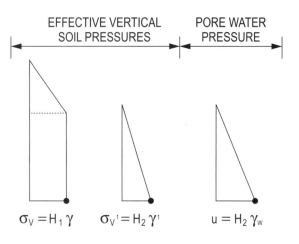

SIMPLE RETAINING WALL
Effective vertical pressures

EFFECTIVE VERTICAL SOIL PRESSURES

PORE WATER PRESSURE

$\sigma_V = H_1 \gamma$ $\sigma_{V'} = H_2 \gamma^1$ $u = H_2 \gamma_w$

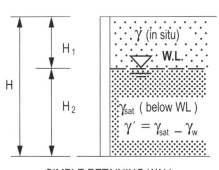

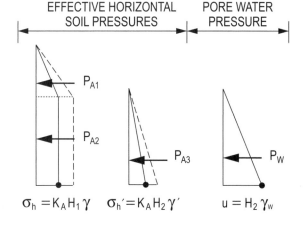

SIMPLE RETAINING WALL
Lateral pressures and forces
(active shown, passive similar)

EFFECTIVE HORIZONTAL SOIL PRESSURES

PORE WATER PRESSURE

$\sigma_h = K_A H_1 \gamma$ $\sigma_h' = K_A H_2 \gamma'$ $u = H_2 \gamma_w$

Active forces on retaining wall per unit wall length (as shown):

K_A = Rankine active earth pressure coefficient (smooth wall, c = 0, level backfill) = $\tan^2 \left(45 - \dfrac{\phi}{2}\right)$

P_{A1} = area of pressure block = $\dfrac{1}{2} K_A H_1^2 \gamma$ [P_W force and remaining P_{Ai} forces found in similar manner]

$P_{A(total)} = P_w + \Sigma P_{Ai}$

Passive forces on retaining wall per unit wall length (similar to the active forces shown):

K_P = Rankine passive earth pressure coefficient (smooth wall, c = 0, level backfill) = $\tan^2 \left(45 + \dfrac{\phi}{2}\right)$

P_{Pi}, P_W, and $P_{P(total)}$ forces computed in similar manner

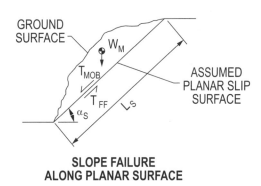

**SLOPE FAILURE
ALONG PLANAR SURFACE**

FS   =   factor of safety against slope instability
       = T_{FF}/T_{MOB}

T_{FF} =   available shearing resistance along slip surface
       = $cL_S + W_M \cos\alpha_S \tan\phi$

T_{MOB} =   mobilized shear force along slip surface
       = $W_M \sin\alpha_S$

L_S =   length of assumed planar slip surface

W_M =   weight of soil above slip surface

α_S =   angle of assumed slip surface with respect to horizontal

UNIFIED SOIL CLASSIFICATION

MAJOR DIVISIONS	FIELD IDENTIFICATION PROCEDURES (EXCLUDING PARTICLES LARGER THAN 3 in. AND BASING FRACTIONS ON ESTIMATED WEIGHTS)	GROUP SYMBOLS a	TYPICAL NAMES	INFORMATION REQUIRED FOR DESCRIBING SOILS
COARSE-GRAINED SOILS (MORE THAN HALF OF MATERIAL IS LARGER THAN NO. 200 SIEVE SIZE) b → GRAVELS (MORE THAN HALF OF COARSE FRACTION IS LARGER THAN NO. 4 SIEVE SIZE) → CLEAN GRAVELS (LITTLE OR NO FINES)	WIDE RANGE IN GRAIN SIZE AND SUBSTANTIAL AMOUNTS OF ALL INTERMEDIATE PARTICLE SIZES	GW	WELL GRADED GRAVELS, GRAVEL-SAND MIXTURES, LITTLE OR NO FINES	GIVE TYPICAL NAME: INDICATE APPROXIMATE PERCENTAGES OF SAND AND GRAVEL; MAXIMUM SIZE; ANGULARITY, SURFACE CONDITION, AND HARDNESS OF THE COARSE GRAINS; LOCAL OR GEOLOGIC NAME AND OTHER PERTINENT DESCRIPTIVE INFORMATION; AND SYMBOLS IN PARENTHESES
CLEAN GRAVELS (LITTLE OR NO FINES)	PREDOMINANTLY ONE SIZE OR A RANGE OF SIZES WITH SOME INTERMEDIATE SIZES MISSING	GP	POORLY GRADED GRAVELS, GRAVEL-SAND MIXTURES, LITTLE OR NO FINES	
GRAVELS WITH FINES (APPRECIABLE AMOUNT OF FINES)	NONPLASTIC FINES (FOR IDENTIFICATION PROCEDURES, SEE ML BELOW)	GM	SILTY GRAVELS, POORLY GRADED GRAVEL-SAND-SILT MIXTURES	FOR UNDISTURBED SOILS ADD INFORMATION ON STRATIFICATION, DEGREE OF COMPACTNESS, CEMENTATION, MOISTURE CONDITIONS AND DRAINAGE CHARACTERISTICS
GRAVELS WITH FINES (APPRECIABLE AMOUNT OF FINES)	PLASTIC FINES (FOR IDENTIFICATION PROCEDURES, SEE CL BELOW)	GC	CLAYEY GRAVELS, POORLY GRADED GRAVEL-SAND-CLAY MIXTURES	
SANDS (MORE THAN HALF OF COARSE FRACTION IS SMALLER THAN NO. 4 SIEVE SIZE) (FOR VISUAL CLASSIFICATION, THE 1/4 IN. SIZE MAY BE USED AS EQUIVALENT TO THE NO. 4 SIEVE SIZE) → CLEAN SANDS (LITTLE OR NO FINES)	WIDE RANGE IN GRAIN SIZES AND SUBSTANTIAL AMOUNTS OF ALL INTERMEDIATE PARTICLE SIZES	SW	WELL GRADED SANDS, GRAVELLY SANDS, LITTLE OR NO FINES	EXAMPLE: SILTY SAND, GRAVELLY; ABOUT 20% HARD, ANGULAR GRAVEL PARTICLES 1/2-IN. MAXIMUM SIZE; ROUNDED AND SUBANGULAR SAND GRAINS COARSE TO FINE, ABOUT 15% NON-PLASTIC FINES WITH LOW DRY STRENGTH; WELL COMPACTED AND MOIST IN PLACE; ALLUVIAL SAND; (SM)
CLEAN SANDS (LITTLE OR NO FINES)	PREDOMINANTLY ONE SIZE OR A RANGE OF SIZES WITH SOME INTERMEDIATE SIZES MISSING	SP	POORLY GRADED SANDS, GRAVELLY SANDS, LITTLE OR NO FINES	
SANDS WITH FINES (APPRECIABLE AMOUNT OF FINES)	NONPLASTIC FINES (FOR IDENTIFICATION PROCEDURES, SEE ML BELOW)	SM	SILTY SANDS, POORLY GRADED SAND-SILT MIXTURES	
SANDS WITH FINES (APPRECIABLE AMOUNT OF FINES)	PLASTIC FINES (FOR IDENTIFICATION PROCEDURES, SEE CL BELOW)	SC	CLAYEY SANDS, POORLY GRADED SAND-CLAY MIXTURES	

IDENTIFICATION PROCEDURES ON FRACTION SMALLER THAN NO. 40 SIEVE SIZE

MAJOR DIVISIONS	DRY STRENGTH (CRUSHING CHARACTERISTICS)	DILATANCY (REACTION TO SHAKING)	TOUGHNESS (CONSISTENCY NEAR PLASTIC LIMIT)	GROUP SYMBOLS a	TYPICAL NAMES	INFORMATION REQUIRED FOR DESCRIBING SOILS
FINE-GRAINED SOILS (MORE THAN HALF OF MATERIAL IS SMALLER THAN NO. 200 SIEVE SIZE) → SILTS AND CLAYS (LIQUID LIMIT LESS THAN 50)	NONE TO SLIGHT	QUICK TO SLOW	NONE	ML	INORGANIC SILTS AND VERY FINE SANDS, ROCK FLOUR, SILTY OR CLAYEY FINE SANDS WITH SLIGHT PLASTICITY	GIVE TYPICAL NAME; INDICATE DEGREE AND CHARACTER OF PLASTICITY, AMOUNT AND MAXIMUM SIZE OF COARSE GRAINS; COLOR IN WET CONDITION, ODOR IF ANY, LOCAL OR GEOLOGIC NAME, AND OTHER PERTINENT DESCRIPTIVE INFORMATION, AND SYMBOL IN PARENTHESES
SILTS AND CLAYS (LIQUID LIMIT LESS THAN 50)	MEDIUM TO HIGH	NONE TO VERY SLOW	MEDIUM	CL	INORGANIC CLAYS OF LOW TO MEDIUM PLASTICITY, GRAVELLY CLAYS, SANDY CLAYS, SILTY CLAYS, LEANS CLAYS	
SILTS AND CLAYS (LIQUID LIMIT LESS THAN 50)	SLIGHT TO MEDIUM	SLOW	SLIGHT	OL	ORGANIC SILTS AND ORGANIC SILT-CLAYS OF LOW PLASTICITY	FOR UNDISTURBED SOILS ADD INFORMATION ON STRUCTURE, STRATIFICATION, CONSISTENCY IN UNDISTURBED AND REMOLDED STATES, MOISTURE AND DRAINAGE CONDITIONS
SILTS AND CLAYS (LIQUID LIMIT GREATER THAN 50)	SLIGHT TO MEDIUM	SLOW TO NONE	SLIGHT TO MEDIUM	MH	INORGANIC SILTS, MICACEOUS OR DIATOMACEOUS FINE SANDY OR SILTY SOILS, ELASTIC SILTS	
SILTS AND CLAYS (LIQUID LIMIT GREATER THAN 50)	HIGH TO VERY HIGH	NONE	HIGH	CH	INORGANIC CLAYS OF HIGH PLASTICITY, FAT CLAYS	EXAMPLE: CLAYEY SILT, BROWN; SLIGHTLY PLASTIC; SMALL PERCENTAGE OF FINE SAND; NUMEROUS VERTICAL ROOT HOLES; FIRM AND DRY IN PLACE; LOESS; (ML)
SILTS AND CLAYS (LIQUID LIMIT GREATER THAN 50)	MEDIUM TO HIGH	NONE TO VERY SLOW	SLIGHT TO MEDIUM	OH	ORGANIC CLAYS OF MEDIUM TO HIGH PLASTICITY	
HIGHLY ORGANIC SOILS	READILY IDENTIFIED BY COLOR, ODOR, SPONGY FEEL AND FREQUENTLY BY FIBROUS TEXTURE			Pt	PEAT AND OTHER HIGHLY ORGANIC SOILS	

(THE NO. 200 SIEVE SIZE IS ABOUT THE SMALLEST PARTICLE VISIBLE TO THE NAKED EYE)

LABORATORY CLASSIFICATION CRITERIA

USE GRAIN SIZE CURVE IN IDENTIFYING THE FRACTIONS AS GIVEN UNDER FIELD IDENTIFICATION

DETERMINE PERCENTAGES OF GRAVEL AND SAND FROM GRAIN SIZE CURVE. DEPENDING ON PERCENTAGE OF FINES (FRACTION SMALLER THAN NO. 200 SIEVE SIZE), COARSE GRAINED SOILS ARE CLASSIFIED AS FOLLOWS:

- LESS THAN 5% — GW, GP, SW, SP
- MORE THAN 12% — GM, GC, SM, SC
- 5% TO 12% — BORDERLINE CASES REQUIRING USE OF DUAL SYMBOLS

GW:
$$C_U = \frac{D_{60}}{D_{10}} \text{ GREATER THAN 4}$$
$$C_C = \frac{(D_{30})^2}{D_{10} \times D_{60}} \text{ BETWEEN 1 AND 3}$$

GP: NOT MEETING ALL GRADATION REQUIREMENT FOR GW

GM: ATTERBERG LIMITS BELOW "A" LINE OR PI LESS THAN 4 / ATTERBERG LIMITS ABOVE "A" LINE OR PI GREATER THAN 7 — ABOVE "A" LINE WITH PI BETWEEN 4 AND 7 ARE BORDERLINE CASES REQUIRING USE OF DUAL SYMBOLS

GC: ATTERBERG LIMITS ABOVE "A" LINE WITH PI GREATER THAN 7

SW:
$$C_U = \frac{D_{60}}{D_{10}} \text{ GREATER THAN 6}$$
$$C_C = \frac{(D_{30})^2}{D_{10} \times D_{60}} \text{ BETWEEN 1 AND 3}$$

SP: NOT MEETING ALL GRADATION REQUIREMENT FOR SW

SM: ATTERBERG LIMITS BELOW "A" LINE OR PI LESS THAN 5 / ATTERBERG LIMITS ABOVE "A" LINE WITH PI BETWEEN 4 AND 7 ARE BORDERLINE CASES REQUIRING USE OF DUAL SYMBOLS

SC: ATTERBERG LIMITS ABOVE "A" LINE WITH PI GREATER THAN 7

PLASTICITY CHART FOR LABORATORY CLASSIFICATION OF FINE GRAINED SOILS

(Plasticity chart: Plasticity Index vs. Liquid Limit, "A" line, zones CL-ML, CL, ML or OL, CH, MH or OH; COMPARING SOILS AT EQUAL LIQUID LIMIT — TOUGHNESS AND DRY STRENGTH INCREASE WITH INCREASING PLASTICITY INDEX)

Lambe, William, and Robert Whitman, *Soil Mechanics*, Wiley, 1969.

AASHTO SOIL CLASSIFICATION

GENERAL CLASSIFICATION	GRANULAR MATERIALS (35% OR LESS PASSING 0.075 SIEVE)							SILT-CLAY MATERIALS (LESS THAN 35% PASSING 0.075 SIEVE)			
GROUP CLASSIFICATION	A–1		A–3	A–2				A–4	A–5	A–6	A–7–5 A–7–6
	A–1–a	A–1–b		A-2-4	A-2-5	A-2-6	A-2-7				
SIEVE ANALYSIS, PERCENT PASSING:											
2.00 mm (No. 10)	≤ 50	–	–	–	–	–	–	–	–	–	–
0.425 mm (No. 40)	≤ 30	≤ 50	≥ 51	–	–	–	–	–	–	–	–
0.075 mm (No. 200)	≤ 15	≤ 25	≤ 10	≤ 35	≤ 35	≤ 35	≤ 35	≥ 36	≥ 36	≥ 36	≥ 36
CHARACTERISTICS OF FRACTION PASSING 0.425 SIEVE (No. 40):											
LIQUID LIMIT	–		–	≤ 40	≥ 41	≤ 40	≥ 41	≤ 40	≥ 41	≤ 40	≥ 41
PLASTICITY INDEX *	6 max		NP	≤ 10	≤ 10	≥ 11	≥ 11	≤ 10	≤ 10	≥ 11	≥ 11
USUAL TYPES OF CONSTITUENT MATERIALS	STONE FRAGM'TS, GRAVEL, SAND		FINE SAND	SILTY OR CLAYEY GRAVEL AND SAND				SILTY SOILS		CLAYEY SOILS	
GENERAL RATING AS A SUBGRADE	EXCELLENT TO GOOD							FAIR TO POOR			

*Plasticity index of A-7-5 subgroup is equal to or less than LL-30. Plasticity index of A-7-6 subgroup is greater than LL-30.
NP = Non-plastic (use "0"). Symbol "–" means that the particular sieve analysis is not considered for that classification.

If the soil classification is A4-A7, then calculate the group index (GI) as shown below and report with classification. The higher the GI, the less suitable the soil. Example: A-6 with GI = 15 is less suitable than A-6 with GI = 10.

$$GI = (F - 35)[\ 0.2 + 0.005(LL - 40)\ ] + 0.01(F - 15)(PI - 10)$$

$$\underbrace{\hphantom{(F-35)}}_{1 \text{ to } 40*} \qquad \underbrace{\hphantom{0.005(LL-40)}}_{1 \text{ to } 20} \qquad \underbrace{\hphantom{(F-15)}}_{1 \text{ to } 40} \underbrace{\hphantom{(PI-10)}}_{1 \text{ to } 20}$$

where: F = Percent passing No. 200 sieve, expressed as a whole number. This percentage is based only on the material passing the No. 200 sieve.

LL = Liquid limit

PI = Plasticity index

*If computed value in (…) falls outside limiting value, then use limiting value.

GENERAL BEARING CAPACITY FACTORS

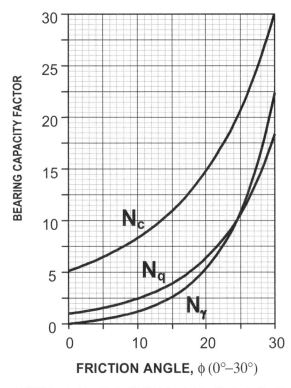

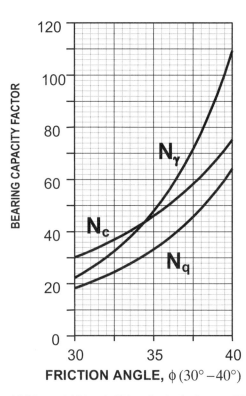

♦ Adapted from AASHTO Standard Specification M145-91, Standard Specification for Classification of Soils and Soil-Aggregate Mixtures for Highway Construction Purposes, 2004.

Vertical Stress Caused by a Point Load

Boussinesq Equation:

$$\sigma_z = \frac{3P}{2\pi} \frac{z^3}{(r^2 + z^2)^{5/2}} = C_r \cdot \frac{P}{z^2}$$

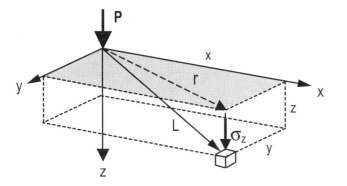

$\frac{r}{z}$	C_r	$\frac{r}{z}$	C_r	$\frac{r}{z}$	C_r
0.0	0.4775	0.32	0.3742	1.00	0.0844
0.02	0.4770	0.34	0.3632	1.20	0.0513
0.04	0.4765	0.36	0.3521	1.40	0.0317
0.06	0.4723	0.38	0.3408	1.60	0.0200
0.08	0.4699	0.40	0.3294	1.80	0.0129
0.10	0.4657	0.45	0.3011	2.00	0.0085
0.12	0.4607	0.50	0.2733	2.20	0.0058
0.14	0.4548	0.55	0.2466	2.40	0.0040
0.16	0.4482	0.60	0.2214	2.60	0.0029
0.18	0.4409	0.65	0.1978	2.80	0.0021
0.20	0.4329	0.70	0.1762	3.00	0.0015
0.22	0.4242	0.75	0.1565	3.20	0.0011
0.24	0.4151	0.80	0.1386	3.40	0.00085
0.26	0.4050	0.85	0.1226	3.60	0.00066
0.28	0.3954	0.90	0.1083	3.80	0.00051
0.30	0.3849	0.95	0.0956	4.00	0.00040

Vertical Stress Beneath a Uniformly Loaded Circular Area

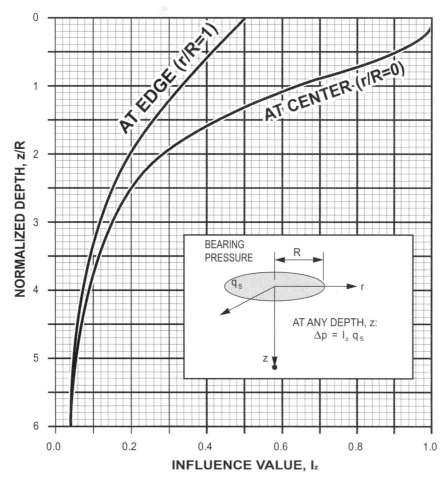

VERTICAL STRESS AT A POINT DIRECTLY BELOW THE CORNER
OF A UNIFORMLY LOADED RECTANGULAR AREA

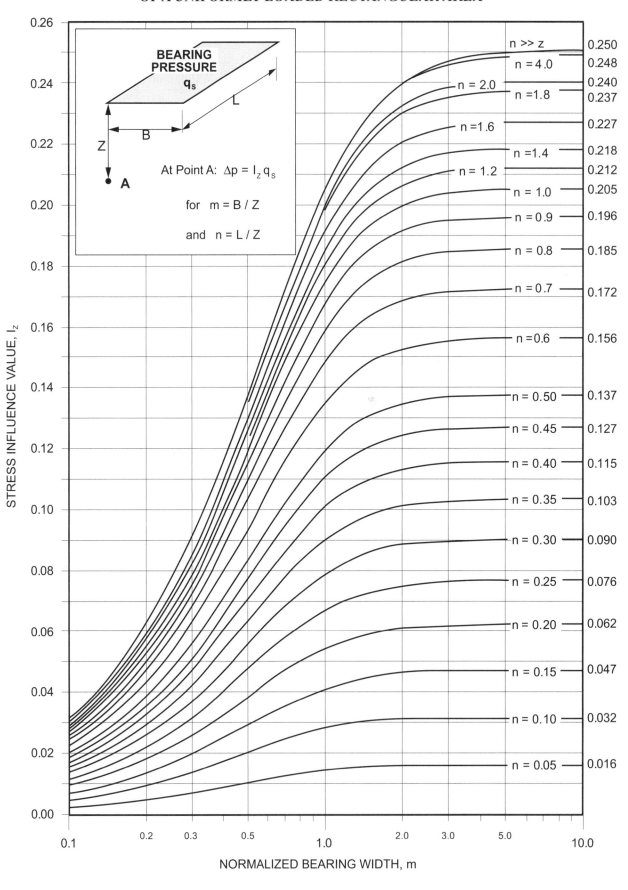

PORE PRESSURE DISTRIBUTION IN A CLAY LAYER BOUNDED AT TOP AND BOTTOM BY PERMEABLE LAYERS

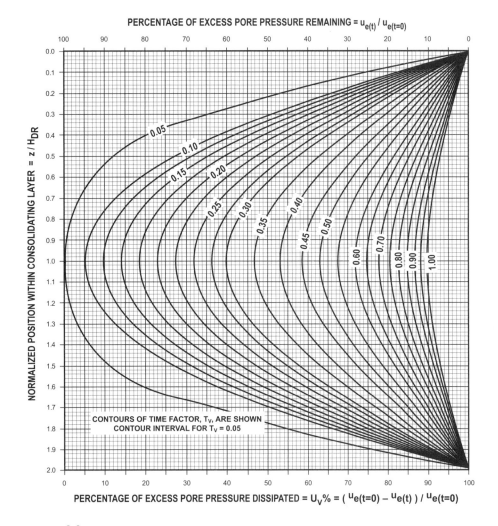

PERCENTAGE OF EXCESS PORE PRESSURE REMAINING = $u_{e(t)} / u_{e(t=0)}$

NORMALIZED POSITION WITHIN CONSOLIDATING LAYER = z / H_{DR}

CONTOURS OF TIME FACTOR, T_V, ARE SHOWN
CONTOUR INTERVAL FOR T_V = 0.05

PERCENTAGE OF EXCESS PORE PRESSURE DISSIPATED = $U_V\% = (u_{e(t=0)} - u_{e(t)}) / u_{e(t=0)}$

DEGREE OF CONSOLIDATION CURVE FOR PORE PRESSURE DISSIPATION WITH TIME

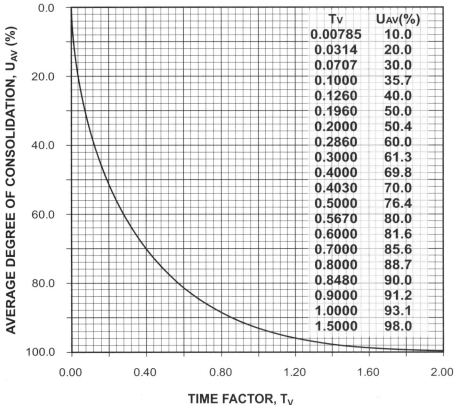

AVERAGE DEGREE OF CONSOLIDATION, U_{AV} (%)

T_V	$U_{AV}(\%)$
0.00785	10.0
0.0314	20.0
0.0707	30.0
0.1000	35.7
0.1260	40.0
0.1960	50.0
0.2000	50.4
0.2860	60.0
0.3000	61.3
0.4000	69.8
0.4030	70.0
0.5000	76.4
0.5670	80.0
0.6000	81.6
0.7000	85.6
0.8000	88.7
0.8480	90.0
0.9000	91.2
1.0000	93.1
1.5000	98.0

TIME FACTOR, T_V

STRUCTURAL ANALYSIS

Influence Lines for Beams and Trusses

An influence line shows the variation of an effect (reaction, shear and moment in beams, bar force in a truss) caused by moving a unit load across the structure. An influence line is used to determine the position of a moveable set of loads that causes the maximum value of the effect.

Moving Concentrated Load Sets

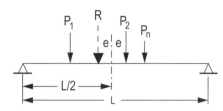

The **absolute maximum moment** produced in a beam by a set of "n" moving loads occurs when the resultant "R" of the load set and an adjacent load are equal distance from the centerline of the beam. In general, two possible load set positions must be considered, one for each adjacent load.

Beam Stiffness and Moment Carryover

$$\theta = \frac{M L}{4 EI} \quad \Rightarrow \quad M = \left(\frac{4 EI}{L}\right)\theta = k_{AB}\,\theta$$

k_{AB} = stiffness $\qquad M_B = M_A/2$ = carryover

Truss Deflection by Unit Load Method

The displacement of a truss joint caused by external effects (truss loads, member temperature change, member misfit) is found by applying a unit load at the point that corresponds to the desired displacement.

$$\Delta_{\text{joint}} = \sum_{i=1}^{\text{members}} f_i (\Delta L)_i$$

where: Δ_{joint} = joint displacement at point of application of unit load (+ in direction of unit load)

f_i = force in member "i" caused by unit load (+ tension)

$(\Delta L)_i$ = change in length caused by external effect (+ for increase in member length):

$= \left(\dfrac{FL}{AE}\right)_i$ for bar force F caused by external load

$= \alpha L_i (\Delta T)_i$ for temperature change in member (α = coefficient of thermal expansion)

$=$ member misfit

L, A = member length and cross-sectional area

E = member elastic modulus

Frame Deflection by Unit Load Method

The displacement of any point on a frame caused by external loads is found by applying a unit load at that point that corresponds to the desired displacement:

$$\Delta = \sum_{i=1}^{\text{members}} \int_{x=0}^{x=L_i} \frac{m_i M_i}{EI_i}\, dx$$

where: Δ = displacement at point of application of unit load (+ in direction of unit load)

m_i = moment equation in member "i" caused by the unit load

M_i = moment equation in member "i" caused by loads applied to frame

L_i = length of member "i"

I_i = moment of inertia of member "i"

If either the real loads or the unit load cause no moment in a member, that member can be omitted from the summation.

Member Fixed-End Moments (magnitudes)

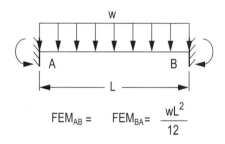

$$FEM_{AB} = \quad FEM_{BA} = \frac{wL^2}{12}$$

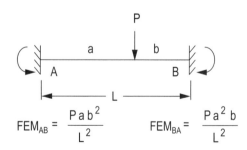

$$FEM_{AB} = \frac{Pab^2}{L^2} \qquad FEM_{BA} = \frac{Pa^2 b}{L^2}$$

STRUCTURAL DESIGN

Definitions (based on ASCE 7-05)

Allowable Stress Design: Method of proportioning structural members such that stresses produced in them by nominal loads do not exceed specified allowable stresses.

Dead Load: Weights of all materials used in a building, including built-in partitions and fixed service equipment ("permanent load").

Design Strength (ϕR_n): The product of the nominal strength and a resistance factor.

Environmental Loads: Loads resulting from acts of nature, such as wind, snow, earthquake (each is a "transient load").

Factored Load (λQ_n): The product of the nominal load and a load factor.

Limit State: A condition beyond which a structure or member becomes unfit for service because it is unsafe (strength limit state) or no longer performs its intended function (serviceability limit state).

Live Load: All loads resulting from the use and occupancy of a building; does not include construction or other roof live loads, and environmental loads such as caused by wind, snow, earthquake, etc. (live loads are "transient loads").

Load Factor (λ): A factor that accounts for deviations of the actual load from the nominal load, or inaccuracies in the load models; also accounts for the probability of two or more extreme transient loads being applied simultaneously.

Nominal Loads (Q_n): The maximum actual loads expected in a building during its life (also called "service loads"). The magnitudes of loads, permanent and transient, computed using load models such as those found in ASCE 7.

Nominal Strength (R_n): The capacity of a structure or member to resist effects of loads. The strength computed using theoretical models or by appropriate laboratory tests.

Resistance Factor (ϕ): A factor that accounts for deviation of the actual strength from the nominal strength due to variations in member or material properties and uncertanties in the nominal load model.

Strength Design: A method of proportioning structural members such that the effects of factored loads on the member do not exceed the design strength of the member: $\Sigma(\lambda Q_n) \leq \phi R_n$. Also called "Load Resistance Factor Design" and "Ultimate Strength Design."

Live Load Reduction

The effect on a building member of nominal occupancy live loads may often be reduced based on the loaded floor area supported by the member. A typical model used for computing reduced live load (as found in ASCE 7 and many building codes) is:

$$L_{\text{reduced}} = L_{\text{nominal}} \left(0.25 + \frac{15}{\sqrt{K_{LL} A_T}} \right) \geq 0.4 L_{\text{nominal}}$$

where:

L_{nominal}	=	nominal live load given in ASCE 7 or a building code
A_T	=	the cumulative floor tributary area supported by the member
$K_{LL} A_T$	=	area of influence supported by the member
K_{LL}	=	ratio of area of influence to the tributary area supported by the member:

$K_{LL} = 4$ (typical columns)

$K_{LL} = 2$ (typical beams and girders)

Load Combinations using Strength Design (LRFD, USD)

Nominal loads used in following combinations:

D = dead loads
E = earthquake loads
L = live loads (floor)
L_r = live loads (roof)
R = rain load
S = snow load
W = wind load

Load factors λ: λ_D (dead load), λ_L (live load), etc.

Basic combinations $L_r/S/R$ = largest of L_r, S, R

L or $0.8W$ = larger of L, $0.8W$

$1.4D$

$1.2D + 1.6L + 0.5 (L_r/S/R)$

$1.2D + 1.6(L_r/S/R) + (L \text{ or } 0.8W)$

$1.2D + 1.6W + L + 0.5(L_r/S/R)$

$1.2D + 1.0E + L + 0.2S$

$0.9D + 1.6W$

$0.9D + 1.0E$

REINFORCED CONCRETE DESIGN ACI 318-05

US Customary units

Definitions

a = depth of equivalent rectangular stress block, in.

A_g = gross area of column, in²

A_s = area of tension reinforcement, in²

A_s' = area of compression reinforcement, in²

A_{st} = total area of longitudinal reinforcement, in²

A_v = area of shear reinforcement within a distance s, in.

b = width of compression face of member, in.

b_e = effective compression flange width, in.

b_w = web width, in.

β_1 = ratio of depth of rectangular stress block, a, to depth to neutral axis, c

$$= 0.85 \geq 0.85 - 0.05\left(\frac{f_c' - 4{,}000}{1{,}000}\right) \geq 0.65$$

c = distance from extreme compression fiber to neutral axis, in.

d = distance from extreme compression fiber to centroid of nonprestressed tension reinforcement, in.

d_t = distance from extreme compression fiber to extreme tension steel, in.

E_c = modulus of elasticity = $33 w_c^{1.5} \sqrt{f_c'}$, psi

ε_t = net tensile strain in extreme tension steel at nominal strength

f_c' = compressive strength of concrete, psi

f_y = yield strength of steel reinforcement, psi

h_f = T-beam flange thickness, in.

M_c = factored column moment, including slenderness effect, in.-lb

M_n = nominal moment strength at section, in.-lb

ϕM_n = design moment strength at section, in.-lb

M_u = factored moment at section, in.-lb

P_n = nominal axial load strength at given eccentricity, lb

ϕP_n = design axial load strength at given eccentricity, lb

P_u = factored axial force at section, lb

ρ_g = ratio of total reinforcement area to cross-sectional area of column = A_{st}/A_g

s = spacing of shear ties measured along longitudinal axis of member, in.

V_c = nominal shear strength provided by concrete, lb

V_n = nominal shear strength at section, lb

ϕV_n = design shear strength at section, lb

V_s = nominal shear strength provided by reinforcement, lb

V_u = factored shear force at section, lb

ASTM STANDARD REINFORCING BARS

BAR SIZE	DIAMETER, IN.	AREA, IN²	WEIGHT, LB/FT
#3	0.375	0.11	0.376
#4	0.500	0.20	0.668
#5	0.625	0.31	1.043
#6	0.750	0.44	1.502
#7	0.875	0.60	2.044
#8	1.000	0.79	2.670
#9	1.128	1.00	3.400
#10	1.270	1.27	4.303
#11	1.410	1.56	5.313
#14	1.693	2.25	7.650
#18	2.257	4.00	13.60

LOAD FACTORS FOR REQUIRED STRENGTH

$U = 1.4\,D$

$U = 1.2\,D + 1.6\,L$

SELECTED ACI MOMENT COEFFICIENTS

Approximate moments in continuous beams of three or more spans, provided:

1. Span lengths approximately equal (length of longer adjacent span within 20% of shorter)
2. Uniformly distributed load
3. Live load not more than three times dead load

M_u = coefficient * w_u * L_n^2

$\quad w_u$ = factored load per unit beam length

$\quad L_n$ = clear span for positive moment; average adjacent clear spans for negative moment

Column

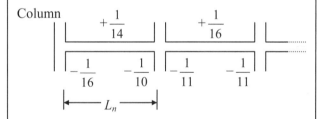

Spandrel beam

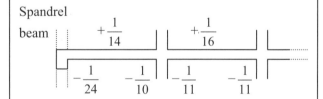

Unrestrained end

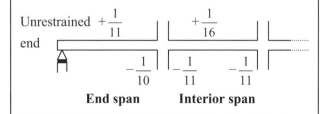

End span   **Interior span**

UNIFIED DESIGN PROVISIONS

Internal Forces and Strains

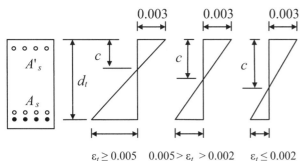

Comp. strain

Net tensile strain: ε_t

Strain Conditions

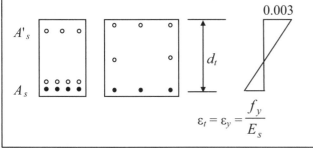

$\varepsilon_t \geq 0.005$ $0.005 > \varepsilon_t > 0.002$ $\varepsilon_t \leq 0.002$

Tension-controlled section: $c \leq 0.375\, d_t$

Transition section

Compression-controlled section: $c \geq 0.6\, d_t$

Balanced Strain: $\varepsilon_t = \varepsilon_y$

$\varepsilon_t = \varepsilon_y = \dfrac{f_y}{E_s}$

RESISTANCE FACTORS, ϕ

Tension-controlled sections ($\varepsilon_t \geq 0.005$): $\phi = 0.9$
Compression-controlled sections ($\varepsilon_t \leq 0.002$):
   Members with spiral reinforcement $\phi = 0.70$
   Members with tied reinforcement $\phi = 0.65$
Transition sections ($0.002 < \varepsilon_t < 0.005$):
   Members with spiral reinforcement $\phi = 0.57 + 67\varepsilon_t$
   Members with tied reinforcement $\phi = 0.48 + 83\varepsilon_t$
Shear and torsion $\phi = 0.75$
Bearing on concrete $\phi = 0.65$

BEAMS − FLEXURE: $\phi M_n \geq M_u$

For all beams

Net tensile strain: $a = \beta_1 c$

$$\varepsilon_t = \frac{0.003(d_t - c)}{c} = \frac{0.003(\beta_1 d_t - a)}{a}$$

Design moment strength: ϕM_n
   where: $\phi = 0.9\ [\varepsilon_t \geq 0.005]$
         $\phi = 0.48 + 83\varepsilon_t\ [0.004 \leq \varepsilon_t < 0.005]$

Reinforcement limits:

$A_{S,max}$ $\varepsilon_t = 0.004$ @ M_n

$$A_{S,min} = \text{larger} \left\{ \frac{3\sqrt{f_c'}\, b_w\, d}{f_y} \text{ or } \frac{200 b_w\, d}{f_y} \right.$$

     $A_{s,min}$ limits need not be applied if
     A_s (provided) $\geq 1.33\, A_s$ (required)

Singly-reinforced beams

$$A_{s,max} = \frac{0.85 f_c'\beta_1 b}{f_y} \left(\frac{3 d_t}{7} \right)$$

$$a = \frac{A_s f_y}{0.85 f_c' b}$$

$$M_n = 0.85 f_c'\, a\, b \left(d - \frac{a}{2} \right) = A_s f_y \left(d - \frac{a}{2} \right)$$

Doubly-reinforced beams

Compression steel yields if:

$$A_s - A_s' \geq \frac{0.85 \beta_1 f_c' d' b}{f_y} \left(\frac{87,000}{87,000 - f_y} \right)$$

If compression steel yields:

$$A_{s,max} = \frac{0.85 f_c'\beta_1 b}{f_y} \left(\frac{3 d_t}{7} \right) + A_s'$$

$$a = \frac{(A_s - A_s') f_y}{0.85 f_c' b}$$

$$M_n = f_y \left[(A_s - A_s')\left(d - \frac{a}{2} \right) + A_s'(d - d') \right]$$

If compression steel does not yield (two steps):
1. Solve for c:

$$c^2 + \left(\frac{(87,000 - 0.85 f_c')\, A_s' - A_s f_y}{0.85 f_c'\beta_1 b} \right) c - \frac{87,000\, A_s' d'}{0.85 f_c'\beta_1 b} = 0$$

Doubly-reinforced beams (continued)

Compression steel does not yield (continued)

2. Compute M_n:

$$M_n = 0.85\, b\, c\, \beta_1 f_c{}' \left(d - \frac{\beta_1 c}{2} \right)$$

$$+ A_s{}' \left(\frac{c - d'}{c} \right)(d - d')\, 87{,}000$$

T-beams − tension reinforcement in stem

Effective flange width:

$$b_e = \begin{cases} 1/4 \bullet \text{ span length} \\ b_w + 16 \bullet h_f \\ \text{beam centerline spacing} \end{cases}$$

smallest

Design moment strength:

$$a = \frac{A_s\, f_y}{0.85\, f_c{}'\, b_e}$$

If $a \leq h_f$:

$$A_{s,max} = \frac{0.85\, f_c{}' \beta_1 b_e}{f_y} \left(\frac{3 d_t}{7} \right)$$

$$M_n = 0.85\, f_c{}'\, a\, b_e \left(d - \frac{a}{2} \right)$$

If $a > h_f$:

$$A_{s,max} = \frac{0.85\, f_c{}' \beta_1 b_e}{f_y} \left(\frac{3 d_t}{7} \right) + \frac{0.85\, f_c{}' (b_e - b_w) h_f}{f_y}$$

Redefine a: $a = \dfrac{A_s\, f_y}{0.85\, f_c{}'\, b_w} - \dfrac{h_f (b_e - b_w)}{b_w}$

$$M_n = 0.85\, f_c{}' \left[h_f\, (b_e - b_w)\left(d - \frac{h_f}{2} \right) + a\, b_w \left(d - \frac{a}{2} \right) \right]$$

Beam width used in shear equations:

$$b_w = \begin{cases} b \text{ (rectangular beams)} \\ b_w \text{ (T-beams)} \end{cases}$$

Nominal shear strength:

$$V_n = V_c + V_s$$

$$V_c = 2\, b_w\, d \sqrt{f_c{}'}$$

$$V_s = \frac{A_v\, f_y\, d}{s} \left(\text{may not exceed } 8\, b_w\, d \sqrt{f_c{}'} \right)$$

Required and maximum-permitted stirrup spacing, s

$$V_u \leq \frac{\phi V_c}{2} : \text{No stirrups required}$$

$$V_u > \frac{\phi V_c}{2} : \text{Use the following table } (A_v \text{ given}):$$

	$\dfrac{\phi V_c}{2} < V_u \leq \phi V_c$	$V_u > \phi V_c$
Required spacing	Smaller of: $s = \dfrac{A_v\, f_y}{50 b_w}$ $s = \dfrac{A_v\, f_y}{0.75\, b_w \sqrt{f_c{}'}}$	$V_s = \dfrac{V_u}{\phi} - V_c$ $s = \dfrac{A_v\, f_y\, d}{V_s}$
Maximum permitted spacing	Smaller of: $s = \dfrac{d}{2}$ OR $s = 24"$	$V_s \leq 4\, b_w\, d \sqrt{f_c{}'}$ Smaller of: $s = \dfrac{d}{2}$ OR $s = 24"$ $V_s > 4\, b_w\, d \sqrt{f_c{}'}$ Smaller of: $s = \dfrac{d}{4}$ $s = 12"$

SHORT COLUMNS
Limits for main reinforcements:

$$\rho_g = \frac{A_{st}}{A_g}$$

$$0.01 \leq \rho_g \leq 0.08$$

Definition of a short column:

$$\frac{KL}{r} \leq 34 - \frac{12 M_1}{M_2}$$

where: $KL = L_{col}$ clear height of column [assume $K = 1.0$]

$r = 0.288h$ rectangular column, h is side length perpendicular to buckling axis (i.e., side length in the plane of buckling)

$r = 0.25h$ circular column, h = diameter

M_1 = smaller end moment
M_2 = larger end moment

$\dfrac{M_1}{M_2}$ $\begin{cases} \text{positive if } M_1, M_2 \text{ cause single curvature} \\ \text{negative if } M_1, M_2 \text{ cause reverse curvature} \end{cases}$

Concentrically-loaded short columns: $\phi P_n \geq P_u$

$$M_1 = M_2 = 0$$

$$\frac{KL}{r} \leq 22$$

Design column strength, spiral columns: $\phi = 0.70$

$$\phi P_n = 0.85\phi \left[0.85 f_c' (A_g - A_{st}) + A_{st} f_y \right]$$

Design column strength, tied columns: $\phi = 0.65$

$$\phi P_n = 0.80\phi \left[0.85 f_c' (A_g - A_{st}) + A_{st} f_y \right]$$

Short columns with end moments:

$$M_u = M_2 \quad \text{or} \quad M_u = P_u e$$

Use *Load-moment strength interaction diagram* to:
1. Obtain ϕP_n at applied moment M_u
2. Obtain ϕP_n at eccentricity e
3. Select A_s for P_u, M_u

LONG COLUMNS − Braced (non-sway) frames
Definition of a long column:

$$\frac{KL}{r} > 34 - \frac{12 M_1}{M_2}$$

Critical load:

$$P_c = \frac{\pi^2 EI}{(KL)^2} = \frac{\pi^2 EI}{(L_{col})^2}$$

where: $EI = 0.25 E_c I_g$

Concentrically-loaded long columns:

$e_{min} = (0.6 + 0.03h)$ minimum eccentricity
$M_1 = M_2 = P_u e_{min}$ (positive curvature)

$$\frac{KL}{r} > 22$$

$$M_c = \frac{M_2}{1 - \dfrac{P_u}{0.75 P_c}}$$

Use *Load-moment strength interaction diagram* to design/analyze column for P_u, M_u

Long columns with end moments:

M_1 = smaller end moment
M_2 = larger end moment

$\dfrac{M_1}{M_2}$ positive if M_1, M_2 produce **single** curvature

$$C_m = 0.6 + \frac{0.4 M_1}{M_2} \geq 0.4$$

$$M_c = \frac{C_m M_2}{1 - \dfrac{P_u}{0.75 P_c}} \geq M_2$$

Use *Load-moment strength interaction diagram* to design/analyze column for P_u, M_u

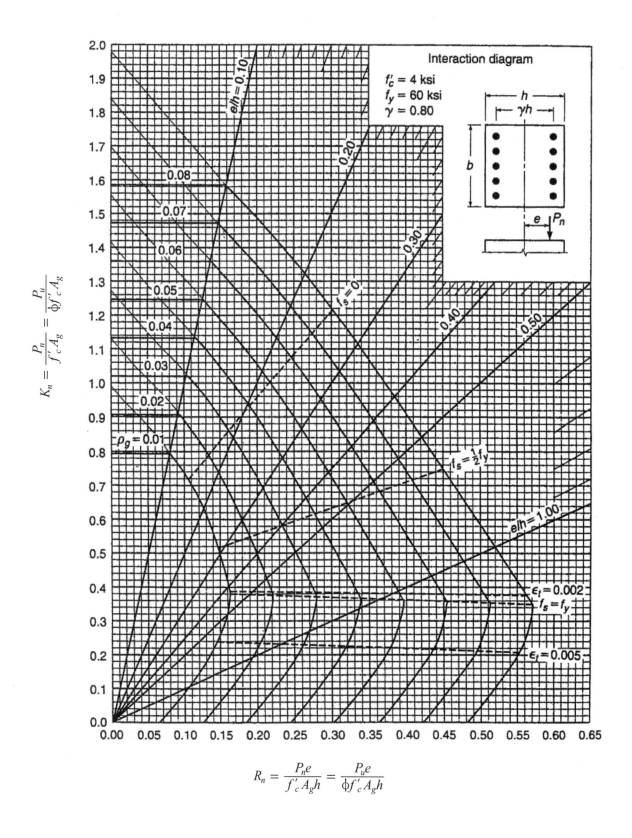

GRAPH A.11

Column strength interaction diagram for rectangular section with bars on end faces and $\gamma = 0.80$ (for instructional use only).

Nilson, Arthur H., David Darwin, and Charles W. Dolan, *Design of Concrete Structures*, 13th ed., McGraw-Hill, New York, 2004.

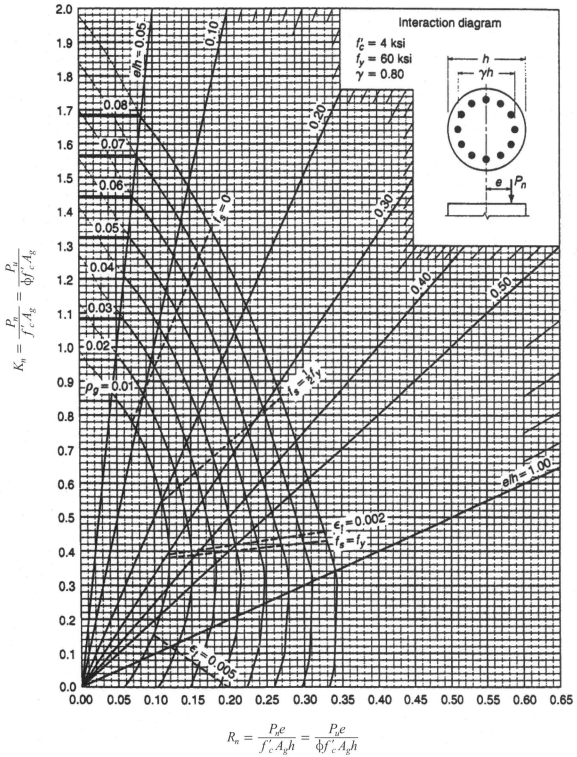

The interaction diagram shows:

Interaction diagram

$f'_c = 4$ ksi
$f_y = 60$ ksi
$\gamma = 0.80$

$$K_n = \frac{P_n}{f'_c A_g} = \frac{P_u}{\phi f'_c A_g}$$

$$R_n = \frac{P_n e}{f'_c A_g h} = \frac{P_u e}{\phi f'_c A_g h}$$

GRAPH A.15

Column strength interaction diagram for circular section $\gamma = 0.80$ (for instructional use only).

Nilson, Arthur H., David Darwin, and Charles W. Dolan, *Design of Concrete Structures*, 13th ed., McGraw-Hill, New York, 2004.

STEEL STRUCTURES (AISC Manual, 13th Edition)
LRFD, E = 29,000 ksi

Definitions (AISC Specifications): LRFD

Available strength: Product of the nominal strength and a resistance factor (same as "design strength" by ASCE 7)

Required strength: The controlling combination of the nominal loads multiplied by load factors (same as critical factored load combination by ASCE 7)

BEAMS
Beam flexure strength of rolled compact sections, flexure about x-axis, F_y = 50 ksi

Compact section criteria:

Flange: $\dfrac{b_f}{2t_f} \leq \lambda_{pf} = \dfrac{64.7}{\sqrt{F_y}}$

Web: $\dfrac{h}{t_w} \leq \lambda_{pw} = \dfrac{640.3}{\sqrt{F_y}}$

For rolled sections, use tabulated values of $\dfrac{b_f}{2t_f}$, $\dfrac{h}{t_w}$

Effect of lateral support on available moment capacity:

L_b = unbraced length of beam segment

ϕ = resistance factor for bending = 0.90

L_p, L_r = length limits

$\left.\begin{array}{l} \phi M_p = \phi F_y Z_x \\ \phi M_r = \phi 0.7 F_x S_x \end{array}\right\}$ AISC Table J3-2

$C_b = \dfrac{12.5\, M_{max}}{2.5 M_{max} + 3 M_A + 4 M_B + 3 M_C} \leq 3.0$

$L_b \leq L_p$: $\phi M_n = \phi M_p$

$L_p < L_b \leq L_r$:

$$\phi M_n = C_b \left[\phi M_p - (\phi M_p - \phi M_r)\left(\frac{L_b - L_p}{L_r - L_p}\right) \right]$$

$$= C_b \left[\phi M_p - BF(L_b - L_p) \right] \leq \phi M_p$$

$L_b > L_r$:

$$\phi M_n = \phi C_b \frac{\pi^2 E}{\left(\frac{L_b}{r_{tx}}\right)^2} \sqrt{1 + 0.078\left(\frac{J}{S_z h_o}\right)\left(\frac{L_b}{r_{ts}}\right)^2}\ S_x \leq \phi M_p$$

where: $r_{ts} = \sqrt[4]{\dfrac{I_y C_w}{S_x^2}}$

h_o = distance between flange centroids

$= d - t_f$

J, C_w = torsion constants (AISC Table 1-1)

See AISC Table 3-10 for ϕM_n vs. L_b curves

Shear – unstiffened beams

$\phi = 0.90$ $\qquad A_w = d t_w$

Rolled W-shapes for $F_y \leq 50$ ksi:

$\phi V_n = \phi(0.6 F_y) A_w$

Built-up I-shaped beams and some rolled W-shapes for $F_y \geq 50$ ksi:

$\dfrac{h}{t_w} \leq \dfrac{418}{\sqrt{F_y}}$: $\qquad \phi V_n = \phi\left(0.6 F_y\right) A_w$

$\dfrac{418}{\sqrt{F_y}} < \dfrac{h}{t_w} \leq \dfrac{522}{\sqrt{F_y}}$:

$$\phi V_n = \phi\left(0.6 F_y\right) A_w \left[\frac{418}{(h/t_w)\sqrt{F_y}} \right]$$

$\dfrac{h}{t_w} > \dfrac{522}{\sqrt{F_y}}$:

$$\phi V_n = \phi\left(0.6 F_y\right) A_w \left[\frac{220,000}{(h/t_w)^2 F_y} \right]$$

COLUMNS
Column effective length KL

AISC Table C-C2.2 – *Approximate Values of Effective Length Factor, K*

AISC Figures C-C2.3 and C-C2.4 – *Alignment Charts*

Column capacity – available strength

$\phi = 0.9$

$\phi P_n = \phi F_{cr} A$

Slenderness ratio: $\dfrac{KL}{r}$

$\dfrac{KL}{r} \leq \dfrac{802.1}{\sqrt{F_y}}$: $\qquad \phi F_{cr} = 0.9 \left[0.658^{\frac{F_y\left(\frac{KL}{r}\right)^2}{286,220}} \right] F_y$

$\dfrac{KL}{r} \leq \dfrac{802.1}{\sqrt{F_y}}$: $\qquad \phi F_{cr} = \dfrac{225,910}{\left(\frac{KL}{r}\right)^2}$

AISC Table 4-22: Available Critical Stress (ϕF_{cr}) for Compression Members

AISC Table 4-1: Available Stress (ϕP_n) in Axial Compression, kips (F_y = 50 ksi)

BEAM-COLUMNS

Sidesway-prevented, x-axis bending, tranverse loading between supports (no moments at ends), ends pinned permitting rotation in plane of bending

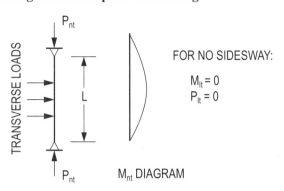

M$_{nt}$ DIAGRAM

FOR NO SIDESWAY:

$M_{lt} = 0$
$P_{lt} = 0$

Required strengths:
$$P_r = P_{nt}$$
$$M_r = B_1 M_{nt}$$

where:
$$B_1 = \frac{C_m}{1 - \dfrac{P_r}{P_{el}}}$$

$C_m = 1.0$ for conditions stated above

$$P_{el} = \frac{\pi^2 EL}{(KL)_x^2} \text{ with respect to bending axis}$$

Strength limit state:

$$\frac{P_r}{\phi P_n} \geq 0.2: \qquad \frac{P_r}{\phi P_n} + \frac{8}{9}\frac{M_r}{\phi M_{nx}} \leq 1.0$$

$$\frac{P_r}{\phi P_n} < 0.2: \qquad \frac{P_r}{0.5(\phi P_n)} + \frac{M_r}{\phi M_{nx}} \leq 1.0$$

where: ϕP_n = compression strength with respect to weak axis (y-axis)

ϕM_n = bending strength with respect to bending axis (x-axis)

TENSION MEMBERS

Flat bars or angles, bolted or welded

Definitions:

Bolt diameter: d_b

Hole diameter: $d_h = d_b - 1/16''$

Gross width of member: b_g

Member thickness: t

Gross area: $A_g = b_g t$ (use tabulated areas for angles)

Net area (parallel holes): $A_n = (b_g - \Sigma d_h) t$

Net area (staggered holes):
$$A_n = \left(b_g - \Sigma d_h + s^2/4g\right)t$$

s = longitudinal spacing of consecutive holes

g = transverse spacing between lines of holes

Effective area (bolted members):
$$A_e = U A_n \begin{cases} U = 1.0\,(\text{flat bars}) \\ U = 1 - \bar{x}/L\,(\text{angles}) \end{cases}$$

Effective area (welded members):

$$A_e = U A_n \begin{cases} \text{Flat bars or angles with transverse welds: U = 1.0} \\ \\ \text{Flat bars of width "w", longitudinal welds of length "L" only:} \\ \qquad \text{U = 1.0 (L} \geq \text{2w)} \\ \qquad \text{U = 0.87 (2w > L} \geq \text{1.5w)} \\ \qquad \text{U = 0.75 (1.5w > L > w)} \\ \\ \text{Angles with longitudinal welds only} \\ \qquad \text{U} = 1 - \bar{x}/L \end{cases}$$

Limit states and available strengths:

Yielding:
$$\phi_y = 0.90$$
$$\phi T_n = \phi_y F_y A_g$$

Fracture:
$$\phi_f = 0.75$$
$$\phi T_n = \phi_f F_u A_e$$

Block shear:
$$\phi = 0.75$$
$$U_{bs} = 1.0 \text{ (flat bars and angles)}$$
A_{gv} = gross area for shear
A_{nv} = net area for shear
A_{nt} = net area for tension

$$\phi T_n = \underset{\text{smaller}}{} \begin{cases} 0.75\,F_u[0.6A_{nv} + U_{bs}A_{nt}] \\ 0.75[0.6F_y A_{gv} + U_{bs}F_u A_{nt}] \end{cases}$$

BOLT STRENGTHS

Definitions:

d_b = bolt diameter

L_e = end distance between center of bolt hole and back edge of member (in direction of applied force)

s = spacing between centers of bolt holes (in direction of applied force)

Bolt tension and shear available strengths (kips/bolt):

Standard size holes. Shear strengths are single shear. Slip-critical connections are Class A faying surface.

Available bolt strength, ϕr_n

BOLT DESIGNATION	BOLT SIZE, in.		
	3/4	7/8	1
TENSION, kips			
A 325	29.8	40.6	53.0
A 307	10.4	20.3	26.5
SHEAR (BEARING-TYPE CONNECTION), kips			
A 325-N	15.9	21.6	28.3
A 325-X	19.9	27.1	35.3
A 306	7.95	10.8	14.1
SHEAR (SLIP-CRITICAL CONNECTION, SLIP IS A SERVICEABILITY LIMIT STATE), kips			
A 325-SC	11.1	15.4	20.2
SHEAR (SLIP-CRITICAL CONNECTION, SLIP IS A STRENGTH LIMIT STATE), kips			
A 325-SC	9.41	13.1	17.1

For slip-critical connections:

Slip is a **serviceability** limit state if load reversals on bolted connection could cause localized failure of the connection (such as by fatigue).

Slip is a **strength** limit state if joint distortion due to slip causes either structure instability or a significant increase of external forces on structure.

Bearing strength of connected member at bolt hole:

The bearing resistance of the connection shall be taken as the sum of the bearing resistances of the individual bolts.

Available strength (kips/bolt/inch thickness):

$$\phi r_n = \phi 1.2 L_c F_u \leq \phi 2.4 d_b F_u$$

$$\phi = 0.75$$

L_c = clear distance between edge of hole and edge of adjacent hole, or edge of member, in direction of force

$$L_c = s - D_h \quad \text{interior holes}$$

$$L_c = L_e - \frac{D_h}{2} \quad \text{end hole}$$

D_b = bolt diameter

D_h = hole diameter
= bolt diameter + clearance

s = center-to-center spacing of interior holes

L_e = end distance (center of end hole to end edge of member)

Available bearing strength, ϕr_n, at bolt holes, kips/bolt/in. thickness *

F_U, ksi CONNECTED MEMBER	BOLT SIZE, in.		
	3/4	7/8	1
$s = 2\,\frac{2}{3}\,d_b$ (MINIMUM PERMITTED)			
58	62.0	72.9	83.7
65	69.5	81.7	93.8
$s = 3"$			
58	78.3	91.3	101
65	87.7	102	113
$L_e = 1\,\frac{1}{4}"$			
58	44.0	40.8	37.5
65	49.4	45.7	42.0
$L_e = 2"$			
58	78.3	79.9	76.7
65	87.7	89.6	85.9

* $D_h = D_b + 1/16\,"$

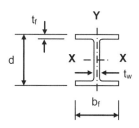

Table 1-1: W-Shapes Dimensions and Properties

Shape	Area A In.^{2}	Depth d In.	Web t_w In.	Flange b_f In.	Flange t_f In.	Compact section $b_f/2t_f$	Compact section h/t_w	r_{ts} in.	h_o in.	Tors. Prop. J in.^{4}	Tors. Prop. C_w in.^{6}	Axis X-X I In.^{4}	Axis X-X S In.^{3}	Axis X-X r In.	Axis X-X Z In.^{3}	Axis Y-Y I In.^{4}	Axis Y-Y r In.
W24X68	20.1	23.7	0.415	8.97	0.585	7.66	52.0	2.30	23.1	1.87	9430	1830	154	9.55	177	70.4	1.87
W24X62	18.2	23.7	0.430	7.04	0.590	5.97	49.7	1.75	23.2	1.71	4620	1550	131	9.23	153	34.5	1.38
W24X55	16.3	23.6	0.395	7.01	0.505	6.94	54.1	1.71	23.1	1.18	3870	1350	114	9.11	134	29.1	1.34
W21X73	21.5	21.2	0.455	8.30	0.740	5.60	41.2	2.19	20.5	3.02	7410	1600	151	8.64	172	70.6	1.81
W21X68	20.0	21.1	0.430	8.27	0.685	6.04	43.6	2.17	20.4	2.45	6760	1480	140	8.60	160	64.7	1.80
W21X62	18.3	21.0	0.400	8.24	0.615	6.70	46.9	2.15	20.4	1.83	5960	1330	127	8.54	144	57.5	1.77
W21X55	16.2	20.8	0.375	8.22	0.522	7.87	50.0	2.11	20.3	1.24	4980	1140	110	8.40	126	48.4	1.73
W21X57	16.7	21.1	0.405	6.56	0.650	5.04	46.3	1.68	20.4	1.77	3190	1170	111	8.36	129	30.6	1.35
W21X50	14.7	20.8	0.380	6.53	0.535	6.10	49.4	1.64	20.3	1.14	2570	984	94.5	8.18	110	24.9	1.30
W21X48	14.1	20.6	0.350	8.14	0.430	9.47	53.6	2.05	20.2	0.803	3950	959	93.0	8.24	107	38.7	1.66
W21X44	13.0	20.7	0.350	6.50	0.450	7.22	53.6	1.60	20.2	0.770	2110	843	81.6	8.06	95.4	20.7	1.26
W18X71	20.8	18.5	0.495	7.64	0.810	4.71	32.4	2.05	17.7	3.49	4700	1170	127	7.50	146	60.3	1.70
W18X65	19.1	18.4	0.450	7.59	0.750	5.06	35.7	2.03	17.6	2.73	4240	1070	117	7.49	133	54.8	1.69
W18X60	17.6	18.2	0.415	7.56	0.695	5.44	38.7	2.02	17.5	2.17	3850	984	108	7.47	123	50.1	1.68
W18X55	16.2	18.1	0.390	7.53	0.630	5.98	41.1	2.00	17.5	1.66	3430	890	98.3	7.41	112	44.9	1.67
W18X50	14.7	18.0	0.355	7.50	0.570	6.57	45.2	1.98	17.4	1.24	3040	800	88.9	7.38	101	40.1	1.65
W18X46	13.5	18.1	0.360	6.06	0.605	5.01	44.6	1.58	17.5	1.22	1720	712	78.8	7.25	90.7	22.5	1.29
W18X40	11.8	17.9	0.315	6.02	0.525	5.73	50.9	1.56	17.4	0.810	1440	612	68.4	7.21	78.4	19.1	1.27
W16X67	19.7	16.3	0.395	10.2	0.67	7.70	35.9	2.82	15.7	2.39	7300	954	117	6.96	130	119	2.46
W16X57	16.8	16.4	0.430	7.12	0.715	4.98	33.0	1.92	15.7	2.22	2660	758	92.2	6.72	105	43.1	1.60
W16X50	14.7	16.3	0.380	7.07	0.630	5.61	37.4	1.89	15.6	1.52	2270	659	81.0	6.68	92.0	37.2	1.59
W16X45	13.3	16.1	0.345	7.04	0.565	6.23	41.1	1.88	15.6	1.11	1990	586	72.7	6.65	82.3	32.8	1.57
W16X40	11.8	16.0	0.305	7.00	0.505	6.93	46.5	1.86	15.5	0.794	1730	518	64.7	6.63	73.0	28.9	1.57
W16X36	10.6	15.9	0.295	6.99	0.430	8.12	48.1	1.83	15.4	0.545	1460	448	56.5	6.51	64.0	24.5	1.52
W14X74	21.8	14.2	0.450	10.1	0.785	6.41	25.4	2.82	13.4	3.87	5990	795	112	6.04	126	134	2.48
W14X68	20.0	14.0	0.415	10.0	0.720	6.97	27.5	2.80	1303	3.01	5380	722	103	6.01	115	121	2.46
W14X61	17.9	13.9	0.375	9.99	0.645	7.75	30.4	2.78	13.2	2.19	4710	640	92.1	5.98	102	107	2.45
W14X53	15.6	13.9	0.370	8.06	0.660	6.11	30.9	2.22	13.3	1.94	2540	541	77.8	5.89	87.1	57.7	1.92
W14X48	14.1	13.8	0.340	8.03	0.595	6.75	33.6	2.20	13.2	1.45	2240	484	70.2	5.85	78.4	51.4	1.91
W12X79	23.2	12.4	0.470	12.1	0.735	8.22	20.7	3.43	11.6	3.84	7330	662	107	5.34	119	216	3.05
W12X72	21.1	12.3	0.430	12.0	0.670	8.99	22.6	3.40	11.6	2.93	6540	597	97.4	5.31	108	195	3.04
W12X65	19.1	12.1	0.390	12.0	0.605	9.92	24.9	3.38	11.5	2.18	5780	533	87.9	5.28	96.8	174	3.02
W12X58	17.0	12.2	0.360	10.0	0.640	7.82	27.0	2.82	11.6	2.10	3570	475	78.0	5.28	86.4	107	2.51
W12X53	15.6	12.1	0.345	9.99	0.575	8.69	28.1	2.79	11.5	1.58	3160	425	70.6	5.23	77.9	95.8	2.48
W12X50	14.6	12.2	0.370	8.08	0.640	6.31	26.8	2.25	11.6	1.71	1880	391	64.2	5.18	71.9	56.3	1.96
W12X45	13.1	12.1	0.335	8.05	0.575	7.00	29.6	2.23	11.5	1.26	1650	348	57.7	5.15	64.2	50.0	1.95
W12X40	11.7	11.9	0.295	8.01	0.515	7.77	33.6	2.21	11.4	0.906	1440	307	51.5	5.13	57.0	44.1	1.94
W10x60	17.6	10.2	0.420	10.1	0.680	7.41	18.7	2.88	9.54	2.48	2640	341	66.7	4.39	74.6	116	2.57
W10x54	15.8	10.1	0.370	10.0	0.615	8.15	21.2	2.86	9.48	1.82	2320	303	60.0	4.37	66.6	103	2.56
W10x49	14.4	10.0	0.340	10.0	0.560	8.93	23.1	2.84	9.42	1.39	2070	272	54.6	4.35	60.4	93.4	2.54
W10x45	13.3	10.1	0.350	8.02	0.620	6.47	22.5	2.27	9.48	1.51	1200	248	49.1	4.32	54.9	53.4	2.01
W10x39	11.5	9.92	0.315	7.99	0.530	7.53	25.0	2.24	9.39	0.976	992	209	42.1	4.27	46.8	45.0	1.98

Adapted from *Steel Construction Manual*, 13th ed., AISC, 2005.

Shape	Z_x in.^{3}	$\phi_b M_{px}$ kip-ft	$\phi_b M_{rx}$ kip-ft	BF kips	L_p ft.	L_r ft.	I_x in.^{4}	$\phi_v V_{nx}$ kips
W24 x 55	**134**	**503**	**299**	**22.2**	**4.73**	**13.9**	**1350**	**251**
W18 x 65	133	499	307	14.9	5.97	18.8	1070	248
W12 x 87	132	495	310	5.76	10.8	43.0	740	194
W16 x 67	130	488	307	10.4	8.69	26.1	954	194
W10 x 100	130	488	294	4.01	9.36	57.7	623	226
W21 x 57	129	484	291	20.1	4.77	14.3	1170	256
W21 x 55	**126**	**473**	**289**	**16.3**	**6.11**	**17.4**	**1140**	**234**
W14 x 74	126	473	294	8.03	8.76	31.0	795	191
W18 x 60	123	461	284	14.5	5.93	18.2	984	227
W12 x 79	119	446	281	5.67	10.8	39.9	662	175
W14 x 68	115	431	270	7.81	8.69	29.3	722	175
W10 x 88	113	424	259	3.95	9.29	51.1	534	197
W18 x 55	**112**	**420**	**258**	**13.9**	**5.90**	**17.5**	**890**	**212**
W21 x 50	**110**	**413**	**248**	**18.3**	**4.59**	**13.6**	**984**	**237**
W12 x 72	108	405	256	5.59	10.7	37.4	597	158
W21 x 48	**107**	**398**	**244**	**14.7**	**6.09**	**16.6**	**959**	**217**
W16 x 57	105	394	242	12.0	5.56	18.3	758	212
W14 x 61	102	383	242	7.46	8.65	27.5	640	156
W18 x 50	101	379	233	13.1	5.83	17.0	800	192
W10 x 77	97.6	366	225	3.90	9.18	45.2	455	169
W12 x 65	96.8	356	231	5.41	11.9	35.1	533	142
W21 x 44	**95.4**	**358**	**214**	**16.8**	**4.45**	**13.0**	**843**	**217**
W16 x 50	92.0	345	213	11.4	5.62	17.2	659	185
W18 x 46	90.7	340	207	14.6	4.56	13.7	712	195
W14 x 53	87.1	327	204	7.93	6.78	22.2	541	155
W12 x 58	86.4	324	205	5.66	8.87	29.9	475	132
W10 x 68	85.3	320	199	3.86	9.15	40.6	394	147
W16 x 45	82.3	309	191	10.8	5.55	16.5	586	167
W18 x 40	**78.4**	**294**	**180**	**13.3**	**4.49**	**13.1**	**612**	**169**
W14 x 48	78.4	294	184	7.66	6.75	21.1	484	141
W12 x 53	77.9	292	185	5.48	8.76	28.2	425	125
W10 x 60	74.6	280	175	3.80	9.08	36.6	341	129
W16 x 40	**73.0**	**274**	**170**	**10.1**	**5.55**	**15.9**	**518**	**146**
W12 x 50	71.9	270	169	5.97	6.92	23.9	391	135
W8 x 67	70.1	263	159	2.60	7.49	47.7	272	154
W14 x 43	69.6	261	164	7.24	6.68	20.0	428	125
W10 x 54	66.6	250	158	3.74	9.04	33.7	303	112
W18 x 35	**66.5**	**249**	**151**	**12.1**	**4.31**	**12.4**	**510**	**159**
W12 x 45	64.2	241	151	5.75	6.89	22.4	348	121
W16 x 36	64.0	240	148	9.31	5.37	15.2	448	140
W14 x 38	61.5	231	143	8.10	5.47	16.2	385	131
W10 x 49	60.4	227	143	3.67	8.97	31.6	272	102
W8 x 58	59.8	224	137	2.56	7.42	41.7	228	134
W12 x 40	57.0	214	135	5.50	6.85	21.1	307	106
W10 x 45	54.9	206	129	3.89	7.10	26.9	248	106

$$\phi_b M_{rx} = \phi_b F_y S_x \qquad BF = \frac{\phi_b M_{px} - \phi_b M_{px}}{L_r - L_p}$$

Adapted from *Steel Construction Manual*, 13th ed., AISC, 2005.

AISC Table 3-2

Z_x

W Shapes – Selection by Z_x

F_y = 50 ksi
ϕ_b = 0.90

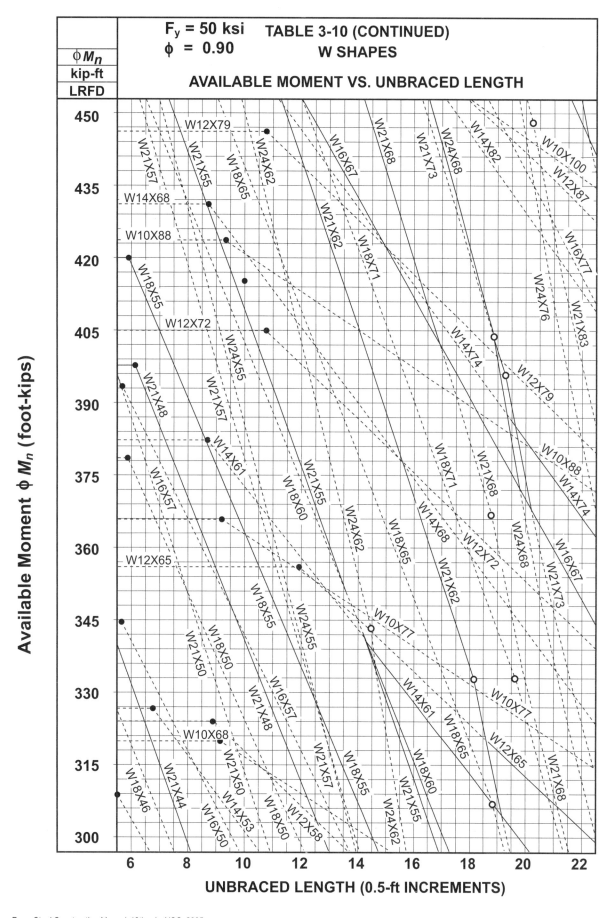

From *Steel Construction Manual*, 13th ed., AISC, 2005.

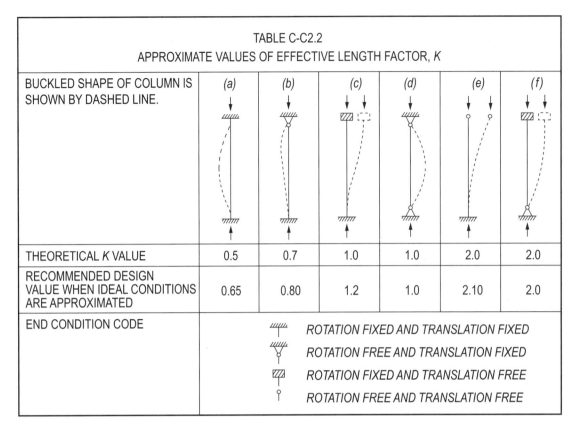

TABLE C-C2.2 APPROXIMATE VALUES OF EFFECTIVE LENGTH FACTOR, *K*						
BUCKLED SHAPE OF COLUMN IS SHOWN BY DASHED LINE.	*(a)*	*(b)*	*(c)*	*(d)*	*(e)*	*(f)*
THEORETICAL *K* VALUE	0.5	0.7	1.0	1.0	2.0	2.0
RECOMMENDED DESIGN VALUE WHEN IDEAL CONDITIONS ARE APPROXIMATED	0.65	0.80	1.2	1.0	2.10	2.0
END CONDITION CODE			ROTATION FIXED AND TRANSLATION FIXED			
			ROTATION FREE AND TRANSLATION FIXED			
			ROTATION FIXED AND TRANSLATION FREE			
			ROTATION FREE AND TRANSLATION FREE			

FOR COLUMN ENDS SUPPORTED BY, BUT NOT RIGIDLY CONNECTED TO, A FOOTING OR FOUNDATION, G IS THEORETICALLY INFINITY BUT UNLESS DESIGNED AS A TRUE FRICTION-FREE PIN, MAY BE TAKEN AS 10 FOR PRACTICAL DESIGNS. IF THE COLUMN END IS RIGIDLY ATTACHED TO A PROPERLY DESIGNED FOOTING, G MAY BE TAKEN AS 1.0. SMALLER VALUES MAY BE USED IF JUSTIFIED BY ANALYSIS.

AISC Figure C-C2.3
Alignment chart, sidesway prevented

AISC Figure C-C2.4
Alignment chart, sidesway **not** prevented

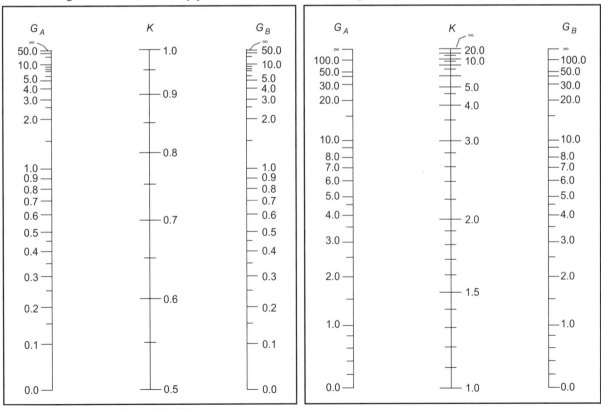

From *Steel Construction Manual*, 13th ed., AISC, 2005.

AISC Table 4-22
Available Critical Stress $\phi_c F_{cr}$ for Compression Members
F_y = 50 ksi $\qquad\qquad$ ϕ_c = 0.90

$\frac{KL}{r}$	ϕF_{cr} ksi	$\frac{KL}{r}$	ϕF_{cr} ksi	$\frac{KL}{r}$	ϕF_{cr} ksi	$\frac{KL}{r}$	ϕF_{cr} ksi	$\frac{KL}{r}$	ϕF_{cr} ksi
1	45.0	41	39.8	81	27.9	121	15.4	161	8.72
2	45.0	42	39.5	82	27.5	122	15.2	162	8.61
3	45.0	43	39.3	83	27.2	123	14.9	163	8.50
4	44.9	44	39.1	84	26.9	124	14.7	164	8.40
5	44.9	45	38.8	85	26.5	125	14.5	165	8.30
6	44.9	46	38.5	86	26.2	126	14.2	166	8.20
7	44.8	47	38.3	87	25.9	127	14.0	167	8.10
8	44.8	48	38.0	88	25.5	128	13.8	168	8.00
9	44.7	49	37.7	89	25.2	129	13.6	169	7.89
10	44.7	50	37.5	90	24.9	130	13.4	170	7.82
11	44.6	51	37.2	91	24.6	131	13.2	171	7.73
12	44.5	52	36.9	92	24.2	132	13.0	172	7.64
13	44.4	53	36.7	93	23.9	133	12.8	173	7.55
14	44.4	54	36.4	94	23.6	134	12.6	174	7.46
15	44.3	55	36.1	95	23.3	135	12.4	175	7.38
16	44.2	56	35.8	96	22.9	136	12.2	176	7.29
17	44.1	57	35.5	97	22.6	137	12.0	177	7.21
18	43.9	58	35.2	98	22.3	138	11.9	178	7.13
19	43.8	59	34.9	99	22.0	139	11.7	179	7.05
20	43.7	60	34.6	100	21.7	140	11.5	180	6.97
21	43.6	61	34.3	101	21.3	141	11.4	181	6.90
22	43.4	62	34.0	102	21.0	142	11.2	182	6.82
23	43.3	63	33.7	103	20.7	143	11.0	183	6.75
24	43.1	64	33.4	104	20.4	144	10.9	184	6.67
25	43.0	65	33.0	105	20.1	145	10.7	185	6.60
26	42.8	66	32.7	106	19.8	146	10.6	186	6.53
27	42.7	67	32.4	107	19.5	147	10.5	187	6.46
28	42.5	68	32.1	108	19.2	148	10.3	188	6.39
29	42.3	69	31.8	109	18.9	149	10.2	189	6.32
30	42.1	70	31.4	110	18.6	150	10.0	190	6.26
31	41.9	71	31.1	111	18.3	151	9.91	191	6.19
32	41.8	72	30.8	112	18.0	152	9.78	192	6.13
33	41.6	73	30.5	113	17.7	153	9.65	193	6.06
34	41.4	74	30.2	114	17.4	154	9.53	194	6.00
35	41.2	75	29.8	115	17.1	155	9.40	195	5.94
36	40.9	76	29.5	116	16.8	156	9.28	196	5.88
37	40.7	77	29.2	117	16.5	157	9.17	197	5.82
38	40.5	78	28.8	118	16.2	158	9.05	198	5.76
39	40.3	79	28.5	119	16.0	159	8.94	199	5.70
40	40.0	80	28.2	120	15.7	160	8.82	200	5.65

Adapted from *Steel Construction Manual*, 13th ed., AISC, 2005.

AISC Table 4-1
Available Strength in Axial Compression, kips—W shapes
LRFD: ϕP_n

$F_y = 50$ ksi
$\phi_c = 0.90$

Selected W14, W12, W10

Shape wt/ft →	W14					W12					W10				
Effective length KL (ft) with respect to least radius of gyration r_y	74	68	61	53	48	58	53	50	45	40	60	54	49	45	39
0	980	899	806	702	636	767	701	657	590	526	794	712	649	597	516
6	922	844	757	633	573	722	659	595	534	475	750	672	612	543	469
7	901	826	740	610	552	707	644	574	516	458	734	658	599	525	452
8	878	804	721	585	529	689	628	551	495	439	717	643	585	505	435
9	853	781	700	557	504	670	610	526	472	419	698	625	569	483	415
10	826	755	677	528	477	649	590	499	448	397	677	607	551	460	395
11	797	728	652	497	449	627	569	471	422	375	655	586	533	435	373
12	766	700	626	465	420	603	547	443	396	351	631	565	513	410	351
13	734	670	599	433	391	578	525	413	370	328	606	543	493	384	328
14	701	639	572	401	361	553	501	384	343	304	581	520	471	358	305
15	667	608	543	369	332	527	477	354	317	280	555	496	450	332	282
16	632	576	515	338	304	500	452	326	291	257	528	472	428	306	260
17	598	544	486	308	276	473	427	297	265	234	501	448	405	281	238
18	563	512	457	278	250	446	402	270	241	212	474	423	383	256	216
19	528	480	428	250	224	420	378	244	217	191	447	399	360	233	195
20	494	448	400	226	202	393	353	220	196	172	420	375	338	210	176
22	428	387	345	186	167	342	306	182	162	142	367	327	295	174	146
24	365	329	293	157	140	293	261	153	136	120	317	282	254	146	122
26	311	281	250	133	120	249	222	130	116	102	270	241	216	124	104
28	268	242	215	115	103	215	192	112	99.8	88.0	233	208	186	107	90.0
30	234	211	187	100	89.9	187	167	97.7	87.0	76.6	203	181	162	93.4	78.4
32	205	185	165	88.1		165	147	82.9	76.4	67.3	179	159	143	82.1	68.9
34	182	164	146			146	130				158	141	126		
36	162	146	130			130	116				141	126	113		
38	146	131	117			117	104				127	113	101		
40	131	119	105			105	93.9				114	102	91.3		

Heavy line indicates KL/r equal to or greater than 200

Adapted from *Steel Construction Manual*, 13th ed., AISC, 2005.

ENVIRONMENTAL ENGINEERING

For information about environmental engineering refer to the **ENVIRONMENTAL ENGINEERING** section.

HYDROLOGY

NRCS (SCS) Rainfall-Runoff

$$Q = \frac{(P - 0.2S)^2}{P + 0.8S}$$

$$S = \frac{1,000}{CN} - 10$$

$$CN = \frac{1,000}{S + 10}$$

P = precipitation (inches),

S = maximum basin retention (inches),

Q = runoff (inches), and

CN = curve number.

Rational Formula

$$Q = CIA, \text{ where}$$

A = watershed area (acres),

C = runoff coefficient,

I = rainfall intensity (in./hr), and

Q = peak discharge (cfs).

Darcy's law

$$Q = -KA(dh/dx), \text{ where}$$

Q = discharge rate (ft^3/sec or m^3/s),

K = hydraulic conductivity (ft/sec or m/s),

h = hydraulic head (ft or m), and

A = cross-sectional area of flow (ft^2 or m^2).

q = $-K(dh/dx)$

   q = specific discharge or Darcy velocity

v = $q/n = -K/n(dh/dx)$

   v = average seepage velocity

   n = effective porosity

Unit hydrograph: The direct runoff hydrograph that would result from one unit of effective rainfall occurring uniformly in space and time over a unit period of time.

Transmissivity, T, is the product of hydraulic conductivity and thickness, b, of the aquifer (L^2T^{-1}).

Storativity or storage coefficient, S, of an aquifer is the volume of water taken into or released from storage per unit surface area per unit change in potentiometric (piezometric) head.

WELL DRAWDOWN:

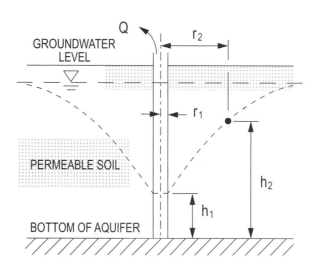

Dupuit's formula

$$Q = \frac{\pi k\left(h_2^2 - h_1^2\right)}{\ln\left(\frac{r_2}{r_1}\right)}$$

where

Q = flow rate of water drawn from well (cfs)

k = permeability of soil (fps)

h_1 = height of water surface above bottom of aquifer at perimeter of well (ft)

h_2 = height of water surface above bottom of aquifer at distance r_2 from well centerline (ft)

r_1 = radius to water surface at perimeter of well, i.e., radius of well (ft)

r_2 = radius to water surface whose height is h_2 above bottom of aquifer (ft)

$\ln$ = natural logarithm

SEWAGE FLOW RATIO CURVES

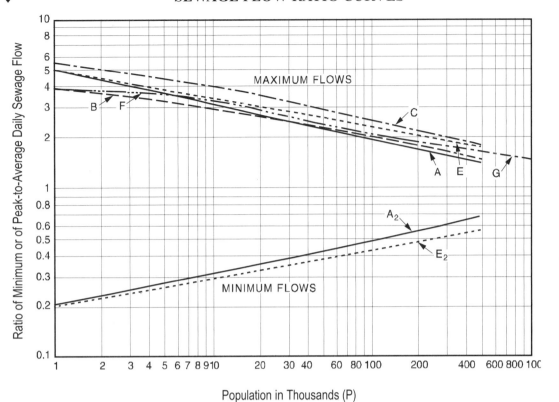

$$\text{Curve } A_2: \frac{P^{0.2}}{5}$$

$$\text{Curve } B: \frac{14}{4 + \sqrt{P}} + 1$$

$$\text{Curve } G: \frac{18 + \sqrt{P}}{4 + \sqrt{P}}$$

Population in Thousands (P)

HYDRAULIC-ELEMENTS GRAPH FOR CIRCULAR SEWERS

Values of: $\frac{f}{f_f}$ and $\frac{n}{n_f}$

- —— n, f variable with depth
- ···· n, f constant
- —·— Independent of n, f

Darcy-Weisbach Friction factor, f

Discharge, Q

Manning's n

Hydraulic radius, R

Velocity, V

Area, A

Ratio of Depth-to-Diameter, $\frac{d}{D}$

Hydraulic Elements: $\frac{V}{V_f}, \frac{Q}{Q_f}, \frac{A}{A_f},$ and $\frac{R}{R_f}$

♦ *Design and Construction of Sanitary and Storm Sewers*, Water Pollution Control Federation and American Society of Civil Engineers, 1970.

Open-Channel Flow
Specific Energy

$$E = \alpha \frac{V^2}{2g} + y = \frac{\alpha Q^2}{2gA^2} + y, \text{ where}$$

E = specific energy,
Q = discharge,
V = velocity,
y = depth of flow,
A = cross-sectional area of flow, and
α = kinetic energy correction factor, usually 1.0.

Critical Depth = that depth in a channel at minimum specific energy

$$\frac{Q^2}{g} = \frac{A^3}{T}$$

where Q and A are as defined above,
g = acceleration due to gravity, and
T = width of the water surface.

For rectangular channels

$$y_c = \left(\frac{q^2}{g}\right)^{1/3}, \text{ where}$$

y_c = critical depth,
q = unit discharge = Q/B,
B = channel width, and
g = acceleration due to gravity.

Froude Number = ratio of inertial forces to gravity forces

$$F_r = \frac{V}{\sqrt{gy_h}}, \text{ where}$$

V = velocity, and
y_h = hydraulic depth = A/T

Specific Energy Diagram

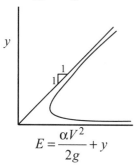

$$E = \frac{\alpha V^2}{2g} + y$$

Alternate depths: depths with the same specific energy.

Uniform flow: a flow condition where depth and velocity do not change along a channel.

Manning's Equation

$$Q = \frac{K}{n} A R^{2/3} S^{1/2}$$

Q = discharge (ft³/sec or m³/s),
K = 1.486 for USCS units, 1.0 for SI units,
A = cross-sectional area of flow (ft² or m²),
R = hydraulic radius = A/P (ft or m),
P = wetted perimeter (ft or m),
S = slope of hydraulic surface (ft/ft or m/m), and
n = Manning's roughness coefficient.

Normal depth (uniform flow depth)

$$AR^{2/3} = \frac{Qn}{KS^{1/2}}$$

Weir Formulas
Fully submerged with no side restrictions

$$Q = CLH^{3/2}$$

V-Notch

$$Q = CH^{5/2}, \text{ where}$$

Q = discharge (cfs or m³/s),
C = 3.33 for submerged rectangular weir (USCS units),
C = 1.84 for submerged rectangular weir (SI units),
C = 2.54 for 90° V-notch weir (USCS units),
C = 1.40 for 90° V-notch weir (SI units),
L = weir length (ft or m), and
H = head (depth of discharge over weir) ft or m.

Hazen-Williams Equation

$$V = k_1 C R^{0.63} S^{0.54}, \text{ where}$$

C = roughness coefficient,
k_1 = 0.849 for SI units, and
k_1 = 1.318 for USCS units,
R = hydraulic radius (ft or m),
S = slope of energy grade line,
  = h_f/L (ft/ft or m/m), and
V = velocity (ft/sec or m/s).

Values of Hazen-Williams Coefficient C

Pipe Material	C
Concrete (regardless of age)	130
Cast iron:	
New	130
5 yr old	120
20 yr old	100
Welded steel, new	120
Wood stave (regardless of age)	120
Vitrified clay	110
Riveted steel, new	110
Brick sewers	100
Asbestos-cement	140
Plastic	150

For additional fluids information, see the **FLUID MECHANICS** section.

TRANSPORTATION

U.S. Customary Units

a = deceleration rate (ft/sec^2)

A = algebraic difference in grades (%)

C = vertical clearance for overhead structure (overpass) located within 200 feet of the midpoint of the curve

e = superelevation (%)

f = side friction factor

$\pm G$ = percent grade divided by 100 (uphill grade "+")

h_1 = height of driver's eyes above the roadway surface (ft)

h_2 = height of object above the roadway surface (ft)

L = length of curve (ft)

L_s = spiral transition length (ft)

R = radius of curve (ft)

S = stopping sight distance (ft)

t = driver reaction time (sec)

V = design speed (mph)

v = vehicle approach speed (fps)

W = width of intersection, curb-to-curb (ft)

l = length of vehicle (ft)

y = length of yellow interval to nearest 0.1 sec (sec)

r = length of red clearance interval to nearest 0.1 sec (sec)

Vehicle Signal Change Interval

$$y = t + \frac{v}{2a \pm 64.4\,G}$$

$$r = \frac{W + l}{v}$$

Stopping Sight Distance

$$S = 1.47Vt + \frac{V^2}{30\left(\left(\dfrac{a}{32.2}\right) \pm G\right)}$$

Transportation Models

See **INDUSTRIAL ENGINEERING** for optimization models and methods, including queueing theory.

Traffic Flow Relationships (q = kv)

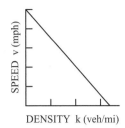

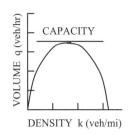

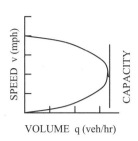

Vertical Curves: Sight Distance Related to Curve Length		
	$S \leq L$	$S > L$
Crest Vertical Curve General equation:	$L = \dfrac{AS^2}{100(\sqrt{2h_1} + \sqrt{2h_2})^2}$	$L = 2S - \dfrac{200\left(\sqrt{h_1} + \sqrt{h_2}\right)^2}{A}$
Standard Criteria: $h_1 = 3.50$ ft and $h_2 = 2.0$ ft:	$L = \dfrac{AS^2}{2{,}158}$	$L = 2S - \dfrac{2{,}158}{A}$
Sag Vertical Curve (based on standard headlight criteria)	$L = \dfrac{AS^2}{400 + 3.5S}$	$L = 2S - \left(\dfrac{400 + 3.5S}{A}\right)$
Sag Vertical Curve (based on riding comfort)	$L = \dfrac{AV^2}{46.5}$	
Sag Vertical Curve (based on adequate sight distance under an overhead structure to see an object beyond a sag vertical curve)	$L = \dfrac{AS^2}{800\left(C - \dfrac{h_1 + h_2}{2}\right)}$	$L = 2S - \dfrac{800}{A}\left(C - \dfrac{h_1 + h_2}{2}\right)$
	$C =$ vertical clearance for overhead structure (overpass) located within 200 feet of the midpoint of the curve	

Horizontal Curves	
Side friction factor (based on superelevation)	$0.01e + f = \dfrac{V^2}{15R}$
Spiral Transition Length	$L_s = \dfrac{3.15V^3}{RC}$ $C =$ rate of increase of lateral acceleration [use 1 ft/sec³ unless otherwise stated]
Sight Distance (to see around obstruction)	$\text{HSO} = R\left[1 - \cos\left(\dfrac{28.65S}{R}\right)\right]$ HSO = Horizontal sight line offset

Horizontal Curve Formulas

D = Degree of Curve, Arc Definition
PC = Point of Curve (also called BC)
PT = Point of Tangent (also called EC)
PI = Point of Intersection
I = Intersection Angle (also called Δ)
      Angle Between Two Tangents
L = Length of Curve, from PC to PT
T = Tangent Distance
E = External Distance
R = Radius
LC = Length of Long Chord
M = Length of Middle Ordinate
c = Length of Sub-Chord
d = Angle of Sub-Chord
l = Curve Length for Sub-Chord

$$R = \frac{5729.58}{D}$$

$$R = \frac{LC}{2\sin(I/2)}$$

$$T = R\tan(I/2) = \frac{LC}{2\cos(I/2)}$$

$$L = RI\frac{\pi}{180} = \frac{I}{D}100$$

$$M = R\left[1 - \cos(I/2)\right]$$

$$\frac{R}{E+R} = \cos(I/2)$$

$$\frac{R-M}{R} = \cos(I/2)$$

$$c = 2R\sin(d/2)$$

$$l = Rd\left(\frac{\pi}{180}\right)$$

$$E = R\left[\frac{1}{\cos(I/2)} - 1\right]$$

Deflection angle per 100 feet of arc length equals $D/2$

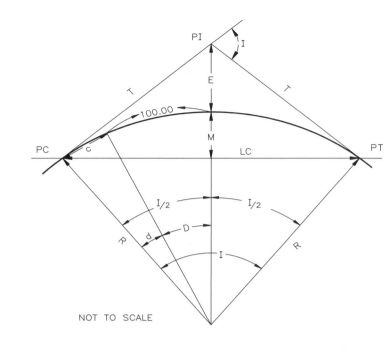

NOT TO SCALE

LATITUDES AND DEPARTURES

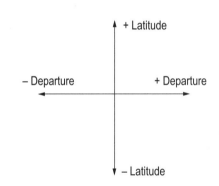

Vertical Curve Formulas

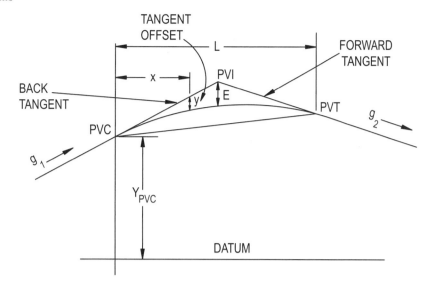

VERTICAL CURVE FORMULAS
NOT TO SCALE

L = Length of Curve (horizontal)

PVC = Point of Vertical Curvature

PVI = Point of Vertical Intersection

PVT = Point of Vertical Tangency

g_1 = Grade of Back Tangent

x = Horizontal Distance from PVC to Point on Curve

g_2 = Grade of Forward Tangent

a = Parabola Constant

y = Tangent Offset

E = Tangent Offset at PVI

r = Rate of Change of Grade

x_m = Horizontal Distance to Min/Max Elevation on Curve = $-\dfrac{g_1}{2a} = \dfrac{g_1 L}{g_1 - g_2}$

Tangent Elevation = $Y_{PVC} + g_1 x$ and = $Y_{PVI} + g_2 (x - L/2)$

Curve Elevation = $Y_{PVC} + g_1 x + ax^2 = Y_{PVC} + g_1 x + [(g_2 - g_1)/(2L)]x^2$

$$y = ax^2 \qquad a = \frac{g_2 - g_1}{2L} \qquad E = a\left(\frac{L}{2}\right)^2 \qquad r = \frac{g_2 - g_1}{L}$$

EARTHWORK FORMULAS

Average End Area Formula, $V = L(A_1 + A_2)/2$

Prismoidal Formula, $V = L(A_1 + 4A_m + A_2)/6$,

where A_m = area of mid-section, and

L = distance between A_1 and A_2

Pyramid or Cone, $V = h$ (Area of Base)/3

AREA FORMULAS

Area by Coordinates: Area = $[X_A(Y_B - Y_N) + X_B(Y_C - Y_A) + X_C(Y_D - Y_B) + ... + X_N(Y_A - Y_{N-1})]/2$

Trapezoidal Rule: Area = $w\left(\dfrac{h_1 + h_n}{2} + h_2 + h_3 + h_4 + ... + h_{n-1}\right)$ $\qquad$ w = common interval

Simpson's 1/3 Rule: Area = $= w\left[h_1 + 2\left(\displaystyle\sum_{k=3,5,...}^{n-2} h_k\right) + 4\left(\displaystyle\sum_{k=2,4,...}^{n-1} h_k\right) + h_n\right]\Big/3$ $\qquad$ n must be odd number of measurements

w = common interval

Highway Pavement Design

AASHTO Structural Number Equation

$$SN = a_1D_1 + a_2D_2 + \ldots + a_nD_n, \text{ where}$$

SN = structural number for the pavement

a_i = layer coefficient and D_i = thickness of layer (inches).

Gross Axle Load		Load Equivalency Factors		Gross Axle Load		Load Equivalency Factors	
kN	lb	Single Axles	Tandem Axles	kN	lb	Single Axles	Tandem Axles
4.45	1,000	0.00002		187.0	42,000	25.64	2.51
8.9	2,000	0.00018		195.7	44,000	31.00	3.00
17.8	4,000	0.00209		200.0	45,000	34.00	3.27
22.25	5,000	0.00500		204.5	46,000	37.24	3.55
26.7	**6,000**	**0.01043**		**213.5**	**48,000**	**44.50**	**4.17**
35.6	8,000	0.0343		222.4	50,000	52.88	4.86
44.5	10,000	0.0877	0.00688	231.3	52,000		5.63
53.4	12,000	0.189	0.0144	240.2	54,000		6.47
62.3	14,000	0.360	0.0270	244.6	55,000		6.93
66.7	**15,000**	**0.478**	**0.0360**	249.0	56,000		**7.41**
71.2	16,000	0.623	0.0472	258.0	58,000		8.45
80.0	18,000	1.000	0.0773	267.0	60,000		9.59
89.0	20,000	1.51	0.1206	275.8	62,000		10.84
97.8	22,000	2.18	0.180	284.5	64,000		12.22
106.8	**24,000**	**3.03**	**0.260**	**289.0**	**65,000**		**12.96**
111.2	25,000	3.53	0.308	293.5	66,000		13.73
115.6	26,000	4.09	0.364	302.5	68,000		15.38
124.5	28,000	5.39	0.495	311.5	70,000		17.19
133.5	30,000	6.97	0.658	320.0	72,000		19.16
142.3	**32,000**	**8.88**	**0.857**	**329.0**	**74,000**		**21.32**
151.2	34,000	11.18	1.095	333.5	75,000		22.47
155.7	35,000	12.50	1.23	338.0	76,000		23.66
160.0	36,000	13.93	1.38	347.0	78,000		26.22
169.0	38,000	17.20	1.70	356.0	80,000		28.99
178.0	**40,000**	**21.08**	**2.08**				
Note: kN converted to lb are within 0.1 percent of lb shown.							

PERFORMANCE-GRADED (PG) BINDER GRADING SYSTEM

PERFORMANCE GRADE	PG 52							PG 58					PG 64				
	−10	−16	−22	−28	−34	−40	−46	−16	−22	−28	−34	−40	−16	−22	−28	−34	−40
AVERAGE 7-DAY MAXIMUM PAVEMENT DESIGN TEMPERATURE, °C[a]	<52							<58					<64				
MINIMUM PAVEMENT DESIGN TEMPERATURE, °C[a]	>−10	>−16	>−22	>−28	>−34	>−40	>−46	>−16	>−22	>−28	>−34	>−40	>−16	>−22	>−28	>−34	>−40
ORIGINAL BINDER																	
FLASH POINT TEMP, T48: MINIMUM °C	230																
VISCOSITY, ASTM D 4402:[b] MAXIMUM, 3 Pa-s (3,000 cP), TEST TEMP, °C	135																
DYNAMIC SHEAR, TP5:[c] $G^*/\sin\delta$, MINIMUM, 1.00 kPa TEST TEMPERATURE @ 10 rad/sec., °C	52							58					64				
ROLLING THIN FILM OVEN (T240) OR THIN FILM OVEN (T179) RESIDUE																	
MASS LOSS, MAXIMUM, %	1.00																
DYNAMIC SHEAR, TP5: $G^*/\sin\delta$, MAXIMUM, 2.20 kPa TEST TEMP @ 10 rad/sec. °C	52							58					64				
PRESSURE AGING VESSEL RESIDUE (PP1)																	
PAV AGING TEMPERATURE, °C[d]	90							100					100				
DYNAMIC SHEAR, TP5: $G^*/\sin\delta$, MINIMUM, 5,000 kPa TEST TEMP @ 10 rad/sec. °C	25	22	19	16	13	10	7	25	22	19	16	13	28	25	22	19	16
PHYSICAL HARDENING [e]	REPORT																
CREEP STIFFNESS, TP1:[f] S, MAXIMUM, 300 MPa M-VALUE, MINIMUM, 0.300 TEST TEMP, @ 60 sec., °C	0	−6	−12	−18	−24	−30	−36	−6	−12	−18	−24	−30	−6	−12	−18	−24	−30
DIRECT TENSION, TP3:[f] FAILURE STRAIN, MINIMUM, 1.0% TEST TEMP @ 1.0 mm/min, °C	0	−6	−12	−18	−24	−30	−36	−6	−12	−18	−24	−30	−6	−12	−18	−24	−30

Federal Highway Administration Report FHWA-SA-95-03, "Background of Superpave Asphalt Mixture Design and Analysis," Nov. 1994.

SUPERPAVE MIXTURE DESIGN: AGGREGATE AND GRADATION REQUIREMENTS

PERCENT PASSING CRITERIA (CONTROL POINTS)					
STANDARD SIEVE, mm	NOMINAL MAXIMUM SIEVE SIZE (mm)				
	9.5	12.5	19.0	25.0	37.5
50					100
37.5				100	90–100
25.0			100	90–100	
19.0		100	90–100		
12.5	100	90–100			
9.5	90–100				
2.36	32–67	28–58	23–49	19–45	15–41
0.075	2–10	2–10	2–8	1–7	0–6

MIXTURE DESIGNATIONS					
SUPERPAVE DESIG. (mm)	9.5	12.5	19.0	25.0	37.5
NOMINAL MAX. SIZE (mm)	9.5	12.5	19.0	25.0	37.5
MAXIMUM SIZE (mm)	12.5	19.0	25.0	37.5	50.0

ADDITIONAL REQUIREMENTS	
FINENESS-TO-EFFECTIVE ASPHALT RATIO, wt./wt.	0.6 – 1.2
SHORT-TERM OVEN AGING AT 135°C , hr.	4
TENSILE STRENGTH RATIO T283, min.	0.80
TRAFFIC (FOR DESIGNATION) BASE ON, yr.	15

TRAFFIC, MILLION EQUIV. SINGLE AXLE LOADS (ESALs)	COARSE AGGREGATE ANGULARITY		FINE AGGREGATE ANGULARITY		FLAT AND ELONGATED PARTICLES	CLAY CONTENT
	DEPTH FROM SURFACE		DEPTH FROM SURFACE		MAXIMUM PERCENT	SAND EQUIVALENT MINIMUM
	≤ 100 mm	> 100 mm	≤ 100 mm	> 100 mm		
< 0.3	55 / –	– / –	–	–	–	40
< 1	65 / –	– / –	40	–	–	40
< 3	75 / –	50 / –	40	40	10	40
< 10	85 / 80	60 / –	45	40	10	45
< 30	95 / 90	80 / 75	45	40	10	45
< 100	100 / 100	95 / 90	45	45	10	50
≥ 100	100 / 100	100 / 100	45	45	10	50

COARSE AGGREGATE ANGULARITY: "85/80" MEANS THAT 85% OF THE COARSE AGGREGATE HAS A MINIMUM OF ONE FRACTURED FACE AND 80% HAS TWO FRACTURED FACES.

FINE AGGREGATE ANGULARITY: CRITERIA ARE PRESENTED AS THE MINIMUM PERCENT AIR VOIDS IN LOOSELY-COMPACTED FINE AGGREGATE.

FLATNESS AND ELONGATED PARTICLES: CRITERIA ARE PRESENTED AS A MAXIMUM PERCENT BY WEIGHT OF FLAT AND ELONGATED PARTICLES.

CLAY CONTENT: CLAY CONTENT IS EXPRESSED AS A PERCENTAGE OF LOCAL SEDIMENT HEIGHT IN A SEDIMENTATION TEST.

MAXIMUM SIZE: ONE SIEVE LARGER THAN THE NOMINAL MAXIMUM SIZE.

NOMINAL MAXIMUM SIZE: ONE SIEVE SIZE LARGER THAT THE FIRST SIEVE TO RETAIN MORE THAN 10% OF THE AGGREGATE.

SUPERPAVE MIXTURE DESIGN: COMPACTION REQUIREMENTS

SUPERPAVE GYRATORY COMPACTION EFFORT												
TRAFFIC, MILLION ESALs	AVERAGE DESIGN HIGH AIR TEMPERATURE											
	< 39°C			39° – 40°C			41° – 42°C			42° – 43°C		
	N_{int}	N_{des}	N_{max}	N_{int}	N_{des}	N_{max}	N_{int}	N_{des}	N_{max}	N_{int}	N_{des}	N_{max}
< 0.3	7	68	104	7	74	114	7	78	121	7	82	127
< 1	7	76	117	7	83	129	7	88	138	8	93	146
< 3	7	86	134	8	95	150	8	100	158	8	105	167
< 10	8	96	152	8	106	169	8	113	181	9	119	192
< 30	8	109	174	9	121	195	9	128	208	9	135	220
< 100	9	126	204	9	139	228	9	146	240	10	153	253
≥ 100	9	142	233	10	158	262	10	165	275	10	177	288

VFA REQUIREMENTS @ 4% AIR VOIDS	
TRAFFIC, MILLION ESALs	DESIGN VFA (%)
< 0.3	70 – 80
< 1	65 – 78
< 3	65 – 78
< 10	65 – 75
< 30	65 – 75
< 100	65 – 75
≥ 100	65 – 75

VMA REQUIREMENTS @ 4% AIR VOIDS					
NOMINAL MAXIMUM AGGREGATE SIZE (mm)	9.5	12.5	19.0	25.0	37.5
MINIMUM VMA (%)	15	14	13	12	11

COMPACTION KEY			
SUPERPAVE GYRATORY COMPACTION	N_{int}	N_{des}	N_{max}
PERCENT OF Gmm	≤ 89%	96%	≤ 98%

Federal Highway Administration Report FHWA-SA-95-03, "Background of Superpave Asphalt Mixture Design and Analysis," Nov. 1994.

CONSTRUCTION

Construction project scheduling and analysis questions may be based on either activity-on-node method or on activity-on-arrow method.

CPM PRECEDENCE RELATIONSHIPS

ACTIVITY-ON-NODE

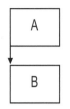

START-TO-START: START OF B
DEPENDS ON THE START OF A

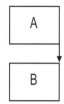

FINISH-TO-FINISH: FINISH OF B
DEPENDS ON THE FINISH OF A

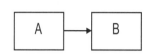

FINISH-TO-START: START OF B
DEPENDS ON THE FINISH OF A

ACTIVITY-ON-ARROW

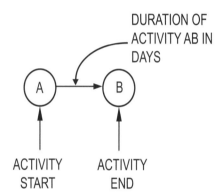

DURATION OF
ACTIVITY AB IN
DAYS

ACTIVITY
START

ACTIVITY
END

ENVIRONMENTAL ENGINEERING

For information about adsorption, chemical thermodynamics, reactors, and mass transfer kinetics, refer to the **CHEMICAL ENGINEERING** and **FLUID MECHANICS** sections.

For information about fluids, refer to the **CIVIL ENGINEERING** and **FLUID MECHANICS** sections.

For information about geohydrology and hydrology, refer to the **CIVIL ENGINEERING** section.

For information about ideal gas law equations, refer to the **THERMODYNAMICS** section.

For information about microbiology (biochemical pathways, cellular biology and organism characteristics), refer to the **BIOLOGY** section.

For information about population growth modeling, refer to the **BIOLOGY** section.

For information about sampling and monitoring (Student's t-Distribution, standard deviation, and confidence intervals), refer to the **MATHEMATICS** section.

AIR POLLUTION

Activated carbon: refer to **WATER TREATMENT** in this section.

Air stripping: refer to **WATER TREATMENT** in this section.

Atmospheric Dispersion Modeling (Gaussian)

σ_y and σ_z as a function of downwind distance and stability class, see following figures.

$$C = \frac{Q}{2\pi u \sigma_y \sigma_z} \exp\left(-\frac{1}{2}\frac{y^2}{\sigma_y^2}\right)\left[\exp\left(-\frac{1}{2}\frac{(z-H)^2}{\sigma_z^2}\right)\right. \\ \left. + \exp\left(-\frac{1}{2}\frac{(z+H)^2}{\sigma_z^2}\right)\right]$$

where

C = steady-state concentration at a point (x, y, z) $(\mu g/m^3)$,

Q = emissions rate $(\mu g/s)$,

σ_y = horizontal dispersion parameter (m),

σ_z = vertical dispersion parameter (m),

u = average wind speed at stack height (m/s),

y = horizontal distance from plume centerline (m),

z = vertical distance from ground level (m),

H = effective stack height (m) = $h + \Delta h$

    where h = physical stack height

Δh = plume rise, and

x = downwind distance along plume centerline (m).

Concentration downwind from elevated source

$$C_{(max)} = \frac{Q}{\pi u \sigma_y \sigma_z} \exp\left(-\frac{1}{2}\frac{(H^2)}{\sigma_z^2}\right)$$

where variables as previous except

$C_{(max)}$ = maximum ground-level concentration

$\sigma_z = \dfrac{H}{\sqrt{2}}$ for neutral atmospheric conditions

Atmospheric Stability Under Various Conditions

Surface Wind Speed[a] (m/s)	Day Solar Insolation			Night Cloudiness[e]	
	Strong[b]	Moderate[c]	Slight[d]	Cloudy (≥4/8)	Clear (≤3/8)
<2	A	A–B[f]	B	E	F
2–3	A–B	B	C	E	F
3–5	B	B–C	C	D	E
5–6	C	C–D	D	D	D
>6	C	D	D	D	D

Notes:

a. Surface wind speed is measured at 10 m above the ground.

b. Corresponds to clear summer day with sun higher than 60° above the horizon.

c. Corresponds to a summer day with a few broken clouds, or a clear day with sun 35-60° above the horizon.

d. Corresponds to a fall afternoon, or a cloudy summer day, or clear summer day with the sun 15-35°.

e. Cloudiness is defined as the fraction of sky covered by the clouds.

f. For A–B, B–C, or C–D conditions, average the values obtained for each.

*   A = Very unstable   D = Neutral

        B = Moderately unstable   E = Slightly stable

        C = Slightly unstable   F = Stable

Regardless of wind speed, Class D should be assumed for overcast conditions, day or night.

Turner, D.B., "Workbook of Atmospheric Dispersion Estimates: An Introduction to Dispersion Modeling," 2nd ed., Lewis Publishing/CRC Press, Florida, 1994.

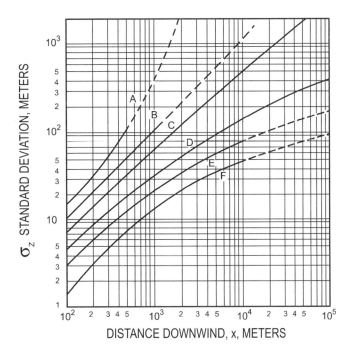

VERTICAL STANDARD DEVIATIONS OF A PLUME

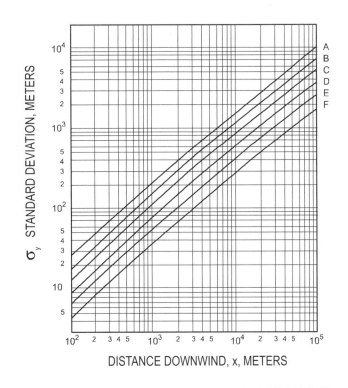

HORIZONTAL STANDARD DEVIATIONS OF A PLUME

A – EXTREMELY UNSTABLE
B – MODERATELY UNSTABLE
C – SLIGHTLY UNSTABLE
D – NEUTRAL
E – SLIGHTLY STABLE
F – MODERATELY STABLE

◆ Turner, D.B., "Workbook of Atmospheric Dispersion Estimates, U.S. Department of Health, Education, and Welfare, Washington, DC, 1970.

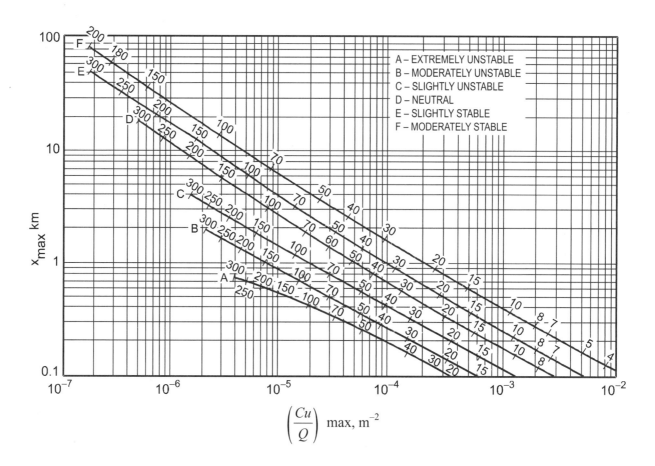

NOTE: Effective stack height shown on curves numerically.

$$\left(\frac{Cu}{Q}\right) \max = e^{[a + b\, lnH + c\,(lnH)^2 + d(lnH)^3]}$$

H = effective stack height, stack height + plume rise, m

Values of Curve-Fit Constants for Estimating $(Cu/Q)_{max}$ from H as a Function of Atmospheric Stability

Stability	Constants			
	a	b	c	d
A	−1.0563	−2.7153	0.1261	0
B	−1.8060	−2.1912	0.0389	0
C	−1.9748	−1.9980	0	0
D	−2.5302	−1.5610	−0.0934	0
E	−1.4496	−2.5910	0.2181	−0.0343
F	−1.0488	−3.2252	0.4977	−0.0765

Adapted from Ranchoux, R.J.P., 1976.

♦ Turner, D.B., "Workbook of Atmospheric Dispersion Estimates: An Introduction to Dispersion Modeling," 2nd ed., Lewis Publishing/CRC Press, Florida, 1994.

Cyclone
Cyclone Collection (Particle Removal) Efficiency

$$\eta = \frac{1}{1 + \left(d_{pc}/d_p\right)^2}, \text{ where}$$

d_{pc} = diameter of particle collected with 50% efficiency,

d_p = diameter of particle of interest, and

η = fractional particle collection efficiency.

◆

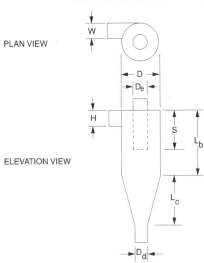

AIR POLLUTION CONTROL

CYCLONE DIMENSIONS

PLAN VIEW

ELEVATION VIEW

Cyclone 50% Collection Efficiency for Particle Diameter

$$d_{pc} = \left[\frac{9\mu W}{2\pi N_e V_i \left(\rho_p - \rho_g\right)}\right]^{0.5}, \text{ where}$$

d_{pc} = diameter of particle that is collected with 50% efficiency (m),

μ = dynamic viscosity of gas (kg/m•s),

W = inlet width of cyclone (m),

N_e = number of effective turns gas makes in cyclone,

V_i = inlet velocity into cyclone (m/s),

ρ_p = density of particle (kg/m^3), and

ρ_g = density of gas (kg/m^3).

◆

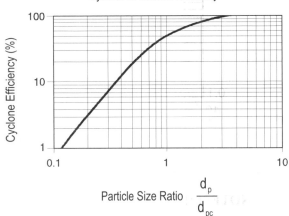

Cyclone Collection Efficiency

Cyclone Effective Number of Turns Approximation

$$N_e = \frac{1}{H}\left[L_b + \frac{L_c}{2}\right], \text{ where}$$

N_e = number of effective turns gas makes in cyclone,

H = inlet height of cyclone (m),

L_b = length of body cyclone (m), and

L_c = length of cone of cyclone (m).

◆

Cyclone Ratio of Dimensions to Body Diameter

Dimension	High Efficiency	Conventional	High Throughput
Inlet height, H	0.44	0.50	0.80
Inlet width, W	0.21	0.25	0.35
Body length, L_b	1.40	1.75	1.70
Cone length, L_c	2.50	2.00	2.00
Vortex finder length, S	0.50	0.60	0.85
Gas exit diameter, D_e	0.40	0.50	0.75
Dust outlet diameter, D_d	0.40	0.40	0.40

◆ Adapted from Cooper, David C. and F.C. Alley, *Air Pollution Control: A Design Approach*, 2nd ed., Waveland Press, Illinois, 1986.

Baghouse

Air-to-Cloth Ratio for Baghouses

Dust	Shaker/Woven Reverse Air/Woven $[m^3/(min \cdot m^2)]$	Pulse Jet/Felt $[m^3/(min \cdot m^2)]$
alumina	0.8	2.4
asbestos	0.9	3.0
bauxite	0.8	2.4
carbon black	0.5	1.5
coal	0.8	2.4
cocoa	0.8	3.7
clay	0.8	2.7
cement	0.6	2.4
cosmetics	0.5	3.0
enamel frit	0.8	2.7
feeds, grain	1.1	4.3
feldspar	0.7	2.7
fertilizer	0.9	2.4
flour	0.9	3.7
fly ash	0.8	1.5
graphite	0.6	1.5
gypsum	0.6	3.0
iron ore	0.9	3.4
iron oxide	0.8	2.1
iron sulfate	0.6	1.8
lead oxide	0.6	1.8
leather dust	1.1	3.7
lime	0.8	3.0
limestone	0.8	2.4
mica	0.8	2.7
paint pigments	0.8	2.1
paper	1.1	3.0
plastics	0.8	2.1
quartz	0.9	2.7
rock dust	0.9	2.7
sand	0.8	3.0
sawdust (wood)	1.1	3.7
silica	0.8	2.1
slate	1.1	3.7
soap detergents	0.6	1.5
spices	0.8	3.0
starch	0.9	2.4
sugar	0.6	2.1
talc	0.8	3.0
tobacco	1.1	4.0
zinc oxide	0.6	1.5

U.S. EPA OAQPS Control Cost Manual, 4th ed., EPA 450/3-90-006 (NTIS PB 90-169954), January 1990.

Electrostatic Precipitator Efficiency

Deutsch-Anderson equation:

$$\eta = 1 - e^{(-WA/Q)}$$

where

η = fractional collection efficiency

W = terminal drift velocity

A = total collection area

Q = volumetric gas flow rate

Note that any consistent set of units can be used for W, A, and Q (for example, ft/min, ft^2, and ft^3/min).

Incineration

$$DRE = \frac{W_{in} - W_{out}}{W_{in}} \times 100\%$$

where

DRE = destruction and removal efficiency (%)

W_{in} = mass feed rate of a particular POHC (kg/h or lb/h)

W_{out} = mass emission rate of the same POHC (kg/h or lb/h)

$$CE = \frac{CO_2}{CO_2 + CO} \times 100\%$$

CO_2 = volume concentration (dry) of CO_2 (parts per million, volume, ppm$_v$)

CO = volume concentration (dry) of CO (ppm$_v$)

CE = combustion efficiency

POHC = principal organic hazardous contaminant

FATE AND TRANSPORT

Microbial Kinetics

BOD Exertion

$$y_t = L(1 - e^{-k_1 t})$$

where

k_1 = deoxygenation rate constant (base e, days^{-1})

L = ultimate BOD (mg/L)

t = time (days)

y_t = the amount of BOD exerted at time t (mg/L)

Stream Modeling: Streeter Phelps

$$D = \frac{k_1 L_0}{k_2 - k_1}\left[\exp(-k_1 t) - \exp(-k_2 t)\right] + D_0 \exp(-k_2 t)$$

$$t_c = \frac{1}{k_2 - k_1}\ln\left[\frac{k_2}{k_1}\left(1 - D_0\frac{(k_2 - k_1)}{k_1 L_0}\right)\right]$$

$$DO = DO_{sat} - D$$

where

D = dissolved oxygen deficit (mg/L)

DO = dissolved oxygen concentration (mg/L)

D_0 = initial dissolved oxygen deficit in mixing zone (mg/L)

DO_{sat} = saturated dissolved oxygen concentration (mg/L)

k_1 = deoxygenation rate constant, base e (days^{-1})

k_2 = reaeration rate constant, base e (days^{-1})

L_0 = initial BOD ultimate in mixing zone (mg/L)

t = time (days)

t_c = time which corresponds with minimum dissolved oxygen (days)

Monod Kinetics—Substrate Limited Growth

Continuous flow systems where growth is limited by one substrate (chemostat):

$$\mu = \frac{Y k_m S}{K_s + S} - k_d = \mu_{max}\frac{S}{K_s + S} - k_d$$

Multiple Limiting Substrates

$$\frac{\mu}{\mu_{max}} = \left[\mu_1(S_1)\right]\left[\mu_2(S_2)\right]\left[\mu_3(S_3)\right]\ldots\left[\mu_n(S_n)\right]$$

$$\text{where } \mu_i = \frac{S_i}{K_{s_i} + S_i} \text{ for } i = 1 \text{ to } n$$

Non-steady State Continuous Flow

$$\frac{dx}{dt} = Dx_0 + (\mu - k_d - D)x$$

Steady State Continuous Flow

$$\mu = D \text{ with } k_d << \mu$$

Product production at steady state, single substrate limiting

$$X_1 = Y_{P/S}(S_0 - S_i)$$

k_d = microbial death rate or endogenous decay rate constant (time^{-1})

k_m = maximum growth rate constant (time^{-1})

K_s = saturation constant or half-velocity constant [= concentration at $\mu_{max}/2$]

S = concentration of substrate in solution (mass/unit volume)

Y = yield coefficient [(mass/L product)/(mass/L food used)]

μ = specific growth rate (time^{-1})

μ_{max} = maximum specific growth rate (time^{-1}) = $Y k_m$

♦ Monod growth rate constant as a function of limiting food concentration.

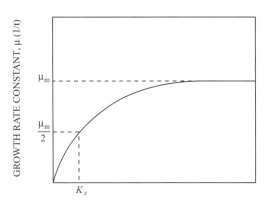

LIMITING FOOD CONCENTRATION, S (mg/L)

X_1 = product (mg/L)

V_r = volume (L)

D = dilution rate (flow f/reactor volume V_r; hr^{-1})

f = flow rate (L/hr)

μ_i = growth rate with one or multiple limiting substrates (hr^{-1})

S_i = substrate i concentration (mass/unit volume)

S_0 = initial substrate concentration (mass/unit volume)

$Y_{P/S}$ = product yield per unit of substrate (mass/mass)

p = product concentration (mass/unit volume)

x = cell concentration (mass/unit volume)

x_0 = initial cell concentration (mass/unit volume)

t = time (time)

♦ Davis, M.L. and D. Cornwell, *Introduction to Environmental Engineering*, 3rd ed., McGraw-Hill, New York, 1998.

Partition Coefficients

Bioconcentration Factor *BCF*

The amount of a chemical to accumulate in aquatic organisms.

$$BCF = C_{org}/C$$

where

C_{org} = equilibrium concentration in organism (mg/kg or ppm)

C = concentration in water (ppm)

Octanol-Water Partition Coefficient

The ratio of a chemical's concentration in the octanol phase to its concentration in the aqueous phase of a two-phase octanol-water system.

$$K_{ow} = C_o/C_w$$

where

C_o = concentration of chemical in octanol phase (mg/L or μg/L)

C_w = concentration of chemical in aqueous phase (mg/L or μg/L)

Organic Carbon Partition Coefficient K_{oc}

$$K_{oc} = C_{soil}/C_{water}$$

where

C_{soil} = concentration of chemical in organic carbon component of soil (μg adsorbed/kg organic *C*, or ppb)

C_{water} = concentration of chemical in water (ppb or μg/kg)

Retardation Factor *R*

$$R = 1 + (\rho/\eta)K_d$$

where

ρ = bulk density

η = porosity

K_d = distribution coefficient

Soil-Water Partition Coefficient $K_{sw} = K_\rho$

$$K_{sw} = X/C$$

where

X = concentration of chemical in soil (ppb or μg/kg)

C = concentration of chemical in water (ppb or μg/kg)

$$K_{sw} = K_{oc}f_{oc}$$

f_{oc} = fraction of organic carbon in the soil (dimensionless)

♦ **Steady-State Reactor Parameters**

Comparison of Steady-State Retention Times (θ) for Decay Reactions of Different Order[a]

Reaction Order	r	Equations for Mean Retention Times (θ)		
		Ideal Batch	**Ideal Plug Flow**	**Ideal CMFR**
Zero[b]	$-k$	$\dfrac{C_o}{k}$	$\dfrac{(C_o - C_t)}{k}$	$\dfrac{(C_o - C_t)}{k}$
First	$-kC$	$\dfrac{1}{k}$	$\dfrac{\ln(C_o/C_t)}{k}$	$\dfrac{(C_o/C_t) - 1}{k}$
Second	$-kC^2$	$\dfrac{1}{kC_o}$	$\dfrac{(C_o/C_t) - 1}{kC_o}$	$\dfrac{(C_o/C_t) - 1}{kC_t}$

[a] C_o = initial concentration or influent concentration; C_t = final condition or effluent concentration.

[b] Expressions are valid for $k\theta \leq C_o$; otherwise $C_t = 0$.

Comparison of Steady-State Performance for Decay Reactions of Different Order[a]

Reaction Order	r	Equations for C_t		
		Ideal Batch	**Ideal Plug Flow**	**Ideal CMFR**
Zero[b] $t \leq C_o/k$	$-k$	$C_o - kt$	$C_o - k\theta$	$C_o - k\theta$
$\quad t > C_o/k$		0		
First	$-kC$	$C_o[\exp(-kt)]$	$C_o[\exp(-k\theta)]$	$\dfrac{C_o}{1 + k\theta}$
Second	$-kC^2$	$\dfrac{C_o}{1 + ktC_o}$	$\dfrac{C_o}{1 + k\theta C_o}$	$\dfrac{(4k\theta C_o + 1)^{1/2} - 1}{2k\theta}$

[a] C_o = initial concentration or influent concentration; C_t = final condition or effluent concentration.

[b] Time conditions are for ideal batch reactor only.

♦ Davis, M.L. and S.J. Masten, *Principles of Environmental Engineering and Science*, McGraw-Hill, New York, 2004.

LANDFILL

Break-Through Time for Leachate to Penetrate a Clay Liner

$$t = \frac{d^2 \eta}{K(d + h)}$$

where

t = breakthrough time (yr)

d = thickness of clay liner (ft)

η = porosity

K = coefficient of permeability (ft/yr)

h = hydraulic head (ft)

Typical porosity values for clays with a coefficient of permeability in the range of 10^{-6} to 10^{-8} cm/s vary from 0.1 to 0.3.

Effect of Overburden Pressure

$$SW_p = SW_i + \frac{p}{a + bp}$$

where

SW_p = specific weight of the waste material at pressure p (lb/yd^3) (typical 1,750 to 2,150)

SW_i = initial compacted specific weight of waste (lb/yd^3) (typical 1,000)

p = overburden pressure (lb/in^2)

a = empirical constant (yd^3/in^2)

b = empirical constant (yd^3/lb)

Gas Flux

$$N_A = \frac{D\eta^{4/3}\left(C_{A_{\mathrm{atm}}} - C_{A_{\mathrm{fill}}}\right)}{L}$$

where

N_A = gas flux of compound A, [g/(cm$^2 \cdot$ s)][lb $\cdot$ mol/(ft$^2 \cdot$ d)]

$C_{A_{\mathrm{atm}}}$ = concentration of compound A at the surface of the landfill cover, g/cm^3 (lb $\cdot$ mol/ft^3)

$C_{A_{\mathrm{fill}}}$ = concentration of compound A at the bottom of the landfill cover, g/cm^3 (lb $\cdot$ mol/ft^3)

L = depth of the landfill cover, cm (ft)

Typical values for the coefficient of diffusion for methane and carbon dioxide are 0.20 cm^2/s (18.6 ft^2/d) and 0.13 cm^2/s (12.1 ft^2/d), respectively.

D = diffusion coefficient, cm^2/s (ft^2/d)

η_{gas} = gas-filled porosity, cm^3/cm^3 (ft^3/ft^3)

η = porosity, cm^3/cm^3 (ft^3/ft^3)

Soil Landfill Cover Water Balance

$$\Delta S_{\mathrm{LC}} = P - R - \mathrm{ET} - \mathrm{PER_{sw}}$$

where

ΔS_{LC} = change in the amount of water held in storage in a unit volume of landfill cover (in.)

P = amount of precipitation per unit area (in.)

R = amount of runoff per unit area (in.)

ET = amount of water lost through evapotranspiration per unit area (in.)

$\mathrm{PER_{sw}}$ = amount of water percolating through the unit area of landfill cover into compacted solid waste (in.)

NOISE POLLUTION

$$\mathrm{SPL\ (dB)} = 10 \log_{10}\left(P^2 / P_0^2\right)$$

$$\mathrm{SPL_{total}} = 10 \log_{10} \Sigma 10^{\mathrm{SPL}/10}$$

Point Source Attenuation

$$\Delta\ \mathrm{SPL\ (dB)} = 10 \log_{10} (r_1/r_2)^2$$

Line Source Attenuation

$$\Delta\ \mathrm{SPL\ (dB)} = 10 \log_{10} (r_1/r_2)$$

where

SPL (dB) = sound pressure level, measured in decibels

P = sound pressure (Pa)

P_0 = reference sound pressure (2×10^{-5} Pa)

$\mathrm{SPL_{total}}$ = sum of multiple sources

Δ SPL (dB) = change in sound pressure level with distance, measured in decibels

r_1 = distance from source to receptor at point 1

r_2 = distance from source to receptor at point 2

POPULATION MODELING

Population Projection Equations

<u>Linear Projection = Algebraic Projection</u>

$$P_t = P_0 + k\Delta t$$

where

P_t = population at time t

P_0 = population at time zero

k = growth rate

Δt = elapsed time in years relative to time zero

<u>Log Growth = Exponential Growth = Geometric Growth</u>

$$P_t = P_0 e^{k\Delta t}$$

$$\ln P_t = \ln P_0 + k\Delta t$$

where

P_t = population at time t

P_0 = population at time zero

k = growth rate

Δt = elapsed time in years relative to time zero

RADIATION

Effective Half-Life

Effective half-life, τ_e, is the combined radioactive and biological half-life.

$$\frac{1}{\tau_e} = \frac{1}{\tau_r} + \frac{1}{\tau_b}$$

where

τ_r = radioactive half-life

τ_b = biological half-life

Half-Life

$$N = N_0 e^{-0.693\, t/\tau}$$

where

N_0 = original number of atoms

N = final number of atoms

t = time

τ = half-life

Flux at distance 2 = (Flux at distance 1) $(r_1/r_2)^2$

The half-life of a biologically degraded contaminant assuming a first-order rate constant is given by:

$$t_{1/2} = \frac{0.693}{k}$$

k = rate constant (time^{-1})

$t_{1/2}$ = half-life (time)

Ionizing Radiation Equations

Daughter Product Activity

$$N_2 = \frac{\lambda_1 N_{10}}{\lambda_2 - \lambda_1}\left(e^{-\lambda_1 t} - e^{-\lambda_2 t}\right)$$

where $\lambda_{1,2}$ = decay constants (time^{-1})

N_{10} = initial activity of parent nuclei

t = time

Daughter Product Maximum Activity Time

$$t' = \frac{\ln\lambda_2 - \ln\lambda_1}{\lambda_2 - \lambda_1}$$

Inverse Square Law

$$\frac{I_1}{I_2} = \frac{(R_2)^2}{(R_1)^2}$$

where $I_{1,2}$ = Radiation intensity at locations 1 and 2

$R_{1,2}$ = Distance from the source at locations 1 and 2

SAMPLING AND MONITORING

Data Quality Objectives (DQO) for Sampling Soils and Solids

Investigation Type	Confidence Level $(1-\alpha)$ (%)	Power $(1-\beta)$ (%)	Minimum Detectable Relative Difference (%)
Preliminary site investigation	70–80	90–95	10–30
Emergency clean-up	80–90	90–95	10–20
Planned removal and remedial response operations	90–95	90–95	10–20

EPA Document "EPA/600/8–89/046" *Soil Sampling Quality Assurance User's Guide*, Chapter 7.
Confidence level: $1-$ (Probability of a Type I error) = $1 - \alpha$ = size probability of not making a Type I error.
Power = $1-$ (Probability of a Type II error) = $1 - \beta$ = probability of not making a Type II error.

CV = $(100 * s)/\overline{x}$

CV = coefficient of variation

s = standard deviation of sample

$\overline{x}$ = sample average

Minimum Detectable Relative Difference = Relative increase over background $[100\,(\mu_s - \mu_B)/\mu_B]$ to be detectable with a probability $(1-\beta)$

Number of samples required in a one-sided one-sample t-test to achieve a minimum detectable relative difference at confidence level $(1-\alpha)$ and power $(1-\beta)$

Coefficient of Variation (%)	Power (%)	Confidence Level (%)	Minimum Detectable Relative Difference (%)				
			5	10	20	30	40
15	95	99	145	39	12	7	5
		95	99	26	8	5	3
		90	78	21	6	3	3
		80	57	15	4	2	2
	90	99	120	32	11	6	5
		95	79	21	7	4	3
		90	60	16	5	3	2
		80	41	11	3	2	1
	80	99	94	26	9	6	5
		95	58	16	5	3	3
		90	42	11	4	2	2
		80	26	7	2	2	1
25	95	99	397	102	28	14	9
		95	272	69	19	9	6
		90	216	55	15	7	5
		80	155	40	11	5	3
	90	99	329	85	24	12	8
		95	272	70	19	9	6
		90	166	42	12	6	4
		80	114	29	8	4	3
	80	99	254	66	19	10	7
		95	156	41	12	6	4
		90	114	30	8	4	3
		80	72	19	5	3	2
35	95	99	775	196	42	25	15
		95	532	134	35	17	10
		90	421	106	28	13	8
		80	304	77	20	9	6
	90	99	641	163	43	21	13
		95	421	107	28	14	8
		90	323	82	21	10	6
		80	222	56	15	7	4
	80	99	495	126	34	17	11
		95	305	78	21	10	7
		90	222	57	15	7	5
		80	140	36	10	5	3

RISK ASSESSMENT/TOXICOLOGY

For information about chemical process safety, refer to the **CHEMICAL ENGINEERING** section.

Dose-Response Curves

The dose-response curve relates toxic response (i.e., percentage of test population exhibiting a specified symptom or dying) to the logarithm of the dosage [i.e., mg/(kg•day) ingested]. A typical dose-response curve is shown below.

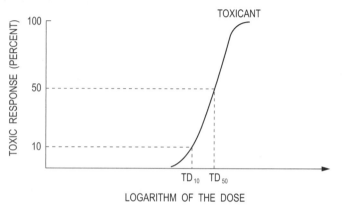

LC_{50}

Median lethal concentration in air that, based on laboratory tests, expected to kill 50% of a group of test animals when administered as a single exposure over one or four hours.

LD_{50}

Median lethal single dose, based on laboratory tests, expected to kill 50% of a group of test animals, usually by oral or skin exposure.

Similar definitions exist for LC_{10} and LD_{10}, where the corresponding percentages are 10%.

♦

Comparative Acutely Lethal Doses

Actual Ranking No.	LD_{50} (mg/kg)	Toxic Chemical
1	15,000	PCBs
2	10,000	Alcohol (ethanol)
3	4,000	Table salt—sodium chloride
4	1,500	Ferrous sulfate—an iron supplement
5	1,375	Malathion—pesticide
6	900	Morphine
7	150	Phenobarbital—a sedative
8	142	Tylenol (acetaminophen)
9	2	Strychnine—a rat poison
10	1	Nicotine
11	0.5	Curare—an arrow poison
12	0.001	2,3,7,8-TCDD (dioxin)
13	0.00001	Botulinum toxin (food poison)

Sequential absorption-disposition-interaction of foreign compounds with humans and animals.

♦

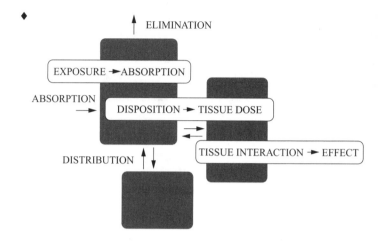

♦ **Selected Chemical Interaction Effects**

Effect	Relative toxicity (hypothetical)	Example
Additive	2 + 3 = 5	Organophosphate pesticides
Synergistic	2 + 3 = 20	Cigarette smoking + asbestos
Antagonistic	6 + 6 = 8	Toluene + benzene or caffeine + alcohol

♦ Williams, P.L., R.C. James, and S.M. Roberts, *Principles of Toxicology: Environmental and Industrial Applications,* 2nd ed., Wiley, New Jersey, 2000.

Hazard Assessment

The fire/hazard diamond below summarizes common hazard data available on the MSDS and is frequently shown on chemical labels.

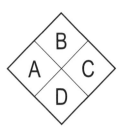

Position A – Hazard (*Blue*)

0 = ordinary combustible hazard

1 = slightly hazardous

2 = hazardous

3 = extreme danger

4 = deadly

Position B – Flammability (*Red*)

0 = will not burn

1 = will ignite if preheated

2 = will ignite if moderately heated

3 = will ignite at most ambient temperature

4 = burns readily at ambient conditions

Position C – Reactivity (*Yellow*)

0 = stable and not reactive with water

1 = unstable if heated

2 = violent chemical change

3 = shock short may detonate

4 = may detonate

Position D – (White)

ALKALI = alkali

OXY = oxidizer

ACID = acid

Cor = corrosive

W = use no water

☢ = radiation hazard

Flammable

Describes any solid, liquid, vapor, or gas that will ignite easily and burn rapidly. A flammable liquid is defined by NFPA and DOT as a liquid with a flash point below 100°F (38°C).

Material Safety Data Sheets (MSDS)

The MSDS indicates chemical source, composition, hazards and health effects, first aid, fire-fighting precautions, accidental-release measures, handling and storage, exposure controls and personal protection, physical and chemical properties, stability and reactivity, toxicological information, ecological hazards, disposal, transport, and other regulatory information.

The MSDS forms for all chemical compounds brought on site should be filed by a designated site safety officer. The MSDS form is provided by the supplier or must be developed when new chemicals are synthesized.

Exposure Limits for Selected Compounds

N	Allowable Workplace Exposure Level (mg/m^3)	Chemical (use)
1	0.1	Iodine
2	5	Aspirin
3	10	Vegetable oil mists (cooking oil)
4	55	1,1,2-Trichloroethane (solvent/degreaser)
5	188	Perchloroethylene (dry-cleaning fluid)
6	170	Toluene (organic solvent)
7	269	Trichloroethylene (solvent/degreaser)
8	590	Tetrahydrofuran (organic solvent)
9	890	Gasoline (fuel)
10	1,590	Naphtha (rubber solvent)
11	1,910	1,1,1-Trichloroethane (solvent/degreaser)

American Conference of Government Industrial Hygienists (ACGIH) 1996 and Williams, P.L., R.C. James, and S.M. Roberts, *Principles of Toxicology: Environmental and Industrial Applications*, 2nd ed., Wiley, New Jersey, 2000.

HAZARDOUS WASTE COMPATIBILITY CHART

KEY

REACTIVITY CODE	CONSEQUENCES
H	HEAT GENERATION
F	FIRE
G	INNOCUOUS & NON-FLAMMABLE GAS
GT	TOXIC GAS GENERATION
GF	FLAMMABLE GAS GENERATION
E	EXPLOSION
P	POLYMERIZATION
S	SOLUBILIZATION OF TOXIC MATERIAL
U	MAY BE HAZARDOUS BUT UNKNOWN

EXAMPLE:

H	HEAT GENERATION,
F	FIRE, AND TOXIC GAS
GT	GENERATION

No.	Reactivity Group Name																					
1	Acid, Minerals, Non-Oxidizing	1																				
2	Acids, Minerals, Oxidizing		2																			
3	Acids, Organic	GF G, GH	H F GT	3																		
4	Alcohols & Glycols	H	H F	H P	4																	
5	Aldehydes	H P	H F P	H P	H	5																
6	Amides	H	H GT		H		6															
7	Amines, Aliphatic & Aromatic	H	H GT	H	H	H		7														
8	Azo Compounds, Diazo Comp., Hydrazines	H, G H GT	H H GT	H G, GT	H H G	H			8													
9	Carbamates	H G, GT	H GT						H G	9												
10	Caustics	H	H	H	H	H			H G	H G	10											
11	Cyanides	GT GF	GT GF	H	GT GF				G			11										
12	Dithiocarbamates	H GF, GF F	H GF GF F	H GF GF GT	H	GF GT	U		H G		G		12									
13	Esters	H	H F	H		H			H G	H G		H	GF GF H	13								
14	Ethers	H	H F						H G	H G			GF GF H	H H F	14							
15	Fluorides, Inorganic	GT	GT	GT												15						
16	Hydrocarbons, Aromatic	H F	H F	H F						H G		H	GF GF H	GF GF H			16					
17	Halogenated Organics	H F, GT	H F, F GT	H F, F GT	H F	GF GF H F		H F GT	H GF, GF	H H F	H	H GF GF H	GF GF H	GF GF H	H H F			17				
18	Isocyanates	H G, G	H G	H G	H P	H P		H P	H G	H G		H, P G	GF GF H	GF GF H	H H F			H G	18			
19	Ketones	H	H F	H F		H			H G	H G		H	GF GF H	GF GF H	H H F			H	H	19		
20	Mercaptans & Other Organic Sulfides	GT GF	H F, F GT	H F				H F GT		H G			GF GF H	F GT H	H H F, F GT		H	GF GF H	H H F, F GT	H	20	
21	Metal, Alkali & Alkaline Earth, Elemental	H GF, GF H F	H GF H F	GF GF H F	GF GF H F	GF GF H F, H F		GF GF H F	H GF, GF H	H GF, GF H	H H	GF GF H	GF GF H	GF GF H	H H F		GF GF H	GF GF H	GF GF H, H H	GF GF H, H H	GF GF H	21
		1	2	3	4	5	6	7	8	9	10	11	12	13	14	15	16	17	18	19	20	21
104	Oxidizing Agents, Strong	H GT	H F, F GT	H GT	H GT	H F, F		H F GT	H GF, GF	H GF GF	H	H E, E GT	GF GF, GT H	H H F, F	H H F, F			H H F, F	H H F, F GT	H H F, F	H H F, F GT	GF GF, H H
105	Reducing Agents, Strong	H GF, GF GT	H F, F GT	H GF	H GF GF F	H GF GF F		H	H H G	H H G	H	GF GF, H	H H GT	H H F	H H F			H H, H T	GF GF, H H	GF GF, H H	GF GF, H H	
106	Water & Mixtures Containing Water	H	H		H				G									H G			GF GF H	
107	Water Reactive Substances																					

104	105	106	107
104			
	105		
	H F, E	106	
		GF GT	107
			107

EXTREMELY REACTIVE! — Do Not Mix With Any Chemical or Waste Material

Risk

Risk characterization estimates the probability of adverse incidence occurring under conditions identified during exposure assessment.

Carcinogens

For carcinogens the added risk of cancer is calculated as follows:

Risk = dose × toxicity = daily dose × CSF

Risk assessment process

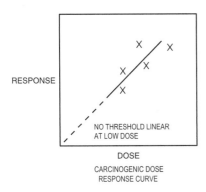

CARCINOGENIC DOSE
RESPONSE CURVE

Noncarcinogens

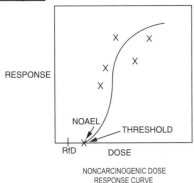

NONCARCINOGENIC DOSE
RESPONSE CURVE

Dose is expressed

$$\left(\frac{\text{mass of chemical}}{\text{body weight} \cdot \text{exposure time}} \right)$$

$NOAEL$ = No Observable Adverse Effect Level. The dose below which there are no harmful effects.

CSF = Cancer Slope Factor. Slope of the dose-response curve for carcinogenic materials.

For noncarcinogens, a hazard index (HI) is calculated as follows:

$$HI = \text{chronic daily intake}/RfD$$

Reference Dose

Reference dose (RfD) is determined from the Noncarcinogenic Dose-Response Curve Using $NOAEL$.

$$RfD = \frac{NOAEL}{UF}$$

and

$$SHD = RfD * W = \frac{NOAEL * W}{UF}$$

where

SHD = safe human dose (mg/day)

$NOAEL$ = threshold dose per kg test animal [mg/(kg•day)] from the dose-response curve

UF = the total uncertainty factor, depending on nature and reliability of the animal test data

W = the weight of the adult male (typically 70 kg)

Exposure

Residential Exposure Equations for Various Pathways

Ingestion in drinking water

$$CDI = \frac{(CW)(IR)(EF)(ED)}{(BW)(AT)}$$

Ingestion while swimming

$$CDI = \frac{(CW)(CR)(ET)(EF)(ED)}{(BW)(AT)}$$

Dermal contact with water

$$AD = \frac{(CW)(SA)(PC)(ET)(EF)(ED)(CF)}{(BW)(AT)}$$

Ingestion of chemicals in soil

$$CDI = \frac{(CS)(IR)(CF)(FI)(EF)(ED)}{(BW)(AT)}$$

Dermal contact with soil

$$AD = \frac{(CS)(CF)(SA)(AF)(ABS)(EF)(ED)}{(BW)(AT)}$$

Inhalation of airborne (vapor phase) chemicals[a]

$$CDI = \frac{(CA)(IR)(ET)(EF)(ED)}{(BW)(AT)}$$

Ingestion of contaminated fruits, vegetables, fish and shellfish

$$CDI = \frac{(CF)(IR)(FI)(EF)(ED)}{(BW)(AT)}$$

where ABS = absorption factor for soil contaminant (unitless)

AD = absorbed dose (mg/[kg•day])

AF = soil-to-skin adherence factor (mg/cm^2)

AT = averaging time (days)

BW = body weight (kg)

CA = contaminant concentration in air (mg/m^3)

CDI = chronic daily intake (mg/[kg•day])

CF = volumetric conversion factor for water
= 1 L/1,000 cm^3

= conversion factor for soil = 10^{-6} kg/mg

CR = contact rate (L/hr)

CS = chemical concentration in soil (mg/kg)

CW = chemical concentration in water (mg/L)

ED = exposure duration (years)

EF = exposure frequency (days/yr or events/year)

ET = exposure time (hr/day or hr/event)

FI = fraction ingested (unitless)

IR = ingestion rate (L/day or mg soil/day or kg/meal)

= inhalation rate (m^3/hr)

PC = chemical-specific dermal permeability constant (cm/hr)

SA = skin surface area available for contact (cm^2)

Risk Assessment Guidance for Superfund. Volume 1, *Human Health Evaluation Manual* (part A). U.S. Environmental Protection Agency, EPA/540/1-89/002, 1989.

[a]For some workplace applications of inhalation exposure, the form of the equation becomes:

$$Dosage = \frac{(\alpha)(BR)(C)(t)}{(BW)}$$

where

Dosage = mg substance per kg body weight

α = fraction of chemical absorbed by the lungs (assume 1.0 unless otherwise specified)

BR = breathing rate of the individual (1.47 m^3/hr for 2 hr or 0.98 m^3/hr for 6 hr; varies some with size of individual)

C = concentration of the substance in the air (mg/m^3)

BW = body weight (kg), usually 70 kg for men and 60 kg for women

t = time (usually taken as 8 hr in these calculations)

Based on animal data, one may use the above relations to calculate the safe air concentration if the safe human dose (*SHD*) is known, using the following relationship:

$$C = \frac{SHD}{(\alpha)(BR)(t)}$$

Intake Rates

EPA Recommended Values for Estimating Intake

Parameter	Standard Value
Average body weight, adult	70 kg
Average body weight, child[a]	
0–1.5 years	10 kg
1.5–5 years	14 kg
5–12 years	26 kg
Amount of water ingested, adult	2 L/day
Amount of water ingested, child	1 L/day
Amount of air breathed, adult	20 m^3/day
Amount of air breathed, child	5 m^3/day
Amount of fish consumed, adult	6.5 g/day
Contact rate, swimming	50 mL/hr
Inhalation rates	
adult (6-hr day)	0.98 m^3/hr
adult (2-hr day)	1.47 m^3/hr
child	0.46 m^3/hr
Skin surface available, adult male	1.94 m^2
Skin surface available, adult female	1.69 m^2
Skin surface available, child	
3–6 years (average for male and female)	0.720 m^2
6–9 years (average for male and female)	0.925 m^2
9–12 years (average for male and female)	1.16 m^2
12–15 years (average for male and female)	1.49 m^2
15–18 years (female)	1.60 m^2
15–18 years (male)	1.75 m^2
Soil ingestion rate, children 1–6 years	200 mg/day
Soil ingestion rate, persons > 6 years	100 mg/day
Skin adherence factor, potting soil to hands	1.45 mg/cm^2
Skin adherence factor, kaolin clay to hands	2.77 mg/cm^2
Exposure duration	
Lifetime (carcinogens, for non-carcinogens use actual exposure duration)	70 years
At one residence, 90th percentile	30 years
National median	5 years
Averaging time	(ED)(365 days/year)
Exposure frequency (EF)	
Swimming	7 days/year
Eating fish and shellfish	48 days/year
Exposure time (ET)	
Shower, 90th percentile	12 min
Shower, 50th percentile	7 min

[a] Data in this category taken from: Copeland, T., A. M. Holbrow, J. M. Otan, et al., "Use of probabilistic methods to understand the conservatism in California's approach to assessing health risks posed by air contaminants," *Journal of the Air and Waste Management Association,* vol. 44, pp. 1399-1413, 1994.

Risk Assessment Guidance for Superfund. Volume 1, *Human Health Evaluation Manual* (part A). U.S. Environmental Protection Agency, EPA/540/1-89/002, 1989.

WASTEWATER TREATMENT AND TECHNOLOGIES

Activated Sludge

$$X_A = \frac{\theta_c Y (S_0 - S_e)}{\theta (1 + k_d \theta_c)}$$

Steady State Mass Balance around Secondary Clarifier:

$$(Q_0 + Q_R) X_A = Q_e X_e + Q_R X_w + Q_w X_w$$

θ_c = Solids residence time = $\dfrac{Vol(X_A)}{Q_w X_w + Q_e X_e}$

Sludge volume/day: $Q_s = \dfrac{M(100)}{\rho_s(\% \text{ solids})}$

$$\text{SVI} = \frac{\text{Sludge volume after settling (mL/L)} * 1,000}{\text{MLSS(mg/L)}}$$

k_d = microbial death ratio; kinetic constant; day^{-1}; typical range 0.1–0.01, typical domestic wastewater value = 0.05 day^{-1}

S_e = effluent BOD or COD concentration (kg/m^3)

S_0 = influent BOD or COD concentration (kg/m^3)

X_A = biomass concentration in aeration tank (MLSS or MLVSS kg/m^3)

Y = yield coefficient (kg biomass/kg BOD or COD consumed); range 0.4–1.2

θ = hydraulic residence time = Vol/Q

Solids loading rate = $Q X/A$

For activated sludge secondary clarifier $Q = Q_0 + Q_R$

Organic loading rate (volumetric) = $Q_0 S_0 / Vol$

Organic loading rate (F:M) = $Q_0 S_0 / (Vol\ X_A)$

Organic loading rate (surface area) = $Q_0 S_0 / A_M$

ρ_s = density of solids (kg/m^3)

A = surface area of unit

A_M = surface area of media in fixed-film reactor

A_x = cross-sectional area of channel

M = sludge production rate (dry weight basis)

Q_0 = influent flow rate

Q_e = effluent flow rate

Q_w = waste sludge flow rate

ρ_s = wet sludge density

R = recycle ratio = Q_R/Q_0

Q_R = recycle flow rate = $Q_0 R$

X_e = effluent suspended solids concentration

X_w = waste sludge suspended solids concentration

Vol = aeration basin volume

Q = flow rate

DESIGN AND OPERATIONAL PARAMETERS FOR ACTIVATED-SLUDGE TREATMENT OF MUNICIPAL WASTEWATER

Type of Process	Mean cell residence time (θ_c, d)	Food-to-mass ratio [(kg BOD$_5$/ (day·kg MLSS)]	Volumetric loading (kg BOD$_5$/m^3)	Hydraulic residence time in aeration basin (θ, h)	Mixed liquor suspended solids (MLSS, mg/L)	Recycle ratio (Q_r/Q)	Flow regime*	BOD$_5$ removal efficiency (%)	Air supplied (m^3/kg BOD$_5$)
Tapered aeration	5–15	0.2–0.4	0.3–0.6	4–8	1,500–3,000	0.25–0.5	PF	85–95	45–90
Conventional	4–15	0.2–0.4	0.3–0.6	4–8	1,500–3,000	0.25–0.5	PF	85–95	45–90
Step aeration	4–15	0.2–0.4	0.6–1.0	3–5	2,000–3,500	0.25–0.75	PF	85–95	45–90
Completely mixed	4–15	0.2–0.4	0.8–2.0	3–5	3,000–6,000	0.25–1.0	CM	85–95	45–90
Contact stabilization	4–15	0.2–0.6	1.0–1.2			0.25–1.0			45–90
Contact basin				0.5–1.0	1,000–3,000		PF	80–90	
Stabilization basin				4–6	4,000–10,000		PF		
High-rate aeration	4–15	0.4–1.5	1.6–16	0.5–2.0	4,000–10,000	1.0–5.0	CM	75–90	25–45
Pure oxygen	8–20	0.2–1.0	1.6–4	1–3	6,000–8,000	0.25–0.5	CM	85–95	
Extended aeration	20–30	0.05–0.15	0.16–0.40	18–24	3,000–6,000	0.75–1.50	CM	75–90	90–125

*PF = plug flow, CM = completely mixed.

Metcalf and Eddy, *Wastewater Engineering: Treatment, Disposal, and Reuse*, 3rd ed., McGraw-Hill, 1991 and McGhee, Terence and E.W. Steel, *Water Supply and Sewerage*, McGraw-Hill, 1991.

♦ Aerobic Digestion

Design criteria for aerobic digesters[a]

Parameter	Value
Hydraulic retention time, 20°C, d[b]	
Waste activated sludge only	10–15
Activated sludge from plant without primary settling	12–18
Primary plus waste activated or trickling-filter sludge[c]	15–20
Solids loading, lb volatile solids/ft³•d	0.1–0.3
Oxygen requirements, lb O_2/lb solids destroyed	
Cell tissue[d]	~2.3
BOD_5 in primary sludge	1.6–1.9
Energy requirements for mixing	
Mechanical aerators, hp/10^3 ft³	0.7–1.50
Diffused-air mixing, ft³/10^3 ft³•min	20–40
Dissolved-oxygen residual in liquid, mg/L	1–2
Reduction in volatile suspended solids, %	40–50

[a] Adapted in part from *Water Pollution Control Federation: Sludge Stabilization, Manual of Practice FD*-9, 1985

[b] Detention times should be increased for operating temperatures below 20°C

[c] Similar detention times are used for primary sludge alone

[d] Ammonia produced during carbonaceous oxidation oxidized to nitrate per equation

$$C_5H_7O_2N + 7O_2 \rightarrow 5CO_2 + N\bar{O}_3 + 3H_2O + H^+$$

Note:
lb/ft³•d × 16.0185	= kg/m³•d
hp/10^3 ft³ × 26.3342	= kW/10^3m³
ft³/10^3ft³•min × 0.001	= m³/m³•min
0.556(°F – 32)	= °C

Tank Volume

$$Vol = \frac{Q_i\left(X_i + FS_i\right)}{X_d\left(k_dP_v + 1/\theta_c\right)}$$

where

Vol = volume of aerobic digester (ft³)

Q_i = influent average flowrate to digester (ft³/d)

X_i = influent suspended solids (mg/L)

F = fraction of the influent BOD_5 consisting of raw primary sludge (expressed as a decimal)

S_i = influent BOD_5 (mg/L)

X_d = digester suspended solids (mg/L)

k_d = reaction-rate constant (d^{-1})

P_v = volatile fraction of digester suspended solids (expressed as a decimal)

θ_c = solids residence time (sludge age) (d)

♦

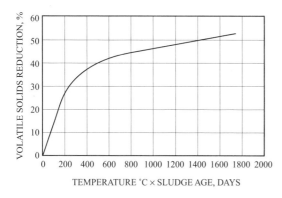

TEMPERATURE °C × SLUDGE AGE, DAYS

VOLATILE SOLIDS REDUCTION IN AN AEROBIC DIGESTER AS A FUNCTION OF DIGESTER LIQUID TEMPERATURE AND DIGESTER SLUDGE AGE

• Anaerobic Digestion

Design parameters for anaerobic digesters

Parameter	Standard-rate	High-rate
Solids residence time, d	30–90	10–20
Volatile solids loading, kg/m³/d	0.5–1.6	1.6–6.4
Digested solids concentration, %	4–6	4–6
Volatile solids reduction, %	35–50	45–55
Gas production (m³/kg VSS added)	0.5–0.55	0.6–0.65
Methane content, %	65	65

Standard Rate

$$\text{Reactor Volume} = \frac{Vol_1 + Vol_2}{2}t_r + Vol_2 t_s$$

High Rate

First stage

$$\text{Reactor Volume} = Vol_1 t_r$$

Second Stage

$$\text{Reactor Volume} = \frac{Vol_1 + Vol_2}{2}t_t + Vol_2 t_s$$

where

Vol_1 = raw sludge input (volume/day)

Vol_2 = digested sludge accumulation (volume/day)

t_r = time to react in a high-rate digester = time to react and thicken in a standard-rate digester

t_t = time to thicken in a high-rate digester

t_s = storage time

Biotower

Fixed-Film Equation without Recycle

$$\frac{S_e}{S_0} = e^{-kD/q^n}$$

Fixed-Film Equation with Recycle

$$\frac{S_e}{S_a} = \frac{e^{-kD/q^n}}{(1 + R) - R\left(e^{-kD/q^n}\right)}$$

where

S_e = effluent BOD_5 (mg/L)

S_0 = influent BOD_5 (mg/L)

R = recycle ratio = Q_0/Q_R

Q_R = recycle flow rate

$$S_a = \frac{S_o + RS_e}{1 + R}$$

D = depth of biotower media (m)

q = hydraulic loading (m³/m² • min)

   = $(Q_0 + RQ_0)/A_{plan}$ (with recycle)

k = treatability constant; functions of wastewater and medium (min^{-1}); range 0.01–0.1; for municipal wastewater and modular plastic media 0.06 min^{-1} @ 20°C

k_T = $k_{20}(1.035)^{T-20}$

n = coefficient relating to media characteristics; modular plastic, n = 0.5

♦ Tchobanoglous, G. and Metcalf and Eddy, *Wastewater Engineering: Treatment, Disposal, and Reuse*, 3rd ed., McGraw-Hill, 1991.

• Peavy, HS, D.R. Rowe and G. Tchobanoglous *Environmental Engineering*, McGraw-Hill, New York, 1985.

Facultative Pond

BOD Loading

Mass (lb/day) = Flow (MGD) × Concentration (mg/L)
$$\times\ 8.34(lb/MGal)/(mg/L)$$

Total System ≤ 35 pounds BOD_5/acre-day

Minimum = 3 ponds

Depth = 3–8 ft

Minimum t = 90–120 days

WATER TREATMENT TECHNOLOGIES

Activated Carbon Adsorption

Freundlich Isotherm

$$\frac{x}{m} = X = KC_e^{1/n}$$

where

x = mass of solute adsorbed

m = mass of adsorbent

X = mass ratio of the solid phase—that is, the mass of adsorbed solute per mass of adsorbent

C_e = equilibrium concentration of solute, mass/volume

K, n = experimental constants

Linearized Form

$$\ln\frac{x}{m} = 1/n\ln C_e + \ln K$$

For linear isotherm, $n = 1$

Langmuir Isotherm

$$\frac{x}{m} = X = \frac{aKC_e}{1 + KC_e}$$

where

a = mass of adsorbed solute required to saturate completely a unit mass of adsorbent

K = experimental constant

Linearized Form

$$\frac{m}{x} = \frac{1}{a} + \frac{1}{aK}\frac{1}{C_e}$$

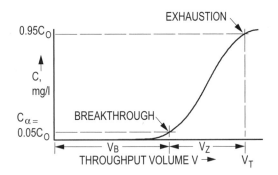

Depth of Sorption Zone

$$Z_s = Z\left[\frac{V_Z}{V_T - 0.5V_Z}\right]$$

where

V_Z = $V_T - V_B$

Z_S = depth of sorption zone

Z = total carbon depth

V_T = total volume treated at exhaustion ($C = 0.95\ C_0$)

V_B = total volume at breakthrough ($C = C_\alpha = 0.05\ C_0$)

C_0 = concentration of contaminant in influent

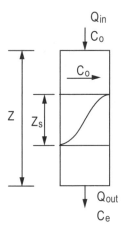

Air Stripping

P_i = HC_i

P_i = partial pressure of component i, atm

H = Henry's Law constant, atm-m^3/kmol

C_i = concentration of component i in solvent, kmol/m^3

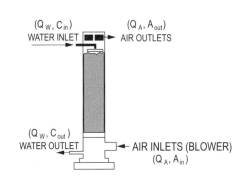

$$A_{out} = H'C_{in}$$
$$Q_W \cdot C_{in} = Q_A H'C_{in}$$
$$Q_W = Q_A H'$$
$$H'(Q_A/Q_W) = 1$$

where

A_{out} = concentration in the effluent air (kmol/m^3)

Q_W = water flow rate (m^3/s)

Q_A = air flow rate (m^3/s)

A_{in} = concentration of contaminant in air (kmol/m^3)

C_{out} = concentration of contaminants in effluent water (kmol/m^3)

C_{in} = concentration of contaminants in influent water (kmol/m^3)

Stripper Packing Height = Z

$$Z = HTU \times NTU$$

Assuming rapid equilibrium:

$$NTU = \left(\frac{R_S}{R_S - 1}\right)\ln\left(\frac{(C_{in}/C_{out})(R_S - 1) + 1}{R_S}\right)$$

where

NTU = number of transfer units

H = Henry's Law constant

H' = H/RT = dimensionless Henry's Law constant

T = temperature in units consistent with R

R = universal gas constant, atm • m^3/(kmol • K)

R_S = stripping factor $H'(Q_A/Q_W)$

C_{in} = concentration in the influent water (kmol/m^3)

C_{out} = concentration in the effluent water (kmol/m^3)

HTU = Height of Transfer Units = $\dfrac{L}{M_W K_L a}$

where

L = liquid molar loading rate [kmol/(s•m^2)]

M_W = molar density of water
(55.6 kmol/m^3) = 3.47 lbmol/ft^3

$K_L a$ = overall transfer rate constant (s^{-1})

Clarifier

Overflow rate = Hydraulic loading rate = $V_o = Q/A_{surface}$

Weir overflow rate = WOR = Q/Weir Length

Horizontal velocity = Approach velocity = V_h
= $Q/A_{cross-section} = Q/A_x$

Hydraulic residence time = $Vol/Q = \theta$

where

Q = flow rate

A_x = cross-sectional area

A = surface area, plan view

Vol = tank volume

Typical Primary Clarifier Efficiency Percent Removal

	Overflow rates			
	1,200 (gpd/ft^2) 48.9 (m/d)	1,000 (gpd/ft^2) 40.7 (m/d)	800 (gpd/ft^2) 32.6 (m/d)	600 (gpd/ft^2) 24.4 (m/d)
Suspended Solids	54%	58%	64%	68%
BOD$_5$	30%	32%	34%	36%

Design Data for Clarifiers for Activated-Sludge Systems

Type of Treatment	Overflow rate, m^3/m^2•d		Loading kg/m^2•h		Depth (m)
	Average	Peak	Average	Peak	
Settling following air-activated sludge (excluding extended aeration)	16–32	40–48	3.0–6.0	9.0	3.5–5
Settling following extended aeration	8–16	24–32	1.0–5.0	7.0	3.5–5

Adapted from Metcalf & Eddy, Inc. [5–36]

Design Criteria for Sedimentation Basins

Type of Basin	Overflow Rate (gpd/ft^2)	Hydraulic Residence Time (hr)
Water Treatment		
Presedimentation	300–500	3–4
Clarification following coagulation and flocculation		
1. Alum coagulation	350–550	4–8
2. Ferric coagulation	550–700	4–8
3. Upflow clarifiers		
a. Ground water	1,500–2,200	1
b. Surface water	1,000–1,500	4
Clarification following lime-soda softening		
1. Conventional	550–1,000	2–4
2. Upflow clarifiers		
a. Ground water	1,000–2,500	1
b. Surface water	1,000–1,800	4
Wastewater Treatment		
Primary clarifiers	600–1,200	2
Fixed film reactors		
1. Intermediate and final clarifiers	400–800	2
Activated sludge	800–1,200	2
Chemical precipitation	800–1,200	2

Weir Loadings
1. Water Treatment—weir overflow rates should not exceed 20,000 gpd/ft
2. Wastewater Treatment
   a. Flow ≤ 1 MGD: weir overflow rates should not exceed 10,000 gpd/ft
   b. Flow > 1 MGD: weir overflow rates should not exceed 15,000 gpd/ft

Horizontal Velocities
1. Water Treatment—horizontal velocities should not exceed 0.5 fpm
2. Wastewater Treatment—no specific requirements (use the same criteria as for water)

Dimensions
1. Rectangular tanks
   a. Length:Width ratio = 3:1 to 5:1
   b. Basin width is determined by the scraper width (or multiples of the scraper width)
   c. Bottom slope is set at 1%
   d. Minimum depth is 10 ft
2. Circular Tanks
   a. Diameters up to 200 ft
   b. Diameters must match the dimensions of the sludge scraping mechanism
   c. Bottom slope is less than 8%
   d. Minimum depth is 10 ft

Length:Width Ratio

Clarifier	3:1 to 5:1
Filter bay	1.2:1 to 1.5:1
Chlorine contact chamber	20:1 to 50:1

Electrodialysis
In n Cells, the Required Current Is:
$$I = (FQN/n) \times (E_1/E_2)$$

where

I = current (amperes)

F = Faraday's constant = 96,487 C/g-equivalent

Q = flow rate (L/s)

N = normality of solution (g-equivalent/L)

n = number of cells between electrodes

E_1 = removal efficiency (fraction)

E_2 = current efficiency (fraction)

Voltage
$$E = IR$$

where

E = voltage requirement (volts)

R = resistance through the unit (ohms)

Required Power
$$P = I^2R \text{ (watts)}$$

Filtration Equations
Effective size = d_{10}

Uniformity coefficient = d_{60}/d_{10}

d_x = diameter of particle class for which $x\%$ of sample is less than (units meters or feet)

Head Loss Through Clean Bed
Rose Equation

Monosized Media

$$h_f = \frac{1.067(V_s)^2 L C_D}{g\eta^4 d}$$

Multisized Media

$$h_f = \frac{1.067(V_s)^2 L}{g\eta^4} \sum \frac{C_{D_{ij}} x_{ij}}{d_{ij}}$$

Carmen-Kozeny Equation

Monosized Media

$$h_f = \frac{f'L(1-\eta)V_s^2}{\eta^3 g d_p}$$

Multisized Media

$$h_f = \frac{L(1-\eta)V_s^2}{\eta^3 g} \sum \frac{f'_{ij} x_{ij}}{d_{ij}}$$

$$f' = \text{friction factor} = 150\left(\frac{1-\eta}{Re}\right) + 1.75$$

where

h_f = head loss through the cleaner bed (m of H_2O)

L = depth of filter media (m)

η = porosity of bed = void volume/total volume

V_s = filtration rate = empty bed approach velocity
  = Q/A_{plan} (m/s)

g = gravitational acceleration (m/s^2)

  Re = Reynolds number = $\dfrac{V_s \rho d}{\mu}$

d_{ij}, d_p, d = diameter of filter media particles; arithmetic average of adjacent screen openings (m); i = filter media (sand, anthracite, garnet); j = filter media particle size

x_{ij} = mass fraction of media retained between adjacent sieves

f'_{ij} = friction factors for each media fraction

C_D = drag coefficient as defined in settling velocity equations

Bed Expansion

Monosized

$$L_{fb} = \frac{L_o(1-\eta_o)}{1 - \left(\dfrac{V_B}{V_t}\right)^{0.22}}$$

Multisized

$$L_{fb} = L_o(1-\eta_o) \sum \frac{x_{ij}}{1 - \left(\dfrac{V_B}{V_{t,i,j}}\right)^{0.22}}$$

$$\eta_{fb} = \left(\frac{V_B}{V_t}\right)^{0.22}$$

where

L_{fb} = depth of fluidized filter media (m)

V_B = backwash velocity (m/s), Q/A_{plan}

V_t = terminal setting velocity

η_{fb} = porosity of fluidized bed

L_o = initial bed depth

η_o = initial bed porosity

Lime-Soda Softening Equations

50 mg/L as $CaCO_3$ equivalent = 1 meq/L

1. Carbon dioxide removal
   $$CO_2 + Ca(OH)_2 \rightarrow CaCO_3(s) + H_2O$$

2. Calcium carbonate hardness removal
   $$Ca(HCO_3)_2 + Ca(OH)_2 \rightarrow 2CaCO_3(s) + 2H_2O$$

3. Calcium non-carbonate hardness removal
   $$CaSO_4 + Na_2CO_3 \rightarrow CaCO_3(s) + 2Na^+ + SO_4^{-2}$$

4. Magnesium carbonate hardness removal
   $$Mg(HCO_3)_2 + 2Ca(OH)_2 \rightarrow 2CaCO_3(s) +$$
   $$Mg(OH)_2(s) + 2H_2O$$

5. Magnesium non-carbonate hardness removal
   $$MgSO_4 + Ca(OH)_2 + Na_2CO_3 \rightarrow CaCO_3(s) +$$
   $$Mg(OH)_2(s) + 2Na^+ + SO_4^{2-}$$

6. Destruction of excess alkalinity
   $$2HCO_3^- + Ca(OH)_2 \rightarrow CaCO_3(s) + CO_3^{2-} + 2H_2O$$

7. Recarbonation
   $$Ca^{2+} + 2OH^- + CO_2 \rightarrow CaCO_3(s) + H_2O$$

Molecular Formulas	Molecular Weight	n # Equiv per mole	Equivalent Weight
CO_3^{2-}	60.0	2	30.0
CO_2	44.0	2	22.0
$Ca(OH)_2$	74.1	2	37.1
$CaCO_3$	100.1	2	50.0
$Ca(HCO_3)_2$	**162.1**	**2**	**81.1**
$CaSO_4$	136.1	2	68.1
Ca^{2+}	40.1	2	20.0
H^+	1.0	1	1.0
HCO_3^-	61.0	1	61.0
$Mg(HCO_3)_2$	**146.3**	**2**	**73.2**
$Mg(OH)_2$	58.3	2	29.2
$MgSO_4$	120.4	2	60.2
Mg^{2+}	24.3	2	12.2
Na^+	23.0	1	23.0
Na_2CO_3	**106.0**	**2**	**53.0**
OH^-	17.0	1	17.0
SO_4^{2-}	96.1	2	48.0

Rapid Mix and Flocculator Design

$$G = \sqrt{\frac{P}{\mu Vol}} = \sqrt{\frac{\gamma H_L}{t\mu}}$$

$Gt = 10^4 - 10^5$

where

G = mixing intensity = root mean square velocity gradient

P = power

Vol = volume

μ = bulk viscosity

γ = specific weight of water

H_L = head loss in mixing zone

t = time in mixing zone

Reel and Paddle

$$P = \frac{C_D A_P \rho_f V_p^3}{2}$$

where

C_D = drag coefficient = 1.8 for flat blade with a L:W > 20:1

A_p = area of blade (m²) perpendicular to the direction of travel through the water

ρ_f = density of H_2O (kg/m³)

V_p = relative velocity of paddle (m/sec)

$V_{mix} = V_p \cdot$ slip coefficient

slip coefficient = 0.5 – 0.75

Turbulent Flow Impeller Mixer

$$P = K_T(n)^3(D_i)^5\rho_f$$

where

K_T = impeller constant (see table)

n = rotational speed (rev/sec)

D_i = impeller diameter (m)

Values of the Impeller Constant K_T
(Assume Turbulent Flow)

Type of Impeller	K_T
Propeller, pitch of 1, 3 blades	0.32
Propeller, pitch of 2, 3 blades	1.00
Turbine, 6 flat blades, vaned disc	6.30
Turbine, 6 curved blades	4.80
Fan turbine, 6 blades at 45°	1.65
Shrouded turbine, 6 curved blades	1.08
Shrouded turbine, with stator, no baffles	1.12

Note: Constant assumes baffled tanks having four baffles at the tank wall with a width equal to 10% of the tank diameter.

Reprinted with permission from *Industrial & Engineering Chemistry,* "Mixing of Liquids in Chemical Processing," J. Henry Rushton, 1952, v. 44, no. 12. p. 2934, American Chemical Society.

Reverse Osmosis

Osmotic Pressure of Solutions of Electrolytes

$$\pi = \phi v \frac{n}{Vol} RT$$

where

π = osmotic pressure, Pa

ϕ = osmotic coefficient

v = number of ions formed from one molecule of electrolyte

n = number of moles of electrolyte

Vol = specific volume of solvent, m³/kmol

R = universal gas constant, Pa • m³/(kmol • K)

T = absolute temperature, K

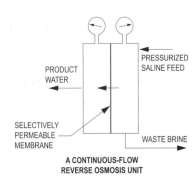

A CONTINUOUS-FLOW REVERSE OSMOSIS UNIT

Salt Flux through the Membrane

$$J_s = (D_s K_s / \Delta Z)(C_{in} - C_{out})$$

where

J_s = salt flux through the membrane [kmol/(m² • s)]

D_s = diffusivity of the solute in the membrane (m²/s)

K_s = solute distribution coefficient (dimensionless)

C = concentration (kmol/m³)

ΔZ = membrane thickness (m)

$$J_s = K_p (C_{in} - C_{out})$$

K_p = membrane solute mass transfer coefficient =

$$\frac{D_s K_s}{\Delta Z} (L/t, m/s)$$

Water Flux

$$J_w = W_p (\Delta P - \Delta \pi)$$

where

J_w = water flux through the membrane [kmol/(m² • s)]

W_p = coefficient of water permeation, a characteristic of the particular membrane [kmol/(m² • s • Pa)]

ΔP = pressure differential across membrane = $P_{in} - P_{out}$ (Pa)

$\Delta \pi$ = osmotic pressure differential across membrane $\pi_{in} - \pi_{out}$ (Pa)

Settling Equations

General Spherical

$$V_t = \sqrt{\frac{4/3 g(\rho_p - \rho_f)d}{C_D \rho_f}}$$

$$C_D = 24/Re \quad (\text{Laminar; } Re \le 1.0)$$

$$= 24/Re + 3/(Re)^{1/2} + 0.34 \quad (\text{Transitional})$$

$$= 0.4 \left(\text{Turbulent; } Re \ge 10^4 \right)$$

$$Re = \text{Reynolds number} = \frac{V_t \rho d}{\mu}$$

where

g = gravitational constant

ρ_p and ρ_f = density of particle and fluid respectively

d = diameter of sphere

C_D = spherical drag coefficient

μ = bulk viscosity of liquid = absolute viscosity

V_t = terminal settling velocity

Stokes' Law

$$V_t = \frac{g(\rho_p - \rho_f)d^2}{18\mu}$$

Approach velocity = horizontal velocity = Q/A_x

Hydraulic loading rate = Q/A

Hydraulic residence time = $Vol/Q = \theta$

where

Q = flow rate

A_x = cross-sectional area

A = surface area, plan view

Vol = tank volume

Ultrafiltration

$$J_w = \frac{\varepsilon r^2 \int \Delta P}{8\mu\delta}$$

where

ε = membrane porosity

r = membrane pore size

ΔP = net transmembrane pressure

μ = viscosity

δ = membrane thickness

J_w = volumetric flux (m/s)

ELECTRICAL AND COMPUTER ENGINEERING

UNITS

The basic electrical units are coulombs for charge, volts for voltage, amperes for current, and ohms for resistance and impedance.

ELECTROSTATICS

$$\mathbf{F}_2 = \frac{Q_1 Q_2}{4\pi\varepsilon r^2}\mathbf{a}_{r12}, \text{ where}$$

$\mathbf{F}_2$ = the force on charge 2 due to charge 1,

Q_i = the ith point charge,

r = the distance between charges 1 and 2,

$\mathbf{a}_{r12}$ = a unit vector directed from 1 to 2, and

ε = the permittivity of the medium.

For free space or air:

$$\varepsilon = \varepsilon_0 = 8.85 \times 10^{-12} \text{ farads/meter}$$

Electrostatic Fields

Electric field intensity $\mathbf{E}$ (volts/meter) at point 2 due to a point charge Q_1 at point 1 is

$$\mathbf{E} = \frac{Q_1}{4\pi\varepsilon r^2}\mathbf{a}_{r12}$$

For a line charge of density ρ_L coulomb/meter on the z-axis, the radial electric field is

$$\mathbf{E}_L = \frac{\rho_L}{2\pi\varepsilon r}\mathbf{a}_r$$

For a sheet charge of density ρ_s coulomb/meter^2 in the x-y plane:

$$\mathbf{E}_s = \frac{\rho_s}{2\varepsilon}\mathbf{a}_z, z > 0$$

Gauss' law states that the integral of the electric flux density $\mathbf{D} = \varepsilon\mathbf{E}$ over a closed surface is equal to the charge enclosed or

$$Q_{encl} = \oiint_s \varepsilon\mathbf{E} \cdot d\mathbf{S}$$

The force on a point charge Q in an electric field with intensity $\mathbf{E}$ is $\mathbf{F} = Q\mathbf{E}$.

The work done by an external agent in moving a charge Q in an electric field from point p_1 to point p_2 is

$$W = -Q\int_{p_1}^{p_2}\mathbf{E} \cdot d\mathbf{l}$$

The energy stored W_E in an electric field $\mathbf{E}$ is

$$W_E = (1/2)\iiint_V \varepsilon|\mathbf{E}|^2 dV$$

Voltage

The potential difference V between two points is the work per unit charge required to move the charge between the points.

For two parallel plates with potential difference V, separated by distance d, the strength of the E field between the plates is

$$E = \frac{V}{d}$$

directed from the + plate to the − plate.

Current

Electric current $i(t)$ through a surface is defined as the rate of charge transport through that surface or

$$i(t) = dq(t)/dt, \text{ which is a function of time } t$$

since $q(t)$ denotes instantaneous charge.

A constant current $i(t)$ is written as I, and the vector current density in amperes/m^2 is defined as $\mathbf{J}$.

Magnetic Fields

For a current carrying wire on the z-axis

$$\mathbf{H} = \frac{\mathbf{B}}{\mu} = \frac{I\mathbf{a}_\phi}{2\pi r}, \text{ where}$$

$\mathbf{H}$ = the magnetic field strength (amperes/meter),

$\mathbf{B}$ = the magnetic flux density (tesla),

$\mathbf{a}_\phi$ = the unit vector in positive ϕ direction in cylindrical coordinates,

I = the current, and

μ = the permeability of the medium.

For air: $\mu = \mu_o = 4\pi \times 10^{-7}$ H/m

Force on a current carrying conductor in a uniform magnetic field is

$$\mathbf{F} = I\mathbf{L} \times \mathbf{B}, \text{ where}$$

$\mathbf{L}$ = the length vector of a conductor.

The energy stored W_H in a magnetic field $\mathbf{H}$ is

$$W_H = (1/2)\iiint_V \mu|\mathbf{H}|^2 dv$$

Induced Voltage

Faraday's Law states for a coil of N turns enclosing flux ϕ:

$$v = -N \, d\phi/dt, \text{ where}$$

v = the induced voltage, and

ϕ = the average flux (webers) enclosed by each turn, and

ϕ = $\int_S \mathbf{B} \cdot d\mathbf{S}$

Resistivity

For a conductor of length L, electrical resistivity ρ, and cross-sectional area A, the resistance is

$$R = \frac{\rho L}{A}$$

For metallic conductors, the resistivity and resistance vary linearly with changes in temperature according to the following relationships:

$$\rho = \rho_0 \left[1 + \alpha(T - T_0) \right], \text{ and}$$
$$R = R_0 \left[1 + \alpha(T - T_0) \right], \text{ where}$$

ρ_0 is resistivity at T_0, R_0 is the resistance at T_0, and α is the temperature coefficient.

Ohm's Law: $V = IR$; $v(t) = i(t)\, R$

Resistors in Series and Parallel

For series connections, the current in all resistors is the same and the equivalent resistance for n resistors in series is

$$R_S = R_1 + R_2 + \ldots + R_n$$

For parallel connections of resistors, the voltage drop across each resistor is the same and the equivalent resistance for n resistors in parallel is

$$R_P = 1/(1/R_1 + 1/R_2 + \ldots + 1/R_n)$$

For two resistors R_1 and R_2 in parallel

$$R_P = \frac{R_1 R_2}{R_1 + R_2}$$

Power Absorbed by a Resistive Element

$$P = VI = \frac{V^2}{R} = I^2 R$$

Kirchhoff's Laws

Kirchhoff's voltage law for a closed path is expressed by

$$\Sigma\, V_{\text{rises}} = \Sigma\, V_{\text{drops}}$$

Kirchhoff's current law for a closed surface is

$$\Sigma\, I_{\text{in}} = \Sigma\, I_{\text{out}}$$

SOURCE EQUIVALENTS

For an arbitrary circuit

The Thévenin equivalent is

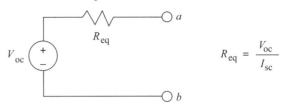

$$R_{\text{eq}} = \frac{V_{\text{oc}}}{I_{\text{sc}}}$$

The open circuit voltage V_{oc} is $V_a - V_b$, and the short circuit current is I_{sc} from a to b.

The Norton equivalent circuit is

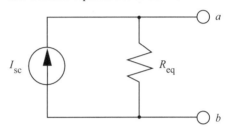

where I_{sc} and R_{eq} are defined above.

A load resistor R_L connected across terminals a and b will draw maximum power when $R_L = R_{\text{eq}}$.

CAPACITORS AND INDUCTORS

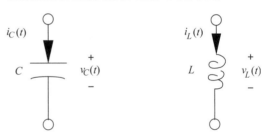

The charge $q_C(t)$ and voltage $v_C(t)$ relationship for a capacitor C in farads is

$$C = q_C(t)/v_C(t) \quad \text{or} \quad q_C(t) = Cv_C(t)$$

A parallel plate capacitor of area A with plates separated a distance d by an insulator with a permittivity ε has a capacitance

$$C = \frac{\varepsilon A}{d}$$

The current-voltage relationships for a capacitor are

$$v_C(t) = v_C(0) + \frac{1}{C} \int_0^t i_C(\tau)\, d\tau$$

and $i_C(t) = C\,(dv_C/dt)$

The energy stored in a capacitor is expressed in joules and given by

$$\text{Energy} = Cv_C^2/2 = q_C^2/2C = q_C v_C/2$$

The inductance L of a coil with N turns is

$$L = N\phi/i_L$$

and using Faraday's law, the voltage-current relations for an inductor are

$$v_L(t) = L\,(di_L/dt)$$

$$i_L(t) = i_L(0) + \frac{1}{L}\int_0^t v_L(\tau)\,d\tau, \text{ where}$$

v_L = inductor voltage,

L = inductance (henrys), and

i_L = inductor current (amperes).

The energy stored in an inductor is expressed in joules and given by

$$\text{Energy} = Li_L^2/2$$

Capacitors and Inductors in Parallel and Series
Capacitors in Parallel

$$C_P = C_1 + C_2 + \ldots + C_n$$

Capacitors in Series

$$C_S = \frac{1}{1/C_1 + 1/C_2 + \ldots + 1/C_n}$$

Inductors in Parallel

$$L_P = \frac{1}{1/L_1 + 1/L_2 + \ldots + 1/L_n}$$

Inductors in Series

$$L_S = L_1 + L_2 + \ldots + L_n$$

AC CIRCUITS
For a sinusoidal voltage or current of frequency f (Hz) and period T (seconds),

$$f = 1/T = \omega/(2\pi), \text{ where}$$

ω = the angular frequency in radians/s.

Average Value
For a periodic waveform (either voltage or current) with period T,

$$X_{ave} = (1/T)\int_0^T x(t)\,dt$$

The average value of a full-wave rectified sinusoid is

$$X_{ave} = (2X_{max})/\pi$$

and half this for half-wave rectification, where

X_{max} = the peak amplitude of the sinusoid.

Effective or RMS Values
For a periodic waveform with period T, the rms or effective value is

$$X_{eff} = X_{rms} = \left[(1/T)\int_0^T x^2(t)\,dt\right]^{1/2}$$

For a sinusoidal waveform and full-wave rectified sine wave,

$$X_{eff} = X_{rms} = X_{max}/\sqrt{2}$$

For a half-wave rectified sine wave,

$$X_{eff} = X_{rms} = X_{max}/2$$

For a periodic signal,

$$X_{rms} = \sqrt{X_{dc}^2 + \sum_{n=1}^{\infty} X_n^2} \text{ where}$$

X_{dc} is the dc component of $x(t)$

X_n is the rms value of the nth harmonic

Sine-Cosine Relations

$$\cos(\omega t) = \sin(\omega t + \pi/2) = -\sin(\omega t - \pi/2)$$

$$\sin(\omega t) = \cos(\omega t - \pi/2) = -\cos(\omega t + \pi/2)$$

Phasor Transforms of Sinusoids

$$P\left[V_{max}\cos(\omega t + \phi)\right] = V_{rms}\angle\phi = \mathbf{V}$$

$$P\left[I_{max}\cos(\omega t + \theta)\right] = I_{rms}\angle\theta = \mathbf{I}$$

For a circuit element, the impedance is defined as the ratio of phasor voltage to phasor current.

$$\mathbf{Z} = \mathbf{V/I}$$

For a Resistor, $\quad \mathbf{Z}_R = R$

For a Capacitor, $\ \mathbf{Z}_C = \dfrac{1}{j\omega C} = jX_C$

For an Inductor,

$$\mathbf{Z}_L = j\omega L = jX_L, \text{ where}$$

X_C and X_L are the capacitive and inductive reactances respectively defined as

$$X_C = -\frac{1}{\omega C} \text{ and } X_L = \omega L$$

Impedances in series combine additively while those in parallel combine according to the reciprocal rule just as in the case of resistors.

ALGEBRA OF COMPLEX NUMBERS

Complex numbers may be designated in rectangular form or polar form. In rectangular form, a complex number is written in terms of its real and imaginary components.

$$z = a + jb, \text{ where}$$

a = the real component,

b = the imaginary component, and

$j = \sqrt{-1}$

In polar form

$$z = c \angle \theta, \text{ where}$$

$c = \sqrt{a^2 + b^2}$,

$\theta = \tan^{-1}(b/a)$,

$a = c \cos \theta$, and

$b = c \sin \theta$.

Complex numbers can be added and subtracted in rectangular form. If

$$z_1 = a_1 + jb_1 = c_1(\cos\theta_1 + j\sin\theta_1)$$
$$= c_1 \angle \theta_1 \text{ and}$$
$$z_2 = a_2 + jb_2 = c_2(\cos\theta_2 + j\sin\theta_2)$$
$$= c_2 \angle \theta_2, \text{ then}$$
$$z_1 + z_2 = (a_1 + a_2) + j(b_1 + b_2) \text{ and}$$
$$z_1 - z_2 = (a_1 - a_2) + j(b_1 - b_2)$$

While complex numbers can be multiplied or divided in rectangular form, it is more convenient to perform these operations in polar form.

$$z_1 \times z_2 = (c_1 \times c_2) \angle \theta_1 + \theta_2$$
$$z_1/z_2 = (c_1/c_2) \angle \theta_1 - \theta_2$$

The complex conjugate of a complex number $z_1 = (a_1 + jb_1)$ is defined as $z_1^* = (a_1 - jb_1)$. The product of a complex number and its complex conjugate is $z_1 z_1^* = a_1^2 + b_1^2$.

RC AND RL TRANSIENTS

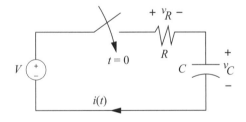

$t \geq 0; v_C(t) = v_C(0)e^{-t/RC} + V(1 - e^{-t/RC})$

$i(t) = \{[V - v_C(0)]/R\}e^{-t/RC}$

$v_R(t) = i(t)R = [V - v_C(0)]e^{-t/RC}$

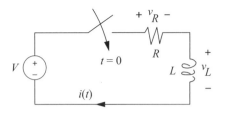

$t \geq 0; i(t) = i(0)e^{-Rt/L} + \dfrac{V}{R}(1 - e^{-Rt/L})$

$v_R(t) = i(t)R = i(0)Re^{-Rt/L} + V(1 - e^{-Rt/L})$

$v_L(t) = L(di/dt) = -i(0)Re^{-Rt/L} + Ve^{-Rt/L}$

where $v(0)$ and $i(0)$ denote the initial conditions and the parameters RC and L/R are termed the respective circuit time constants.

RESONANCE

The radian resonant frequency for both parallel and series resonance situations is

$$\omega_0 = \frac{1}{\sqrt{LC}} = 2\pi f_0 \,(\text{rad/s})$$

Series Resonance

$$\omega_0 L = \frac{1}{\omega_0 C}$$

$Z = R$ at resonance.

$$Q = \frac{\omega_0 L}{R} = \frac{1}{\omega_0 CR}$$

$$BW = \omega_0/Q \,(\text{rad/s})$$

Parallel Resonance

$$\omega_0 L = \frac{1}{\omega_0 C} \text{ and}$$

$Z = R$ at resonance.

$$Q = \omega_0 RC = \frac{R}{\omega_0 L}$$

$$BW = \omega_0/Q \,(\text{rad/s})$$

TWO-PORT PARAMETERS

A two-port network consists of two input and two output terminals as shown.

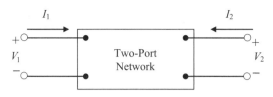

A two-port network may be represented by an equivalent circuit using a set of two-port parameters. Three commonly used sets of parameters are impedance, admittance, and hybrid parameters. The following table describes the equations used for each of these sets of parameters.

Parameter Type	Equations	Definitions
Impedance (z)	$V_1 = z_{11}I_1 + z_{12}I_2$ $V_2 = z_{21}I_1 + z_{22}I_2$	$z_{11} = \left.\dfrac{V_1}{I_1}\right|_{I_2=0} \quad z_{12} = \left.\dfrac{V_1}{I_2}\right|_{I_1=0} \quad z_{21} = \left.\dfrac{V_2}{I_1}\right|_{I_2=0} \quad z_{22} = \left.\dfrac{V_2}{I_2}\right|_{I_1=0}$
Admittance (y)	$I_1 = y_{11}V_1 + y_{12}V_2$ $I_2 = y_{21}V_1 + y_{22}V_2$	$y_{11} = \left.\dfrac{I_1}{V_1}\right|_{V_2=0} \quad y_{12} = \left.\dfrac{I_1}{V_2}\right|_{V_1=0} \quad y_{21} = \left.\dfrac{I_2}{V_1}\right|_{V_2=0} \quad y_{22} = \left.\dfrac{I_2}{V_2}\right|_{V_1=0}$
Hybrid (h)	$V_1 = h_{11}I_1 + h_{12}V_2$ $I_2 = h_{21}I_1 + h_{22}V_2$	$h_{11} = \left.\dfrac{V_1}{I_1}\right|_{V_2=0} \quad h_{12} = \left.\dfrac{V_1}{V_2}\right|_{I_1=0} \quad h_{21} = \left.\dfrac{I_2}{I_1}\right|_{V_2=0} \quad h_{22} = \left.\dfrac{I_2}{V_2}\right|_{I_1=0}$

AC POWER

Complex Power

Real power P (watts) is defined by

$$P = (\tfrac{1}{2})V_{max}I_{max}\cos\theta$$
$$= V_{rms}I_{rms}\cos\theta$$

where θ is the angle measured from $\mathbf{V}$ to $\mathbf{I}$. If $\mathbf{I}$ leads (lags) $\mathbf{V}$, then the power factor (*p.f.*),

$$p.f. = \cos\theta$$

is said to be a leading (lagging) *p.f.*

Reactive power Q (vars) is defined by

$$Q = (\tfrac{1}{2})V_{max}I_{max}\sin\theta$$
$$= V_{rms}I_{rms}\sin\theta$$

Complex power $\mathbf{S}$ (volt-amperes) is defined by

$$\mathbf{S} = \mathbf{VI}^* = P + jQ,$$

where $\mathbf{I}^*$ is the complex conjugate of the phasor current.

$S = VI$ $Q = VI\sin\theta$ $P = VI\cos\theta$ θ

Complex Power Triangle (Inductive Load)

For resistors, $\theta = 0$, so the real power is

$$P = V_{rms}I_{rms} = V_{rms}^2/R = I_{rms}^2 R$$

Balanced Three-Phase (3-ϕ) Systems

The 3-ϕ line-phase relations are

for a delta
$$V_L = V_P$$
$$I_L = \sqrt{3}\,I_P$$

for a wye
$$V_L = \sqrt{3}\,V_P = \sqrt{3}\,V_{LN}$$
$$I_L = I_P$$

where subscripts L and P denote line and phase respectively.

A balanced 3-ϕ delta-connected load impedance can be converted to an equivalent wye-connected load impedance using the following relationship

$$\mathbf{Z}_\Delta = 3\mathbf{Z}_Y$$

The following formulas can be used to determine 3-ϕ power for balanced systems.

$$\mathbf{S} = P + jQ$$
$$|\mathbf{S}| = 3V_P I_P = \sqrt{3}\,V_L I_L$$
$$\mathbf{S} = 3\mathbf{V}_P\mathbf{I}_P^* = \sqrt{3}\,V_L I_L\left(\cos\theta_P + j\sin\theta_P\right)$$

For balanced 3-ϕ wye- and delta-connected loads

$$\mathbf{S} = \frac{V_L^2}{Z_Y^*} \qquad \mathbf{S} = 3\frac{V_L^2}{Z_\Delta^*}$$

where

S = total 3-ϕ complex power (VA)

$|S|$ = total 3-ϕ apparent power (VA)

P = total 3-ϕ real power (W)

Q = total 3-ϕ reactive power (var)

θ_P = power factor angle of each phase

V_L = rms value of the line-to-line voltage

V_{LN} = rms value of the line-to-neutral voltage

I_L = rms value of the line current

I_P = rms value of the phase current

For a 3-ϕ wye-connected source or load with line-to-neutral voltages

$$\mathbf{V}_{an} = V_P \angle 0°$$
$$\mathbf{V}_{bn} = V_P \angle -120°$$
$$\mathbf{V}_{cn} = V_P \angle 120°$$

The corresponding line-to-line voltages are

$$\mathbf{V}_{ab} = \sqrt{3}\, V_P \angle 30°$$
$$\mathbf{V}_{bc} = \sqrt{3}\, V_P \angle -90°$$
$$\mathbf{V}_{ca} = \sqrt{3}\, V_P \angle 150°$$

Transformers (Ideal)

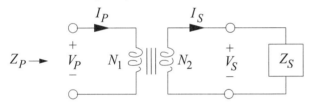

Turns Ratio

$$a = N_1/N_2$$
$$a = \left|\frac{\mathbf{V}_P}{\mathbf{V}_S}\right| = \left|\frac{\mathbf{I}_S}{\mathbf{I}_P}\right|$$

The impedance seen at the input is
$$\mathbf{Z}_P = a^2 \mathbf{Z}_S$$

AC Machines

The synchronous speed n_s for ac motors is given by

$$n_s = 120f/p, \text{ where}$$

f = the line voltage frequency in Hz and

p = the number of poles.

The slip for an induction motor is

$$\text{slip} = (n_s - n)/n_s, \text{ where}$$

n = the rotational speed (rpm).

DC Machines

The armature circuit of a dc machine is approximated by a series connection of the armature resistance R_a, the armature inductance L_a, and a dependent voltage source of value

$$V_a = K_a n\phi \text{ volts, where}$$

K_a = constant depending on the design,

n = is armature speed in rpm, and

ϕ = the magnetic flux generated by the field.

The field circuit is approximated by the field resistance R_f in series with the field inductance L_f. Neglecting saturation, the magnetic flux generated by the field current I_f is

$$\phi = K_f I_f \qquad \text{webers}$$

The mechanical power generated by the armature is

$$P_m = V_a I_a \qquad \text{watts}$$

where I_a is the armature current. The mechanical torque produced is

$$T_m = (60/2\pi)K_a \phi I_a \text{ newton-meters.}$$

ELECTROMAGNETIC DYNAMIC FIELDS

The integral and point form of Maxwell's equations are

$$\oint \mathbf{E} \cdot d\ell = - \iint_S (\partial \mathbf{B}/\partial t) \cdot d\mathbf{S}$$
$$\oint \mathbf{H} \cdot d\ell = I_{\text{enc}} + \iint_S (\partial \mathbf{D}/\partial t) \cdot d\mathbf{S}$$
$$\oiint_{S_V} \mathbf{D} \cdot d\mathbf{S} = \iint_V \rho dv$$
$$\oiint_{S_V} \mathbf{B} \cdot d\mathbf{S} = 0$$
$$\nabla \times \mathbf{E} = -\partial \mathbf{B}/\partial t$$
$$\nabla \times \mathbf{H} = \mathbf{J} + \partial \mathbf{D}/\partial t$$
$$\nabla \cdot \mathbf{D} = \rho$$
$$\nabla \cdot \mathbf{B} = 0$$

The sinusoidal wave equation in $\mathbf{E}$ for an isotropic homogeneous medium is given by

$$\nabla^2 \mathbf{E} = -\omega^2 \mu\varepsilon \mathbf{E}$$

The *EM* energy flow of a volume V enclosed by the surface S_V can be expressed in terms of Poynting's Theorem

$$- \oiint_{S_V} (\mathbf{E} \times \mathbf{H}) \cdot d\mathbf{S} = \iiint_V \mathbf{J} \cdot \mathbf{E}\, dv$$
$$+ \frac{\partial}{\partial t}\left\{ \iiint_V \left(\varepsilon E^2/2 + \mu H^2/2\right) dv \right\}$$

where the left-side term represents the energy flow per unit time or power flow into the volume V, whereas the $\mathbf{J}\cdot\mathbf{E}$ represents the loss in V and the last term represents the rate of change of the energy stored in the $\mathbf{E}$ and $\mathbf{H}$ fields.

LOSSLESS TRANSMISSION LINES

The wavelength, λ, of a sinusoidal signal is defined as the distance the signal will travel in one period.

$$\lambda = \frac{U}{f}$$

where U is the velocity of propagation and f is the frequency of the sinusoid.

The characteristic impedance, $\mathbf{Z}_0$, of a transmission line is the input impedance of an infinite length of the line and is given by

$$\mathbf{Z}_0 = \sqrt{L/C}$$

where L and C are the per unit length inductance and capacitance of the line.

The reflection coefficient at the load is defined as

$$\Gamma = \frac{\mathbf{Z}_L - \mathbf{Z}_0}{\mathbf{Z}_L + \mathbf{Z}_0}$$

and the standing wave ratio SWR is

$$\text{SWR} = \frac{1 + |\Gamma|}{1 - |\Gamma|}$$

$\beta = $ Propagation constant $= \dfrac{2\pi}{\lambda}$

For sinusoidal voltages and currents:

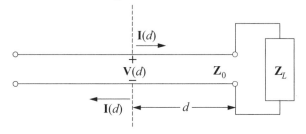

Voltage across the transmission line:

$$\mathbf{V}(d) = \mathbf{V}^+ e^{j\beta d} + \mathbf{V}^- e^{-j\beta d}$$

Current along the transmission line:

$$\mathbf{I}(d) = \mathbf{I}^+ e^{j\beta d} + \mathbf{I}^- e^{-j\beta d}$$

where $\mathbf{I}^+ = \mathbf{V}^+/\mathbf{Z}_0$ and $\mathbf{I}^- = -\mathbf{V}^-/\mathbf{Z}_0$

Input impedance at d

$$\mathbf{Z}_{\text{in}}(d) = \mathbf{Z}_L \frac{\mathbf{Z}_{0L} + j\mathbf{Z}_0 \tan(\beta d)}{\mathbf{Z}_0 + j\mathbf{Z}_L \tan(\beta d)}$$

FOURIER SERIES

Every periodic function $f(t)$ which has the period $T = 2\pi/\omega_0$ and has certain continuity conditions can be represented by a series plus a constant

$$f(t) = a_0 + \sum_{n=1}^{\infty} \left[a_n \cos(n\omega_0 t) + b_n \sin(n\omega_0 t) \right]$$

The above holds if $f(t)$ has a continuous derivative $f'(t)$ for all t. It should be noted that the various sinusoids present in the series are orthogonal on the interval 0 to T and as a result the coefficients are given by

$$a_0 = (1/T) \int_0^T f(t)\, dt$$
$$a_n = (2/T) \int_0^T f(t) \cos(n\omega_0)\, dt \qquad n = 1, 2, \ldots$$
$$b_n = (2/T) \int_0^T f(t) \sin(n\omega_0 t)\, dt \qquad n = 1, 2, \ldots$$

The constants a_n and b_n are the *Fourier coefficients* of $f(t)$ for the interval 0 to T and the corresponding series is called the *Fourier series of $f(t)$* over the same interval.

The integrals have the same value when evaluated over any interval of length T.

If a Fourier series representing a periodic function is truncated after term $n = N$ the mean square value F_N^2 of the truncated series is given by the Parseval relation. This relation says that the mean-square value is the sum of the mean-square values of the Fourier components, or

$$F_N^2 = a_0^2 + (1/2) \sum_{n=1}^{N} \left(a_n^2 + b_n^2 \right)$$

and the RMS value is then defined to be the square root of this quantity or F_N.

Three useful and common Fourier series forms are defined in terms of the following graphs (with $\omega_0 = 2\pi/T$). Given:

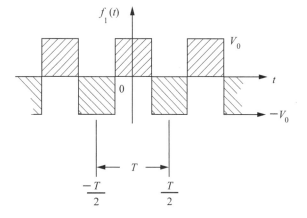

then

$$f_1(t) = \sum_{\substack{n=1 \\ (n\ \text{odd})}}^{\infty} (-1)^{(n-1)/2} \left(4V_0/n\pi \right) \cos(n\omega_0 t)$$

Given:

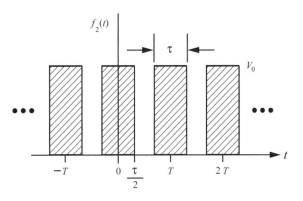

then

$$f_2(t) = \frac{V_0\tau}{T} + \frac{2V_0\tau}{T} \sum_{n=1}^{\infty} \frac{\sin(n\pi\tau/T)}{(n\pi\tau/T)} \cos(n\omega_0 t)$$

$$f_2(t) = \frac{V_0\tau}{T} \sum_{n=-\infty}^{\infty} \frac{\sin(n\pi\tau/T)}{(n\pi\tau/T)} e^{jn\omega_0 t}$$

Given:

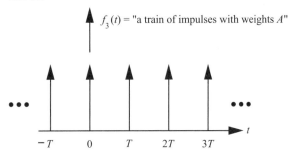

$f_3(t) =$ "a train of impulses with weights A"

then

$$f_3(t) = \sum_{n=-\infty}^{\infty} A\delta(t - nT)$$

$$f_3(t) = (A/T) + (2A/T) \sum_{n=1}^{\infty} \cos(n\omega_0 t)$$

$$f_3(t) = (A/T) \sum_{n=-\infty}^{\infty} e^{jn\omega_0 t}$$

LAPLACE TRANSFORMS

The unilateral Laplace transform pair

$$F(s) = \int_{0^-}^{\infty} f(t)e^{-st}dt$$

$$f(t) = \frac{1}{2\pi j} \int_{\sigma - j\infty}^{\sigma + j\infty} F(s)e^{st}dt$$

represents a powerful tool for the transient and frequency response of linear time invariant systems. Some useful Laplace transform pairs are:

$f(t)$	$F(s)$
$\delta(t)$, Impulse at $t = 0$	1
$u(t)$, Step at $t = 0$	$1/s$
$t[u(t)]$, Ramp at $t = 0$	$1/s^2$
$e^{-\alpha t}$	$1/(s + \alpha)$
$te^{-\alpha t}$	$1/(s + \alpha)^2$
$e^{-\alpha t} \sin \beta t$	$\beta/[(s + \alpha)^2 + \beta^2]$
$e^{-\alpha t} \cos \beta t$	$(s + \alpha)/[(s + \alpha)^2 + \beta^2]$
$\dfrac{d^n f(t)}{dt^n}$	$s^n F(s) - \sum\limits_{m=0}^{n-1} s^{n-m-1} \dfrac{d^m f(0)}{dt^m}$
$\int_0^t f(\tau)d\tau$	$(1/s)F(s)$
$\int_0^t x(t - \tau)h(t)d\tau$	$H(s)X(s)$
$f(t - \tau) u(t - \tau)$	$e^{-\tau s}F(s)$
$\lim\limits_{t\to\infty} f(t)$	$\lim\limits_{s\to 0} sF(s)$
$\lim\limits_{t\to 0} f(t)$	$\lim\limits_{s\to\infty} sF(s)$

[Note: The last two transforms represent the Final Value Theorem (F.V.T.) and Initial Value Theorem (I.V.T.) respectively. It is assumed that the limits exist.]

DIFFERENCE EQUATIONS

Difference equations are used to model discrete systems. Systems which can be described by difference equations include computer program variables iteratively evaluated in a loop, sequential circuits, cash flows, recursive processes, systems with time-delay components, etc. Any system whose input $x(t)$ and output $y(t)$ are defined only at the equally spaced intervals $t = kT$ can be described by a difference equation.

First-Order Linear Difference Equation

A first-order difference equation

$$y[k] + a_1 y[k - 1] = x[k]$$

Second-Order Linear Difference Equation

A second-order difference equation is

$$y[k] + a_1 y[k-1] + a_2 y[k-2] = x[k]$$

z-Transforms

The transform definition is

$$F(z) = \sum_{k=0}^{\infty} f[k] z^{-k}$$

The inverse transform is given by the contour integral

$$f(k) = \frac{1}{2\pi j} \oint_{\Gamma} F(z) z^{k-1} dz$$

and it represents a powerful tool for solving linear shift invariant difference equations. A limited unilateral list of z-transform pairs follows:

$f[k]$	$F(z)$
$\delta[k]$, Impulse at $k = 0$	1
$u[k]$, Step at $k = 0$	$1/(1 - z^{-1})$
β^k	$1/(1 - \beta z^{-1})$
$y[k-1]$	$z^{-1} Y(z) + y(-1)$
$y[k-2]$	$z^{-2} Y(z) + y(-2) + y(-1) z^{-1}$
$y[k+1]$	$zY(z) - zy(0)$
$y[k+2]$	$z^2 Y(z) - z^2 y(0) - zy(1)$
$\sum_{m=0}^{\infty} X[k-m] h[m]$	$H(z)X(z)$
$\lim_{k \to 0} f[k]$	$\lim_{z \to \infty} F(z)$
$\lim_{k \to \infty} f[k]$	$\lim_{z \to 1}(1 - z^{-1}) F(z)$

[Note: The last two transform pairs represent the Initial Value Theorem (I.V.T.) and the Final Value Theorem (F.V.T.) respectively.]

CONVOLUTION

Continuous-time convolution:

$$v(t) = x(t) * y(t) = \int_{-\infty}^{\infty} x(\tau) y(t - \tau) d\tau$$

Discrete-time convolution:

$$v[n] = x[n] * y[n] = \sum_{k=-\infty}^{\infty} x[k] y[n - k]$$

DIGITAL SIGNAL PROCESSING

A discrete-time, linear, time-invariant (DTLTI) system with a single input $x[n]$ and a single output $y[n]$ can be described by a linear difference equation with constant coefficients of the form

$$y[n] + \sum_{i=1}^{k} b_i y[n - i] = \sum_{i=0}^{l} a_i x[n - i]$$

If all initial conditions are zero, taking a z-transform yields a transfer function

$$H(z) = \frac{Y(z)}{X(z)} = \frac{\sum_{i=0}^{l} a_i z^{k-1}}{z^k + \sum_{i=1}^{k} b_i z^{k-1}}$$

Two common discrete inputs are the unit-step function $u[n]$ and the unit impulse function $\delta[n]$, where

$$u[n] = \begin{Bmatrix} 0 & n < 0 \\ 1 & n \geq 0 \end{Bmatrix} \text{ and } \delta[n] = \begin{Bmatrix} 1 & n = 0 \\ 0 & n \neq 0 \end{Bmatrix}$$

The impulse response $h[n]$ is the response of a discrete-time system to $x[n] = \delta[n]$.

A finite impulse response (FIR) filter is one in which the impulse response $h[n]$ is limited to a finite number of points:

$$h[n] = \sum_{i=0}^{k} a_i \delta[n - i]$$

The corresponding transfer function is given by

$$H(z) = \sum_{i=0}^{k} a_i z^{-i}$$

where k is the order of the filter.

An infinite impulse response (IIR) filter is one in which the impulse response $h[n]$ has an infinite number of points:

$$h[n] = \sum_{i=0}^{\infty} a_i \delta[n - i]$$

COMMUNICATION THEORY AND CONCEPTS

The following concepts and definitions are useful for communications systems analysis.

Functions

Unit step, $u(t)$	$u(t) = \begin{cases} 0 & t < 0 \\ 1 & t > 0 \end{cases}$
Rectangular pulse, $\Pi(t/\tau)$	$\Pi(t/\tau) = \begin{cases} 1 & \lvert t/\tau \rvert < \dfrac{1}{2} \\ 0 & \lvert t/\tau \rvert > \dfrac{1}{2} \end{cases}$
Triangular pulse, $\Lambda(t/\tau)$	$\Lambda(t/\tau) = \begin{cases} 1 - \lvert t/\tau \rvert & \lvert t/\tau \rvert < 1 \\ 0 & \lvert t/\tau \rvert > 1 \end{cases}$
Sinc, $\mathrm{sinc}(at)$	$\mathrm{sinc}(at) = \dfrac{\sin(a\pi t)}{a\pi t}$
Unit impulse, $\delta(t)$	$\displaystyle\int_{-\infty}^{+\infty} x(t+t_0)\delta(t)\,dt = x(t_0)$ for every $x(t)$ defined and continuous at $t = t_0$. This is equivalent to $\displaystyle\int_{-\infty}^{+\infty} x(t)\delta(t-t_0)\,dt = x(t_0)$

The Convolution Integral

$$x(t) * h(t) = \int_{-\infty}^{+\infty} x(\lambda)h(t-\lambda)\,d\lambda$$
$$= h(t) * x(t) = \int_{-\infty}^{+\infty} h(\lambda)x(t-\lambda)\,d\lambda$$

In particular,

$$x(t)*\delta(t-t_0) = x(t-t_0)$$

The Fourier Transform and its Inverse

$$X(f) = \int_{-\infty}^{+\infty} x(t)e^{-j2\pi ft}\,dt$$
$$x(t) = \int_{-\infty}^{+\infty} X(f)e^{j2\pi ft}\,df$$

We say that $x(t)$ and $X(f)$ form a *Fourier transform pair*:

$$x(t) \leftrightarrow X(f)$$

Fourier Transform Pairs

$x(t)$	$X(f)$
1	$\delta(f)$
$\delta(t)$	1
$u(t)$	$\dfrac{1}{2}\delta(f) + \dfrac{1}{j2\pi f}$
$\Pi(t/\tau)$	$\tau f\,\mathrm{sinc}(\tau f)$
$\mathrm{sinc}(Bt)$	$\dfrac{1}{B}\Pi(f/B)$
$\Lambda(t/\tau)$	$\tau f\,\mathrm{sinc}^2(\tau f)$
$e^{-at}u(t)$	$\dfrac{1}{a+j2\pi f} \quad a > 0$
$te^{-at}u(t)$	$\dfrac{1}{(a+j2\pi f)^2} \quad a > 0$
$e^{-a\lvert t\rvert}$	$\dfrac{2a}{a^2+(2\pi f)^2} \quad a > 0$
$e^{-(at)^2}$	$\dfrac{\sqrt{\pi}}{a}e^{-(\pi f/a)^2}$
$\cos(2\pi f_0 t + \theta)$	$\dfrac{1}{2}[e^{j\theta}\delta(f-f_0) + e^{-j\theta}\delta(f+f_0)]$
$\sin(2\pi f_0 t + \theta)$	$\dfrac{1}{2j}[e^{j\theta}\delta(f-f_0) - e^{-j\theta}\delta(f+f_0)]$
$\displaystyle\sum_{n=-\infty}^{n=+\infty} \delta(t-nT_s)$	$\displaystyle f_s\sum_{k=-\infty}^{k=+\infty} \delta(f-kf_s) \quad f_s = \dfrac{1}{T_s}$

Fourier Transform Theorems

Linearity	$ax(t) + by(t)$	$aX(f) + bY(f)$
Scale change	$x(at)$	$\dfrac{1}{\lvert a \rvert} X\left(\dfrac{f}{a}\right)$
Time reversal	$x(-t)$	$X(-f)$
Duality	$X(t)$	$x(-f)$
Time shift	$x(t - t_0)$	$X(f)e^{-j2\pi f t_0}$
Frequency shift	$x(t)e^{j2\pi f_0 t}$	$X(f - f_0)$
Modulation	$x(t)\cos 2\pi f_0 t$	$\dfrac{1}{2} X(f - f_0)$ $+\dfrac{1}{2} X(f + f_0)$
Multiplication	$x(t)y(t)$	$X(f) * Y(f)$
Convolution	$x(t) * y(t)$	$X(f)Y(f)$
Differentiation	$\dfrac{d^n x(t)}{dt^n}$	$(j2\pi f)^n X(f)$
Integration	$\displaystyle\int_{-\infty}^{t} x(\lambda)d\lambda$	$\dfrac{1}{j2\pi f} X(f)$ $+\dfrac{1}{2} X(0)\delta(f)$

Frequency Response and Impulse Response

The *frequency response* $H(f)$ of a system with input $x(t)$ and output $y(t)$ is given by

$$H(f) = \frac{Y(f)}{X(f)}$$

This gives

$$Y(f) = H(f)X(f)$$

The response $h(t)$ of a linear time-invariant system to a unit-impulse input $\delta(t)$ is called the *impulse response* of the system. The response $y(t)$ of the system to any input $x(t)$ is the convolution of the input $x(t)$ with the impulse response $h(t)$:

$$y(t) = x(t) * h(t) = \int_{-\infty}^{+\infty} x(\lambda)h(t - \lambda)d\lambda$$
$$= h(t) * x(t) = \int_{-\infty}^{+\infty} h(\lambda)x(t - \lambda)d\lambda$$

Therefore, the impulse response $h(t)$ and frequency response $H(f)$ form a Fourier transform pair:

$$h(t) \leftrightarrow H(f)$$

Parseval's Theorem

The total energy in an energy signal (finite energy) $x(t)$ is given by

$$E = \int_{-\infty}^{+\infty} \lvert x(t) \rvert^2 dt = \int_{-\infty}^{+\infty} \lvert X(f) \rvert^2 df$$

Parseval's Theorem for Fourier Series

As described in the following section, a periodic signal $x(t)$ with period T_0 and fundamental frequency $f_0 = 1/T_0 = \omega_0/2\pi$ can be represented by a complex-exponential Fourier series

$$x(t) = \sum_{n=-\infty}^{n=+\infty} X_n e^{jn2\pi f_0 t}$$

The average power in the dc component and the first N harmonics is

$$P = \sum_{n=-N}^{n=+N} \lvert X_n \rvert^2 = X_0^2 + 2 \sum_{n=0}^{n=N} \lvert X_n \rvert^2$$

The total average power in the periodic signal $x(t)$ is given by *Parseval's theorem*:

$$P = \frac{1}{T_0} \int_{t_0}^{t_0 + T_0} \lvert x(t) \rvert^2 dt = \sum_{n=-\infty}^{n=+\infty} \lvert X_n \rvert^2$$

AM (Amplitude Modulation)

$$x_{AM}(t) = A_c\left[A + m(t)\right]\cos(2\pi f_c t)$$
$$= A'_c\left[1 + am_n(t)\right]\cos(2\pi f_c t)$$

The modulation index is a, and the normalized message is

$$m_n(t) = \frac{m(t)}{\max\lvert m(t) \rvert}$$

The efficiency η is the percent of the total transmitted power that contains the message.

$$\eta = \frac{a^2 <m_n^2(t)>}{1 + a^2 <m_n^2(t)>} 100 \text{ percent}$$

where the mean-squared value or normalized average power in $m_n(t)$ is

$$<m_n^2(t)> = \lim_{T \to \infty} \frac{1}{2T} \int_{-T}^{+T} \lvert m_n(t) \rvert^2 dt$$

If $M(f) = 0$ for $\lvert f \rvert > W$, then the bandwidth of $x_{AM}(t)$ is $2W$. AM signals can be demodulated with an envelope detector or a synchronous demodulator.

DSB (Double-Sideband Modulation)

$$x_{DSB}(t) = A_c m(t)\cos(2\pi f_c t)$$

If $M(f) = 0$ for $|f| > W$, then the bandwidth of $m(t)$ is W and the bandwidth of $x_{DSB}(t)$ is $2W$. DSB signals must be demodulated with a synchronous demodulator. A Costas loop is often used.

SSB (Single-Sideband Modulation)

Lower Sideband:

$$x_{LSB}(t) \longleftrightarrow X_{LSB}(f) = X_{DSB}(f)\Pi\left(\frac{f}{2f_c}\right)$$

Upper Sideband:

$$x_{USB}(t) \longleftrightarrow X_{USB}(f) = X_{DSB}(f)\left[1 - \Pi\left(\frac{f}{2f_c}\right)\right]$$

In either case, if $M(f) = 0$ for $|f| > W$, then the bandwidth of $x_{LSB}(t)$ or of $x_{USB}(t)$ is W. SSB signals can be demodulated with a synchronous demodulator or by carrier reinsertion and envelope detection.

Angle Modulation

$$x_{Ang}(t) = A_c\cos\left[2\pi f_c t + \phi(t)\right]$$

The *phase deviation* $\phi(t)$ is a function of the message $m(t)$. The *instantaneous phase* is

$$\phi_i(t) = 2\pi f_c t + \phi(t) \quad \text{radians}$$

The *instantaneous frequency* is

$$\omega_i(t) = \frac{d}{dt}\phi_i(t) = 2\pi f_c + \frac{d}{dt}\phi(t) \quad \text{radians/s}$$

The *frequency deviation* is

$$\Delta\omega(t) = \frac{d}{dt}\phi(t) \quad \text{radians/s}$$

PM (Phase Modulation)

The *phase deviation* is

$$\phi(t) = k_P m(t) \quad \text{radians}$$

FM (Frequency Modulation)

The *phase deviation* is

$$\phi(t) = k_F \int_{-\infty}^{t} m(\lambda)\, d\lambda \quad \text{radians.}$$

The *frequency-deviation ratio* is

$$D = \frac{k_F \max|m(t)|}{2\pi W}$$

where W is the message bandwidth. If $D \ll 1$ (narrowband FM), the 98% power bandwidth B is

$$B \cong 2W$$

If $D > 1$, (wideband FM) the 98% power bandwidth B is given by *Carson's rule*:

$$B \cong 2(D + 1)W$$

The *complete* bandwidth of an angle-modulated signal is infinite.

A discriminator or a phase-lock loop can demodulate angle-modulated signals.

Sampled Messages

A lowpass message $m(t)$ can be exactly reconstructed from uniformly spaced samples taken at a sampling frequency of $f_s = 1/T_s$

$$f_s \geq 2W \text{ where } M(f) = 0 \text{ for } f > W$$

The frequency $2W$ is called the *Nyquist frequency*. Sampled messages are typically transmitted by some form of pulse modulation. The minimum bandwidth B required for transmission of the modulated message is inversely proportional to the pulse length τ.

$$B \propto \frac{1}{\tau}$$

Frequently, for approximate analysis

$$B \cong \frac{1}{2\tau}$$

is used as the *minimum* bandwidth of a pulse of length τ.

Ideal-Impulse Sampling

$$x_\delta(t) = m(t) \sum_{n=-\infty}^{n=+\infty} \delta(t - nT_s) = \sum_{n=-\infty}^{n=+\infty} m(nT_s)\delta(t - nT_s)$$

$$X_\delta(f) = M(f) * f_s \sum_{k=-\infty}^{k=+\infty} \delta(f - kf_s)$$

$$= f_s \sum_{k=-\infty}^{k=+\infty} M(f - kf_s)$$

The message $m(t)$ can be recovered from $x_\delta(t)$ with an ideal lowpass filter of bandwidth W.

PAM (Pulse-Amplitude Modulation)
Natural Sampling:
A PAM signal can be generated by multiplying a message by a pulse train with pulses having duration τ and period $T_s = 1/f_s$

$$x_N(t) = m(t) \sum_{n=-\infty}^{n=+\infty} \Pi\left[\frac{t - nT_s}{\tau}\right] = \sum_{n=-\infty}^{n=+\infty} m(t)\Pi\left[\frac{t - nT_s}{\tau}\right]$$

$$X_N(f) = \tau f_s \sum_{k=-\infty}^{k=+\infty} \text{sinc}(k\tau f_s)M(f - kf_s)$$

The message $m(t)$ can be recovered from $x_N(t)$ with an ideal lowpass filter of bandwidth W.

PCM (Pulse-Code Modulation)

PCM is formed by sampling a message $m(t)$ and digitizing the sample values with an A/D converter. For an n-bit binary word length, transmission of a pulse-code-modulated lowpass message $m(t)$, with $M(f) = 0$ for $f > W$, requires the transmission of at least $2nW$ binary pulses per second. A binary word of length n bits can represent q quantization levels:

$$q = 2^n$$

The minimum bandwidth required to transmit the PCM message will be

$$B \propto nW = 2W \log_2 q$$

ANALOG FILTER CIRCUITS

Analog filters are used to separate signals with different frequency content. The following circuits represent simple analog filters used in communications and signal processing.

First-Order Low-Pass Filters

$$|\mathbf{H}(j\omega)|$$

$$|\mathbf{H}(j\omega_c)| = \frac{1}{\sqrt{2}}|\mathbf{H}(0)|$$

Frequency Response

$$\mathbf{H}(s) = \frac{\mathbf{V}_2}{\mathbf{V}_1} = \frac{R_P}{R_1} \cdot \frac{1}{1 + sR_PC}$$

$$R_P = \frac{R_1R_2}{R_1 + R_2} \qquad \omega_c = \frac{1}{R_PC}$$

$$\mathbf{H}(s) = \frac{\mathbf{V}_2}{\mathbf{V}_1} = \frac{R_2}{R_S} \cdot \frac{1}{1 + sL/R_S}$$

$$R_S = R_1 + R_2 \qquad \omega_c = \frac{R_S}{L}$$

$$\mathbf{H}(s) = \frac{\mathbf{I}_2}{\mathbf{I}_1} = \frac{R_P}{R_2} \cdot \frac{1}{1 + sR_PC}$$

$$R_P = \frac{R_1R_2}{R_1 + R_2} \qquad \omega_c = \frac{1}{R_PC}$$

First-Order High-Pass Filters

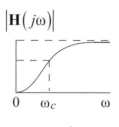

$$|\mathbf{H}(j\omega)|$$

$$|\mathbf{H}(j\omega_c)| = \frac{1}{\sqrt{2}}|\mathbf{H}(j\infty)|$$

Frequency Response

$$\mathbf{H}(s) = \frac{\mathbf{V}_2}{\mathbf{V}_1} = \frac{R_2}{R_S} \cdot \frac{sR_SC}{1 + sR_SC}$$

$$R_S = R_1 + R_2 \qquad \omega_c = \frac{1}{R_SC}$$

$$\mathbf{H}(s) = \frac{\mathbf{V}_2}{\mathbf{V}_1} = \frac{R_P}{R_1} \cdot \frac{sL/R_P}{1 + sL/R_P}$$

$$R_P = \frac{R_1R_2}{R_1 + R_2} \qquad \omega_c = \frac{R_P}{L}$$

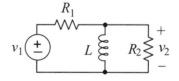

$$\mathbf{H}(s) = \frac{\mathbf{I}_2}{\mathbf{I}_1} = \frac{R_P}{R_2} \cdot \frac{sL/R_P}{1 + sL/R_P}$$

$$R_P = \frac{R_1R_2}{R_1 + R_2} \qquad \omega_c = \frac{R_P}{L}$$

Band-Pass Filters

$$\left|\mathbf{H}(j\omega)\right|$$

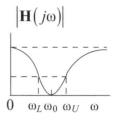

$$\left|\mathbf{H}(j\omega_L)\right| = \left|\mathbf{H}(j\omega_U)\right| = \frac{1}{\sqrt{2}}\left|\mathbf{H}(j\omega_0)\right|$$

3-dB Bandwidth $= BW = \omega_U - \omega_L$

Frequency Response

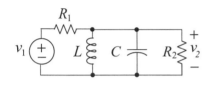

$$\mathbf{H}(s) = \frac{\mathbf{V}_2}{\mathbf{V}_1} = \frac{1}{R_1 C} \bullet \frac{s}{s^2 + s/R_P C + 1/LC}$$

$$R_P = \frac{R_1 R_2}{R_1 + R_2} \qquad \omega_0 = \frac{1}{\sqrt{LC}}$$

$$\left|\mathbf{H}(j\omega_0)\right| = \frac{R_2}{R_1 + R_2} = \frac{R_P}{R_1} \qquad BW = \frac{1}{R_P C}$$

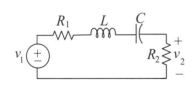

$$\mathbf{H}(s) = \frac{\mathbf{V}_2}{\mathbf{V}_1} = \frac{R_2}{L} \bullet \frac{s}{s^2 + sR_S/L + 1/LC}$$

$$R_S = R_1 + R_2 \qquad \omega_0 = \frac{1}{\sqrt{LC}}$$

$$\left|\mathbf{H}(j\omega_0)\right| = \frac{R_2}{R_1 + R_2} = \frac{R_2}{R_S} \qquad BW = \frac{R_S}{L}$$

Band-Reject Filters

$$\left|\mathbf{H}(j\omega)\right|$$

$$\left|\mathbf{H}(j\omega_L)\right| = \left|\mathbf{H}(j\omega_U)\right| = \left[1 - \frac{1}{\sqrt{2}}\right]\left|\mathbf{H}(0)\right|$$

3-dB Bandwidth $= BW = \omega_U - \omega_L$

Frequency Response

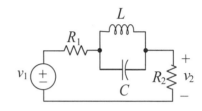

$$\mathbf{H}(s) = \frac{\mathbf{V}_2}{\mathbf{V}_1} = \frac{R_2}{R_S} \bullet \frac{s^2 + 1/LC}{s^2 + s/R_S C + 1/LC}$$

$$R_S = R_1 + R_2 \qquad \omega_0 = \frac{1}{\sqrt{LC}}$$

$$\left|\mathbf{H}(0)\right| = \frac{R_2}{R_1 + R_2} = \frac{R_2}{R_S} \qquad BW = \frac{1}{R_S C}$$

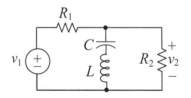

$$\mathbf{H}(s) = \frac{\mathbf{V}_2}{\mathbf{V}_1} = \frac{R_P}{R_1} \bullet \frac{s^2 + 1/LC}{s^2 + sR_P/L + 1/LC}$$

$$R_P = \frac{R_1 R_2}{R_1 + R_2} \qquad \omega_0 = \frac{1}{\sqrt{LC}}$$

$$\left|\mathbf{H}(0)\right| = \frac{R_2}{R_1 + R_2} = \frac{R_P}{R_1} \qquad BW = \frac{R_P}{L}$$

Phase-Lead Filter

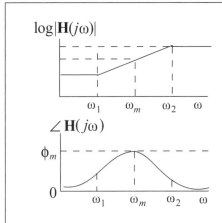

Frequency Response

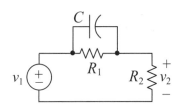

$$\mathbf{H}(s) = \frac{\mathbf{V}_2}{\mathbf{V}_1} = \frac{R_P}{R_1} \cdot \frac{1 + sR_1C}{1 + sR_PC}$$

$$= \frac{\omega_1}{\omega_2} \cdot \frac{1 + s/\omega_1}{1 + s/\omega_2}$$

$$R_P = \frac{R_1 R_2}{R_1 + R_2} \qquad \omega_1 = \frac{1}{R_1 C} \qquad \omega_2 = \frac{1}{R_P C}$$

$$\omega_m = \sqrt{\omega_1 \omega_2} \qquad \max\{\angle\mathbf{H}(j\omega_m)\} = \phi_m$$

$$\phi_m = \arctan\sqrt{\frac{\omega_2}{\omega_1}} - \arctan\sqrt{\frac{\omega_1}{\omega_2}}$$

$$= \arctan\frac{\omega_2 - \omega_1}{2\omega_m}$$

$$\mathbf{H}(0) = \frac{R_P}{R_1} = \frac{\omega_1}{\omega_2}$$

$$|\mathbf{H}(j\omega_m)| = \sqrt{\frac{\omega_1}{\omega_2}}$$

$$\mathbf{H}(j\infty) = 1$$

Phase-Lag Filter

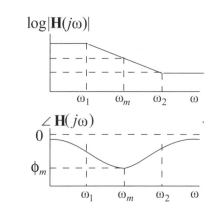

Frequency Response

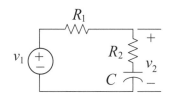

$$\mathbf{H}(s) = \frac{\mathbf{V}_2}{\mathbf{V}_1} = \frac{1 + sR_2C}{1 + sR_SC}$$

$$= \frac{1 + s/\omega_2}{1 + s/\omega_1}$$

$$R_S = R_1 + R_2 \qquad \omega_1 = \frac{1}{R_S C} \qquad \omega_2 = \frac{1}{R_2 C}$$

$$\omega_m = \sqrt{\omega_1 \omega_2} \qquad \min\{\angle\mathbf{H}(j\omega_m)\} = \phi_m$$

$$\phi_m = \arctan\sqrt{\frac{\omega_1}{\omega_2}} - \arctan\sqrt{\frac{\omega_2}{\omega_1}}$$

$$= \arctan\frac{\omega_1 - \omega_2}{2\omega_m}$$

$$\mathbf{H}(0) = 1$$

$$|\mathbf{H}(j\omega_m)| = \sqrt{\frac{\omega_1}{\omega_2}}$$

$$\mathbf{H}(j\infty) = \frac{R_2}{R_S} = \frac{\omega_1}{\omega_2}$$

OPERATIONAL AMPLIFIERS

Ideal

$v_0 = A(v_1 - v_2)$

where

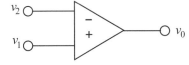

A is large ($> 10^4$), and

$v_1 - v_2$ is small enough so as not to saturate the amplifier.

For the ideal operational amplifier, assume that the input currents are zero and that the gain A is infinite so when operating linearly $v_2 - v_1 = 0$.

For the two-source configuration with an ideal operational amplifier,

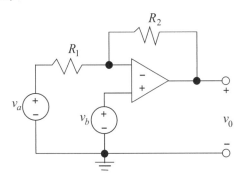

$$v_0 = -\frac{R_2}{R_1} v_a + \left(1 + \frac{R_2}{R_1}\right) v_b$$

If $v_a = 0$, we have a non-inverting amplifier with

$$v_0 = \left(1 + \frac{R_2}{R_1}\right) v_a$$

If $v_b = 0$, we have an inverting amplifier with

$$v_0 = -\frac{R_2}{R_1} v_a$$

SOLID-STATE ELECTRONICS AND DEVICES

Conductivity of a semiconductor material:

$$\sigma = q\,(n\mu_n + p\mu_p), \text{ where}$$

$\mu_n \equiv$ electron mobility,

$\mu_p \equiv$ hole mobility,

$n \equiv$ electron concentration,

$p \equiv$ hole concentration, and

$q \equiv$ charge on an electron (1.6×10^{-19}C).

Doped material:

p-type material; $p_p \approx N_a$

n-type material; $n_n \approx N_d$

Carrier concentrations at equilibrium

$(p)(n) = n_i^2$, where

$n_i \equiv$ intrinsic concentration.

Built-in potential (contact potential) of a p-n junction:

$$V_0 = \frac{kT}{q} \ln \frac{N_a N_d}{n_i^2}$$

Thermal voltage

$$V_T = \frac{kT}{q} \approx 0.026\text{V at 300K}$$

N_a = acceptor concentration,

N_d = donor concentration,

T = temperature (K), and

k = Boltzmann's Constant = 1.38×10^{-23} J/K

Capacitance of abrupt $p - n$ junction diode

$$C(V) = C_0 \big/ \sqrt{1 - V/V_{bi}}, \text{ where:}$$

C_0 = junction capacitance at $V = 0$,

V = potential of anode with respect to cathode, and

V_{bi} = junction contact potential.

Resistance of a diffused layer is

$R = R_s (L/W)$, where:

R_s = sheet resistance = ρ/d in ohms per square

ρ = resistivity,

d = thickness,

L = length of diffusion, and

W = width of diffusion.

TABULATED CHARACTERISTICS FOR:
Diodes
Bipolar Junction Transistors (BJT)
N-Channel JFET and MOSFETs
Enhancement MOSFETs
are on the following pages.

DIODES

Device and Schematic Symbol	Ideal $I-V$ Relationship	Piecewise-Linear Approximation of the $I-V$ Relationship	Mathematical $I-V$ Relationship
(Junction Diode) 		 (0.5 to 0.7)V v_B = breakdown voltage	Shockley Equation $i_D \approx I_s\left[e^{(v_D/\eta V_T)} - 1\right]$ where I_s = saturation current η = emission coefficient, typically 1 for Si V_T = thermal voltage = $\dfrac{kT}{q}$
(Zener Diode) 		 (0.5 to 0.7)V v_Z = Zener voltage	Same as above.

NPN Bipolar Junction Transistor (BJT)

Schematic Symbol	Mathematical Relationships	Large-Signal (DC) Equivalent Circuit	Low-Frequency Small-Signal (AC) Equivalent Circuit
  NPN – Transistor	$i_E = i_B + i_C$ $i_C = \beta i_B$ $i_C = \alpha i_E$ $\alpha = \beta/(\beta+1)$ $i_C \approx I_S e^{(V_{BE}/V_T)}$ I_S = emitter saturation current V_T = thermal voltage <u>Note</u>: These relationships are valid in the active mode of operation.	<u>Active Region:</u> base emitter junction forward biased; base collector juction reverse biased <u>Saturation Region:</u> both junctions forward biased 	Low Frequency: $g_m \approx I_{CQ}/V_T$ $r_\pi \approx \beta/g_m$, $r_o = \left[\dfrac{\partial v_{CE}}{\partial i_c}\right]_{Q_{point}} \approx \dfrac{V_A}{I_{CQ}}$ where I_{CQ} = dc collector current at the Q_{point} V_A = Early voltage
  PNP – Transistor	Same as for NPN with current directions and voltage polarities reversed.	<u>Cutoff Region:</u> both junctions reverse biased Same as NPN with current directions and voltage polarities reversed	Same as for NPN.

	N-Channel Junction Field Effect Transistors (JFETs) and Depletion MOSFETs (Low and Medium Frequency)	
Schematic Symbol	**Mathematical Relationships**	**Small-Signal (AC) Equivalent Circuit**
N-CHANNEL JFET P-CHANNEL JFET N-CHANNEL DEPLETION MOSFET (NMOS) SIMPLIFIED SYMBOL	<u>Cutoff Region:</u> $v_{GS} < V_p$ $i_D = 0$ <u>Triode Region:</u> $v_{GS} > V_p$ and $v_{GD} > V_p$ $i_D = (I_{DSS}/V_p^2)[2v_{DS}(v_{GS} - V_p) - v_{DS}^2]$ <u>Saturation Region:</u> $v_{GS} > V_p$ and $v_{GD} < V_p$ $i_D = I_{DSS}(1 - v_{GS}/V_p)^2$ where I_{DSS} = drain current with $v_{GS} = 0$ (in the saturation region) $= KV_p^2$, K = conductivity factor, and V_p = pinch-off voltage.	$g_m = \dfrac{2\sqrt{I_{DSS}I_D}}{|V_p|}$ in saturation region where $r_d = \left|\dfrac{\partial v_{ds}}{\partial i_d}\right|_{Q_{point}}$
P-CHANNEL DEPLETION MOSFET (PMOS) SIMPLIFIED SYMBOL	Same as for N-Channel with current directions and voltage polarities reversed.	Same as for N-Channel.

Enhancement MOSFET (Low and Medium Frequency)		
Schematic Symbol	**Mathematical Relationships**	**Small-Signal (AC) Equivalent Circuit**
N-CHANNEL ENHANCEMENT MOSFET (NMOS) SIMPLIFIED SYMBOL	<u>Cutoff Region</u>: $v_{GS} < V_t$ $i_D = 0$ <u>Triode Region</u>: $v_{GS} > V_t$ and $v_{GD} > V_t$ $i_D = K\,[2v_{DS}\,(v_{GS} - V_t) - v_{DS}^2\,]$ <u>Saturation Region</u>: $v_{GS} > V_t$ and $v_{GD} < V_t$ $i_D = K\,(v_{GS} - V_t)^2$ where $K =$ conductivity factor $V_t =$ threshold voltage	$g_m = 2K(v_{GS} - V_t)$ in saturation region where $$r_d = \left|\frac{\partial v_{ds}}{\partial i_d}\right|_{Q_{\text{point}}}$$
P-CHANNEL ENHANCEMENT MOSFET (PMOS) SIMPLIFIED SYMBOL	Same as for N-channel with current directions and voltage polarities reversed.	Same as for N-channel.

NUMBER SYSTEMS AND CODES

An unsigned number of base-r has a decimal equivalent D defined by

$$D = \sum_{k=0}^{n} a_k r^k + \sum_{i=1}^{m} a_i r^{-i}, \text{ where}$$

a_k = the $(k+1)$ digit to the left of the radix point and
a_i = the ith digit to the right of the radix point.

Binary Number System

In digital computers, the base-2, or binary, number system is normally used. Thus the decimal equivalent, D, of a binary number is given by

$$D = a_k 2^k + a_{k-1} 2^{k-1} + \ldots + a_0 + a_{-1} 2^{-1} + \ldots$$

Since this number system is so widely used in the design of digital systems, we use a short-hand notation for some powers of two:

$2^{10} = 1{,}024$ is abbreviated "k" or "kilo"

$2^{20} = 1{,}048{,}576$ is abbreviated "M" or "mega"

Signed numbers of base-r are often represented by the radix complement operation. If M is an N-digit value of base-r, the radix complement $R(M)$ is defined by

$$R(M) = r^N - M$$

The 2's complement of an N-bit binary integer can be written

2's Complement $(M) = 2^N - M$

This operation is equivalent to taking the 1's complement (inverting each bit of M) and adding one.

The following table contains equivalent codes for a four-bit binary value.

Binary Base-2	Decimal Base-10	Hexa-decimal Base-16	Octal Base-8	BCD Code	Gray Code
0000	0	0	0	0	0000
0001	1	1	1	1	0001
0010	2	2	2	2	0011
0011	3	3	3	3	0010
0100	4	4	4	4	0110
0101	5	5	5	5	0111
0110	6	6	6	6	0101
0111	7	7	7	7	0100
1000	8	8	10	8	1100
1001	9	9	11	9	1101
1010	10	A	12	---	1111
1011	11	B	13	---	1110
1100	12	C	14	---	1010
1101	13	D	15	---	1011
1110	14	E	16	---	1001
1111	15	F	17	---	1000

LOGIC OPERATIONS AND BOOLEAN ALGEBRA

Three basic logic operations are the "AND ($\cdot$)," "OR (+)," and "Exclusive-OR $\oplus$" functions. The definition of each function, its logic symbol, and its Boolean expression are given in the following table.

Function / Inputs	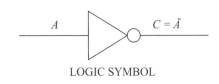 AND	OR	XOR
$A\ B$	$C = A \cdot B$	$C = A + B$	$C = A \oplus B$
0 0	0	0	0
0 1	0	1	1
1 0	0	1	1
1 1	1	1	0

As commonly used, A AND B is often written AB or $A \bullet B$.

The not operator inverts the sense of a binary value $(0 \to 1, 1 \to 0)$

NOT OPERATOR

Input	Output
A	$C = \bar{A}$
0	1
1	0

LOGIC SYMBOL

$A \longrightarrow C = \bar{A}$

DeMorgan's Theorems

first theorem: $\overline{A + B} = \bar{A} \cdot \bar{B}$

second theorem: $\overline{A \cdot B} = \bar{A} + \bar{B}$

These theorems define the NAND gate and the NOR gate. Logic symbols for these gates are shown below.

NAND Gates: $\overline{A \bullet B} = \bar{A} + \bar{B}$

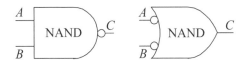

NOR Gates: $\overline{A + B} = \bar{A} \bullet \bar{B}$

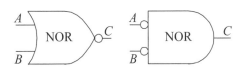

FLIP-FLOPS

A flip-flop is a device whose output can be placed in one of two states, 0 or 1. The flip-flop output is synchronized with a clock (CLK) signal. Q_n represents the value of the flip-flop output before CLK is applied, and Q_{n+1} represents the output after CLK has been applied. Three basic flip-flops are described below.

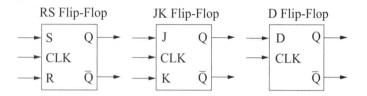

SR	Q_{n+1}		JK	Q_{n+1}		D	Q_{n+1}
00	Q_n no change		00	Q_n no change		0	0
01	0		01	0		1	1
10	1		10	1			
11	x invalid		11	$\overline{Q_n}$ toggle			

Composite Flip-Flop State Transition						
Q_n	Q_{n+1}	S	R	J	K	D
0	0	0	x	0	x	0
0	1	1	0	1	x	1
1	0	0	1	x	1	0
1	1	x	0	x	0	1

Switching Function Terminology

Minterm, m_i – A product term which contains an occurrence of every variable in the function.

Maxterm, M_i – A sum term which contains an occurrence of every variable in the function.

Implicant – A Boolean algebra term, either in sum or product form, which contains one or more minterms or maxterms of a function.

Prime Implicant – An implicant which is not entirely contained in any other implicant.

Essential Prime Implicant – A prime implicant which contains a minterm or maxterm which is not contained in any other prime implicant.

A function can be described as a sum of minterms using the notation

$$F(ABCD) = \Sigma m(h, i, j, \ldots)$$
$$= m_h + m_i + m_j + \ldots$$

A function can be described as a product of maxterms using the notation

$$G(ABCD) = \Pi M(h, i, j, \ldots)$$
$$= M_h \bullet M_i \bullet M_j \bullet \ldots$$

A function represented as a sum of minterms only is said to be in *canonical sum of products* (SOP) form. A function represented as a product of maxterms only is said to be in *canonical product of sums* (POS) form. A function in canonical SOP form is often represented as a *minterm list*, while a function in canonical POS form is often represented as a *maxterm list*.

A *Karnaugh Map* (K-Map) is a graphical technique used to represent a truth table. Each square in the K-Map represents one minterm, and the squares of the K-Map are arranged so that the adjacent squares differ by a change in exactly one variable. A four-variable K-Map with its corresponding minterms is shown below. K-Maps are used to simplify switching functions by visually identifying all essential prime implicants.

Four-variable Karnaugh Map

AB \ CD	00	01	11	10
00	m_0	m_1	m_3	m_2
01	m_4	m_5	m_7	m_6
11	m_{12}	m_{13}	m_{15}	m_{14}
10	m_8	m_9	m_{11}	m_{10}

INDUSTRIAL ENGINEERING

LINEAR PROGRAMMING

The general linear programming (LP) problem is:

$$\text{Maximize } Z = c_1 x_1 + c_2 x_2 + \ldots + c_n x_n$$

Subject to:

$$a_{11} x_1 + a_{12} x_2 + \ldots + a_{1n} x_n \leq b_1$$
$$a_{21} x_1 + a_{22} x_2 + \ldots + a_{2n} x_n \leq b_2$$
$$\vdots$$
$$a_{m1} x_1 + a_{m2} x_2 + \ldots + a_{mn} x_n \leq b_m$$
$$x_1, \ldots, x_n \geq 0$$

An LP problem is frequently reformulated by inserting non-negative slack and surplus variables. Although these variables usually have zero costs (depending on the application), they can have non-zero cost coefficients in the objective function. A slack variable is used with a "less than" inequality and transforms it into an equality. For example, the inequality $5x_1 + 3x_2 + 2x_3 \leq 5$ could be changed to $5x_1 + 3x_2 + 2x_3 + s_1 = 5$ if s_1 were chosen as a slack variable. The inequality $3x_1 + x_2 - 4x_3 \geq 10$ might be transformed into $3x_1 + x_2 - 4x_3 - s_2 = 10$ by the addition of the surplus variable s_2. Computer printouts of the results of processing an LP usually include values for all slack and surplus variables, the dual prices, and the reduced costs for each variable.

Dual Linear Program

Associated with the above linear programming problem is another problem called the dual linear programming problem. If we take the previous problem and call it the primal problem, then in matrix form the primal and dual problems are respectively:

Primal	Dual
Maximize $Z = cx$	Minimize $W = yb$
Subject to: $Ax \leq b$	Subject to: $yA \geq c$
$x \geq 0$	$y \geq 0$

It is assumed that if A is a matrix of size $[m \times n]$, then y is a $[1 \times m]$ vector, c is a $[1 \times n]$ vector, b is an $[m \times 1]$ vector, and x is an $[n \times 1]$ vector.

Network Optimization

Assume we have a graph $G(N, A)$ with a finite set of nodes N and a finite set of arcs A. Furthermore, let

$N = \{1, 2, \ldots, n\}$

x_{ij} = flow from node i to node j

c_{ij} = cost per unit flow from i to j

u_{ij} = capacity of arc (i, j)

b_i = net flow generated at node i

We wish to minimize the total cost of sending the available supply through the network to satisfy the given demand. The minimal cost flow model is formulated as follows:

$$\text{Minimize } Z = \sum_{i=1}^{n} \sum_{j=1}^{n} c_{ij} x_{ij}$$

subject to

$$\sum_{j=1}^{n} x_{ij} - \sum_{j=1}^{n} x_{ji} = b_i \text{ for each node } i \in N$$

and

$$0 \leq x_{ij} \leq u_{ij} \text{ for each arc } (i, j) \in A$$

The constraints on the nodes represent a conservation of flow relationship. The first summation represents total flow out of node i, and the second summation represents total flow into node i. The net difference generated at node i is equal to b_i.

Many models, such as shortest-path, maximal-flow, assignment and transportation models can be reformulated as minimal-cost network flow models.

STATISTICAL QUALITY CONTROL

Average and Range Charts

n	A_2	D_3	D_4
2	1.880	0	3.268
3	1.023	0	2.574
4	0.729	0	2.282
5	0.577	0	2.114
6	0.483	0	2.004
7	0.419	0.076	1.924
8	0.373	0.136	1.864
9	0.337	0.184	1.816
10	0.308	0.223	1.777

X_i = an individual observation

n = the sample size of a group

k = the number of groups

R = (range) the difference between the largest and smallest observations in a sample of size n.

$$\overline{X} = \frac{X_1 + X_2 + \ldots + X_n}{n}$$

$$\overline{\overline{X}} = \frac{\overline{X}_1 + \overline{X}_2 + \ldots + \overline{X}_n}{k}$$

$$\overline{R} = \frac{R_1 + R_2 + \ldots + R_k}{k}$$

The R Chart formulas are:

$$CL_R = \overline{R}$$
$$UCL_R = D_4 \overline{R}$$
$$LCL_R = D_3 \overline{R}$$

The $\overline{X}$ Chart formulas are:

$$CL_X = \overline{\overline{X}}$$

$$UCL_X = \overline{\overline{X}} + A_2\overline{R}$$

$$LCL_X = \overline{\overline{X}} - A_2\overline{R}$$

Standard Deviation Charts

n	A_3	B_3	B_4
2	2.659	0	3.267
3	1.954	0	2.568
4	1.628	0	2.266
5	1.427	0	2.089
6	1.287	0.030	1.970
7	1.182	0.119	1.882
8	1.099	0.185	1.815
9	1.032	0.239	1.761
10	0.975	0.284	1.716

$$UCL_X = \overline{\overline{X}} + A_3\overline{S}$$

$$CL_X = \overline{\overline{X}}$$

$$LCL_X = \overline{\overline{X}} - A_3\overline{S}$$

$$UCL_S = B_4\overline{S}$$

$$CL_S = \overline{S}$$

$$LCL_S = B_3\overline{S}$$

Approximations

The following table and equations may be used to generate initial approximations of the items indicated.

n	c_4	d_2	d_3
2	0.7979	1.128	0.853
3	0.8862	1.693	0.888
4	0.9213	2.059	0.880
5	0.9400	2.326	0.864
6	0.9515	2.534	0.848
7	0.9594	2.704	0.833
8	0.9650	2.847	0.820
9	0.9693	2.970	0.808
10	0.9727	3.078	0.797

$$\hat{\sigma} = \overline{R}/d_2$$

$$\hat{\sigma} = \overline{S}/c_4$$

$$\sigma_R = d_3\hat{\sigma}$$

$$\sigma_S = \hat{\sigma}\sqrt{1 - c_4^2}, \text{ where}$$

$\hat{\sigma}$ = an estimate of σ,

σ_R = an estimate of the standard deviation of the ranges of the samples, and

σ_S = an estimate of the standard deviation of the standard deviations of the samples.

Tests for Out of Control

1. A single point falls outside the (three sigma) control limits.
2. Two out of three successive points fall on the same side of and more than two sigma units from the center line.
3. Four out of five successive points fall on the same side of and more than one sigma unit from the center line.
4. Eight successive points fall on the same side of the center line.

PROCESS CAPABILITY

Actual Capability

$$PCR_k = C_{pk} = \min\left(\frac{\mu - LSL}{3\sigma}, \frac{USL - \mu}{3\sigma}\right)$$

Potential Capability (i.e., Centered Process)

$$PCR = C_p = \frac{USL - LSL}{6\sigma}, \text{ where}$$

μ and σ are the process mean and standard deviation, respectively, and LSL and USL are the lower and upper specification limits, respectively.

QUEUEING MODELS

Definitions

P_n = probability of n units in system,

L = expected number of units in the system,

L_q = expected number of units in the queue,

W = expected waiting time in system,

W_q = expected waiting time in queue,

λ = mean arrival rate (constant),

$\tilde{\lambda}$ = effective arrival rate,

μ = mean service rate (constant),

ρ = server utilization factor, and

s = number of servers.

Kendall notation for describing a queueing system:
$A / B / s / M$

A = the arrival process,

B = the service time distribution,

s = the number of servers, and

M = the total number of customers including those in service.

Fundamental Relationships

$$L = \lambda W$$

$$L_q = \lambda W_q$$

$$W = W_q + 1/\mu$$

$$\rho = \lambda/(s\mu)$$

Single Server Models ($s = 1$)

Poisson Input—Exponential Service Time: $M = \infty$

$$P_0 = 1 - \lambda/\mu = 1 - \rho$$
$$P_n = (1 - \rho)\rho^n = P_0\rho^n$$
$$L = \rho/(1 - \rho) = \lambda/(\mu - \lambda)$$
$$L_q = \lambda^2/[\mu(\mu - \lambda)]$$
$$W = 1/[\mu(1 - \rho)] = 1/(\mu - \lambda)$$
$$W_q = W - 1/\mu = \lambda/[\mu(\mu - \lambda)]$$

Finite queue: $M < \infty$

$$\tilde{\lambda} = \lambda(1 - P_n)$$
$$P_0 = (1 - \rho)/(1 - \rho^{M+1})$$
$$P_n = [(1 - \rho)/(1 - \rho^{M+1})]\rho^n$$
$$L = \rho/(1 - \rho) - (M + 1)\rho^{M+1}/(1 - \rho^{M+1})$$
$$L_q = L - (1 - P_0)$$

Poisson Input—Arbitrary Service Time

Variance σ^2 is known. For constant service time, $\sigma^2 = 0$.

$$P_0 = 1 - \rho$$
$$L_q = (\lambda^2\sigma^2 + \rho^2)/[2(1 - \rho)]$$
$$L = \rho + L_q$$
$$W_q = L_q/\lambda$$
$$W = W_q + 1/\mu$$

Poisson Input—Erlang Service Times, $\sigma^2 = 1/(k\mu^2)$

$$L_q = [(1 + k)/(2k)][(\lambda^2)/(\mu(\mu - \lambda))]$$
$$= [\lambda^2/(k\mu^2) + \rho^2]/[2(1 - \rho)]$$
$$W_q = [(1 + k)/(2k)]\{\lambda/[\mu(\mu - \lambda)]\}$$
$$W = W_q + 1/\mu$$

Multiple Server Model ($s > 1$)

Poisson Input—Exponential Service Times

$$P_0 = \left[\sum_{n=0}^{s-1}\frac{\left(\frac{\lambda}{\mu}\right)^n}{n!} + \frac{\left(\frac{\lambda}{\mu}\right)^s}{s!}\left(\frac{1}{1 - \frac{\lambda}{s\mu}}\right)\right]^{-1}$$

$$= 1\left/\left[\sum_{n=0}^{s-1}\frac{(s\rho)^n}{n!} + \frac{(s\rho)^s}{s!(1 - \rho)}\right]\right.$$

$$L_q = \frac{P_0\left(\frac{\lambda}{\mu}\right)^s\rho}{s!(1 - \rho)^2}$$

$$= \frac{P_0 s^s\rho^{s+1}}{s!(1 - \rho)^2}$$

$$P_n = P_0(\lambda/\mu)^n/n! \qquad 0 \le n \le s$$
$$P_n = P_0(\lambda/\mu)^n/(s!s^{n-s}) \quad n \ge s$$
$$W_q = L_q/\lambda$$
$$W = W_q + 1/\mu$$
$$L = L_q + \lambda/\mu$$

Calculations for P_0 and L_q can be time consuming; however, the following table gives formulas for 1, 2, and 3 servers.

s	P_0	L_q
1	$1 - \rho$	$\rho^2/(1 - \rho)$
2	$(1 - \rho)/(1 + \rho)$	$2\rho^3/(1 - \rho^2)$
3	$\dfrac{2(1 - \rho)}{2 + 4\rho + 3\rho^2}$	$\dfrac{9\rho^4}{2 + 2\rho - \rho^2 - 3\rho^3}$

SIMULATION

1. Random Variate Generation

The linear congruential method of generating pseudorandom numbers U_i between 0 and 1 is obtained using $Z_n = (aZ_{n-1} + C) \pmod{m}$ where a, C, m, and Z_0 are given nonnegative integers and where $U_i = Z_i/m$. Two integers are equal (mod m) if their remainders are the same when divided by m.

2. Inverse Transform Method

If X is a continuous random variable with cumulative distribution function $F(x)$, and U_i is a random number between 0 and 1, then the value of X_i corresponding to U_i can be calculated by solving $U_i = F(x_i)$ for x_i. The solution obtained is $x_i = F^{-1}(U_i)$, where F^{-1} is the inverse function of $F(x)$.

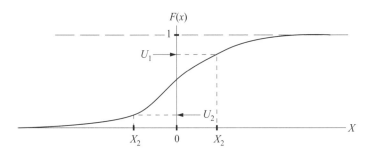

Inverse Transform Method for Continuous Random Variables

FORECASTING

Moving Average

$$\hat{d}_t = \frac{\sum_{i=1}^{n}d_{t-i}}{n}, \text{ where}$$

$\hat{d}_t$ = forecasted demand for period t,

d_{t-i} = actual demand for ith period preceding t, and

n = number of time periods to include in the moving average.

Exponentially Weighted Moving Average

$$\hat{d}_t = \alpha d_{t-1} + (1 - \alpha)\hat{d}_{t-1}, \text{ where}$$

$\hat{d}_t$ = forecasted demand for t, and

α = smoothing constant, $0 \le \alpha \le 1$

LINEAR REGRESSION

Least Squares

$y = \hat{a} + \hat{b}x$, where

$y - \text{intercept: } \hat{a} = \bar{y} - \hat{b}\bar{x}$,

and slope: $\hat{b} = SS_{xy}/SS_{xx}$,

$S_{xy} = \sum\limits_{i=1}^{n} x_i y_i - (1/n)\left(\sum\limits_{i=1}^{n} x_i\right)\left(\sum\limits_{i=1}^{n} y_i\right)$,

$S_{xx} = \sum\limits_{i=1}^{n} x_i^2 - (1/n)\left(\sum\limits_{i=1}^{n} x_i\right)^2$,

$n = \text{sample size}$,

$\bar{y} = (1/n)\left(\sum\limits_{i=1}^{n} y_i\right)$, and

$\bar{x} = (1/n)\left(\sum\limits_{i=1}^{n} x_i\right)$.

Standard Error of Estimate

$S_e^2 = \dfrac{S_{xx}S_{yy} - S_{xy}^2}{S_{xx}(n-2)} = MSE$, where

$S_{yy} = \sum\limits_{i=1}^{n} y_i^2 - (1/n)\left(\sum\limits_{i=1}^{n} y_i\right)^2$

Confidence Interval for a

$\hat{a} \pm t_{\alpha/2, n-2} \sqrt{\left(\dfrac{1}{n} + \dfrac{\bar{x}^2}{S_{xx}}\right) MSE}$

Confidence Interval for b

$\hat{b} \pm t_{\alpha/2, n-2} \sqrt{\dfrac{MSE}{S_{xx}}}$

Sample Correlation Coefficient

$r = \dfrac{S_{xy}}{\sqrt{S_{xx}S_{yy}}}$

ONE-WAY ANALYSIS OF VARIANCE (ANOVA)

Given independent random samples of size n_i from k populations, then:

$\sum\limits_{i=1}^{k} \sum\limits_{j=1}^{n_i} \left(x_{ij} - \bar{x}\right)^2$

$= \sum\limits_{i=1}^{k} \sum\limits_{j=1}^{n_i} \left(x_{ij} - \bar{x}_i\right)^2 + \sum\limits_{i=1}^{k} n_i \left(\bar{x}_i - \bar{x}\right)^2$ or

$SS_{\text{total}} = SS_{\text{error}} + SS_{\text{treatments}}$

Let T be the grand total of all $N = \Sigma_i n_i$ observations and T_i be the total of the n_i observations of the ith sample.

$C = T^2/N$

$SS_{\text{total}} = \sum\limits_{i=1}^{k} \sum\limits_{j=1}^{n_i} x_{ij}^2 - C$

$SS_{\text{treatments}} = \sum\limits_{i=1}^{k} \left(T_i^2/n_i\right) - C$

$SS_{\text{error}} = SS_{\text{total}} - SS_{\text{treatments}}$

See One-Way ANOVA table later in this chapter.

RANDOMIZED BLOCK DESIGN

The experimental material is divided into n randomized blocks. One observation is taken at random for every treatment within the same block. The total number of observations is $N = nk$. The total value of these observations is equal to T. The total value of observations for treatment i is T_i. The total value of observations in block j is B_j.

$C = T^2/N$

$SS_{\text{total}} = \sum\limits_{i=1}^{k} \sum\limits_{j=1}^{n} x_{ij}^2 - C$

$SS_{\text{blocks}} = \sum\limits_{j=1}^{n} \left(B_j^2/k\right) - C$

$SS_{\text{treatments}} = \sum\limits_{i=1}^{k} \left(T_i^2/n\right) - C$

$SS_{\text{error}} = SS_{\text{total}} - SS_{\text{blocks}} - SS_{\text{treatments}}$

See Two-Way ANOVA table later in this chapter.

2^n FACTORIAL EXPERIMENTS

Factors: $X_1, X_2, \ldots, X_n$

Levels of each factor: 1, 2 (sometimes these levels are represented by the symbols – and +, respectively)

r = number of observations for each experimental condition (treatment),

E_i = estimate of the effect of factor X_i, $i = 1, 2, \ldots, n$,

E_{ij} = estimate of the effect of the interaction between factors X_i and X_j,

$\bar{Y}_{ik}$ = average response value for all $r2^{n-1}$ observations having X_i set at level k, $k = 1, 2$, and

$\bar{Y}_{ij}^{km}$ = average response value for all $r2^{n-2}$ observations having X_i set at level k, $k = 1, 2$, and X_j set at level m, $m = 1, 2$.

$E_i = \bar{Y}_{i2} - \bar{Y}_{i1}$

$E_{ij} = \dfrac{\left(\bar{Y}_{ij}^{22} - \bar{Y}_{ij}^{21}\right) - \left(\bar{Y}_{ij}^{12} - \bar{Y}_{ij}^{11}\right)}{2}$

ANALYSIS OF VARIANCE FOR 2^n FACTORIAL DESIGNS

Main Effects

Let E be the estimate of the effect of a given factor, let L be the orthogonal contrast belonging to this effect. It can be proved that

$$E = \frac{L}{2^{n-1}}$$

$$L = \sum_{c=1}^{m} a_{(c)} \overline{Y}_{(c)}$$

$$SS_L = \frac{rL^2}{2^n}, \text{ where}$$

m = number of experimental conditions ($m = 2^n$ for n factors),

$a_{(c)} = -1$ if the factor is set at its low level (level 1) in experimental condition c,

$a_{(c)} = +1$ if the factor is set at its high level (level 2) in experimental condition c,

r = number of replications for each experimental condition

$\overline{Y}_{(c)}$ = average response value for experimental condition c, and

SS_L = sum of squares associated with the factor.

Interaction Effects

Consider any group of two or more factors.

$a_{(c)} = +1$ if there is an even number (or zero) of factors in the group set at the low level (level 1) in experimental condition $c = 1, 2, ..., m$

$a_{(c)} = -1$ if there is an odd number of factors in the group set at the low level (level 1) in experimental condition $c = 1, 2, ..., m$

It can be proved that the interaction effect E for the factors in the group and the corresponding sum of squares SS_L can be determined as follows:

$$E = \frac{L}{2^{n-1}}$$

$$L = \sum_{c=1}^{m} a_{(c)} \overline{Y}_{(c)}$$

$$SS_L = \frac{rL^2}{2^n}$$

Sum of Squares of Random Error

The sum of the squares due to the random error can be computed as

$$SS_{error} = SS_{total} - \Sigma_i SS_i - \Sigma_i \Sigma_j SS_{ij} - ... - SS_{12...n}$$

where SS_i is the sum of squares due to factor X_i, SS_{ij} is the sum of squares due to the interaction of factors X_i and X_j, and so on. The total sum of squares is equal to

$$SS_{total} = \sum_{c=1}^{m} \sum_{k=1}^{r} Y_{ck}^2 - \frac{T^2}{N}$$

where Y_{ck} is the kth observation taken for the cth experimental condition, $m = 2^n$, T is the grand total of all observations, and $N = r2^n$.

RELIABILITY

If P_i is the probability that component i is functioning, a reliability function $R(P_1, P_2, ..., P_n)$ represents the probability that a system consisting of n components will work.

For n independent components connected in series,

$$R(P_1, P_2, ...P_n) = \prod_{i=1}^{n} P_i$$

For n independent components connected in parallel,

$$R(P_1, P_2, ...P_n) = 1 - \prod_{i=1}^{n} (1 - P_i)$$

LEARNING CURVES

The time to do the repetition N of a task is given by

$$T_N = KN^s, \text{ where}$$

K = constant, and

s = ln (learning rate, as a decimal)/ln 2; or, learning rate = 2^s.

If N units are to be produced, the average time per unit is given by

$$T_{avg} = \frac{K}{N(1+s)} \left[(N + 0.5)^{(1+s)} - 0.5^{(1+s)} \right]$$

INVENTORY MODELS

For instantaneous replenishment (with constant demand rate, known holding and ordering costs, and an infinite stockout cost), the economic order quantity is given by

$$EOQ = \sqrt{\frac{2AD}{h}}, \text{ where}$$

A = cost to place one order,

D = number of units used per year, and

h = holding cost per unit per year.

Under the same conditions as above with a finite replenishment rate, the economic manufacturing quantity is given by

$$EMQ = \sqrt{\frac{2AD}{h(1 - D/R)}}, \text{ where}$$

R = the replenishment rate.

ERGONOMICS

NIOSH Formula

Recommended Weight Limit (pounds)

$$= 51(10/H)(1 - 0.0075|V - 30|)(0.82 + 1.8/D)(1 - 0.0032A)(FM)(CM)$$

where

H = horizontal distance of the hand from the midpoint of the line joining the inner ankle bones to a point projected on the floor directly below the load center, in inches

V = vertical distance of the hands from the floor, in inches

D = vertical travel distance of the hands between the origin and destination of the lift, in inches

A = asymmetry angle, in degrees

FM = frequency multiplier (see table)

CM = coupling multiplier (see table)

Frequency Multiplier Table

F, min^{-1}	≤ 8 hr/day		≤ 2 hr/day		≤ 1 hr/day	
	V < 30 in.	V ≥ 30 in.	V < 30 in.	V ≥ 30 in.	V < 30 in.	V ≥ 30 in.
0.2	0.85		0.95		1.00	
0.5	0.81		0.92		0.97	
1	0.75		0.88		0.94	
2	0.65		0.84		0.91	
3	0.55		0.79		0.88	
4	0.45		0.72		0.84	
5	0.35		0.60		0.80	
6	0.27		0.50		0.75	
7	0.22		0.42		0.70	
8	0.18		0.35		0.60	
9		0.15	0.30		0.52	
10		0.13	0.26		0.45	
11				0.23	0.41	
12				0.21	0.37	
13	0.00					0.34
14						0.31
15						0.28

Coupling Multiplier (CM) Table
(Function of Coupling of Hands to Load)

Container			Loose Part / Irreg. Object	
Optimal Design		Not	Comfort Grip	Not
Opt. Handles or Cut-outs	Not	POOR	GOOD	
GOOD	Flex Fingers	90 Degrees	Not	
	FAIR		POOR	

Coupling	V < 30 in. or 75 cm	V ≥ 30 in. or 75 cm
GOOD	1.00	
FAIR	0.95	
POOR	0.90	

Biomechanics of the Human Body

Basic Equations

$$H_x + F_x = 0$$
$$H_y + F_y = 0$$
$$H_z + W + F_z = 0$$
$$T_{Hxz} + T_{Wxz} + T_{Fxz} = 0$$
$$T_{Hyz} + T_{Wyz} + T_{Fyz} = 0$$
$$T_{Hxy} + T_{Fxy} = 0$$

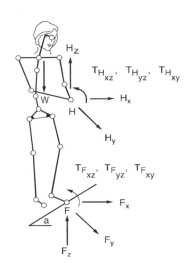

The coefficient of friction μ and the angle α at which the floor is inclined determine the equations at the foot.

$$F_x = \mu F_z$$

With the slope angle α

$$F_x = \alpha F_z \cos \alpha$$

Of course, when motion must be considered, dynamic conditions come into play according to Newton's Second Law. Force transmitted with the hands is counteracted at the foot. Further, the body must also react with internal forces at all points between the hand and the foot.

PERMISSIBLE NOISE EXPOSURE (OSHA)

Noise dose (D) should not exceed 100%.

$$D = 100\% \times \sum \frac{C_i}{T_i}$$

where C_i = time spent at specified sound pressure level, SPL, (hours)

T_i = time permitted at SPL (hours)

$\sum C_i$ = 8 (hours)

For $80 \leq SPL \leq 130$ dBA, $T_i = 2^{\left(\frac{105 - SPL}{2}\right)}$ (hours)

If D > 100%, noise abatement required.

If $50\% \leq D \leq 100\%$, hearing conservation program required.

Note: D = 100% is equivalent to 90 dBA time-weighted average (TWA). D = 50% equivalent to TWA of 85 dBA.

Hearing conservation program requires: (1) testing employee hearing, (2) providing hearing protection at employee's request, and (3) monitoring noise exposure.

Exposure to impulsive or impact noise should not exceed 140 dB sound pressure level (SPL).

FACILITY PLANNING

Equipment Requirements

$$M_j = \sum_{i=1}^{n} \frac{P_{ij}T_{ij}}{C_{ij}} \text{ where}$$

M_j = number of machines of type j required per production period,

P_{ij} = desired production rate for product i on machine j, measured in pieces per production period,

T_{ij} = production time for product i on machine j, measured in hours per piece,

C_{ij} = number of hours in the production period available for the production of product i on machine j, and

n = number of products.

People Requirements

$$A_j = \sum_{i=1}^{n} \frac{P_{ij}T_{ij}}{C_{ij}}, \text{ where}$$

A_j = number of crews required for assembly operation j,

P_{ij} = desired production rate for product i and assembly operation j (pieces per day),

T_{ij} = standard time to perform operation j on product i (minutes per piece),

C_{ij} = number of minutes available per day for assembly operation j on product i, and

n = number of products.

Standard Time Determination

$$ST = NT \times AF$$

where

NT = normal time, and

AF = allowance factor.

Case 1: Allowances are based on the *job time*.

$AF_{job} = 1 + A_{job}$

A_{job} = allowance fraction (percentage/100) based on *job time*.

Case 2: Allowances are based on *workday*.

$AF_{time} = 1/(1 - A_{day})$

A_{day} = allowance fraction (percentage/100) based on *workday*.

Plant Location

The following is one formulation of a discrete plant location problem.

Minimize

$$z = \sum_{i=1}^{m} \sum_{j=1}^{n} c_{ij}y_{ij} + \sum_{j=1}^{n} f_j x_j$$

subject to

$$\sum_{i=1}^{m} y_{ij} \leq mx_j, \qquad j = 1, \ldots, n$$

$$\sum_{j=1}^{n} y_{ij} = 1, \qquad i = 1, \ldots, m$$

$$y_{ij} \geq 0, \text{ for all } i, j$$

$$x_j = (0, 1), \text{ for all } j, \text{ where}$$

m = number of customers,

n = number of possible plant sites,

y_{ij} = fraction or proportion of the demand of customer i which is satisfied by a plant located at site j; $i = 1, \ldots, m$; $j = 1, \ldots, n$,

x_j = 1, if a plant is located at site j,

x_j = 0, otherwise,

c_{ij} = cost of supplying the entire demand of customer i from a plant located at site j, and

f_j = fixed cost resulting from locating a plant at site j.

Material Handling

Distances between two points (x_1, y_1) and (x_2, y_2) under different metrics:

Euclidean:

$$D = \sqrt{(x_1 - x_2)^2 + (y_1 - y_2)^2}$$

Rectilinear (or Manhattan):

$$D = |x_1 - x_2| + |y_1 - y_2|$$

Chebyshev (simultaneous x and y movement):
$$D = \max\left(|x_1 - x_2|, |y_1 - y_2|\right)$$

Line Balancing

$$N_{\min} = \left(OR \times \sum_i t_i \middle/ OT\right)$$
$$= \text{Theoretical minimum number of stations}$$

$$\text{Idle Time/Station} = CT - ST$$

$$\text{Idle Time/Cycle} = \Sigma(CT - ST)$$

$$\text{Percent Idle Time} = \frac{\text{Idle Time/Cycle}}{N_{\text{actual}} \times CT} \times 100, \text{ where}$$

CT = cycle time (time between units),

OT = operating time/period,

OR = output rate/period,

ST = station time (time to complete task at each station),

t_i = individual task times, and

N = number of stations.

Job Sequencing

Two Work Centers—Johnson's Rule

1. Select the job with the shortest time, from the list of jobs, and its time at each work center.

2. If the shortest job time is the time at the first work center, schedule it first, otherwise schedule it last. Break ties arbitrarily.

3. Eliminate that job from consideration.

4. Repeat 1, 2, and 3 until all jobs have been scheduled.

CRITICAL PATH METHOD (CPM)

d_{ij} = duration of activity (i, j),

CP = critical path (longest path),

T = duration of project, and

$$T = \sum_{(i,j) \in CP} d_{ij}$$

PERT

(a_{ij}, b_{ij}, c_{ij}) = (optimistic, most likely, pessimistic) durations for activity (i, j),

μ_{ij} = mean duration of activity (i, j),

σ_{ij} = standard deviation of the duration of activity (i, j),

μ = project mean duration, and

σ = standard deviation of project duration.

$$\mu_{ij} = \frac{a_{ij} + 4b_{ij} + c_{ij}}{6}$$

$$\sigma_{ij} = \frac{c_{ij} - a_{ij}}{6}$$

$$\mu = \sum_{(i,j) \in CP} \mu_{ij}$$

$$\sigma^2 = \sum_{(i,j) \in CP} \sigma_{ij}^2$$

TAYLOR TOOL LIFE FORMULA

$$VT^n = C, \text{ where}$$

V = speed in surface feet per minute,

T = tool life in minutes, and

C, n = constants that depend on the material and on the tool.

WORK SAMPLING FORMULAS

$$D = Z_{\alpha/2} \sqrt{\frac{p(1 - p)}{n}} \text{ and } R = Z_{\alpha/2} \sqrt{\frac{1 - p}{pn}}, \text{ where}$$

p = proportion of observed time in an activity,

D = absolute error,

R = relative error $(R = D/p)$, and

n = sample size.

ONE-WAY ANOVA TABLE

Source of Variation	Degrees of Freedom	Sum of Squares	Mean Square	F
Between Treatments	$k-1$	$SS_{\text{treatments}}$	$MST = \dfrac{SS_{\text{treatments}}}{k-1}$	$\dfrac{MST}{MSE}$
Error	$N-k$	SS_{error}	$MSE = \dfrac{SS_{\text{error}}}{N-k}$	
Total	$N-1$	SS_{total}		

TWO-WAY ANOVA TABLE

Source of Variation	Degrees of Freedom	Sum of Squares	Mean Square	F
Between Treatments	$k-1$	$SS_{\text{treatments}}$	$MST = \dfrac{SS_{\text{treatments}}}{k-1}$	$\dfrac{MST}{MSE}$
Between Blocks	$n-1$	SS_{blocks}	$MSB = \dfrac{SS_{\text{blocks}}}{n-1}$	$\dfrac{MSB}{MSE}$
Error	$(k-1)(n-1)$	SS_{error}	$MSE = \dfrac{SS_{\text{error}}}{(k-1)(n-1)}$	
Total	$N-1$	SS_{total}		

PROBABILITY AND DENSITY FUNCTIONS: MEANS AND VARIANCES

Variable	Equation	Mean	Variance
Binomial Coefficient	$\binom{n}{x} = \dfrac{n!}{x!(n-x)!}$		
Binomial	$b(x;n,p) = \binom{n}{x} p^x (1-p)^{n-x}$	np	$np(1-p)$
Hyper Geometric	$h(x;n,r,N) = \binom{r}{x} \dfrac{\binom{N-r}{n-x}}{\binom{N}{n}}$	$\dfrac{nr}{N}$	$\dfrac{r(N-r)n(N-n)}{N^2(N-1)}$
Poisson	$f(x;\lambda) = \dfrac{\lambda^x e^{-\lambda}}{x!}$	λ	λ
Geometric	$g(x;p) = p\,(1-p)^{x-1}$	$1/p$	$(1-p)/p^2$
Negative Binomial	$f(y;r,p) = \binom{y+r-1}{r-1} p^r (1-p)^y$	r/p	$r\,(1-p)/p^2$
Multinomial	$f(x_1,\ldots,x_k) = \dfrac{n!}{x_1!,\ldots,x_k!} p_1^{x_1} \cdots p_k^{x_k}$	np_i	$np_i\,(1-p_i)$
Uniform	$f(x) = 1/(b-a)$	$(a+b)/2$	$(b-a)^2/12$
Gamma	$f(x) = \dfrac{x^{\alpha-1} e^{-x/\beta}}{\beta^\alpha \Gamma(\alpha)};\quad \alpha > 0, \beta > 0$	$\alpha\beta$	$\alpha\beta^2$
Exponential	$f(x) = \dfrac{1}{\beta} e^{-x/\beta}$	β	β^2
Weibull	$f(x) = \dfrac{\alpha}{\beta} x^{\alpha-1} e^{-x^\alpha/\beta}$	$\beta^{1/\alpha}\Gamma[(\alpha+1)/\alpha]$	$\beta^{2/\alpha}\left[\Gamma\left(\dfrac{\alpha+1}{\alpha}\right) - \Gamma^2\left(\dfrac{\alpha+1}{\alpha}\right)\right]$
Normal	$f(x) = \dfrac{1}{\sigma\sqrt{2\pi}} e^{-\frac{1}{2}\left(\frac{x-\mu}{\sigma}\right)^2}$	μ	σ^2
Triangular	$f(x) = \begin{cases} \dfrac{2(x-a)}{(b-a)(m-a)} & \text{if } a \le x \le m \\[2mm] \dfrac{2(b-x)}{(b-a)(b-m)} & \text{if } m < x \le b \end{cases}$	$\dfrac{a+b+m}{3}$	$\dfrac{a^2+b^2+m^2-ab-am-bm}{18}$

HYPOTHESIS TESTING

Table A. Tests on means of normal distribution—variance known.

Hypothesis	Test Statistic	Criteria for Rejection
H_0: $\mu = \mu_0$ H_1: $\mu \neq \mu_0$		$\lvert Z_0 \rvert > Z_{\alpha/2}$
H_0: $\mu = \mu_0$ H_0: $\mu < \mu_0$	$Z_0 \equiv \dfrac{\overline{X} - \mu_0}{\sigma/\sqrt{n}}$	$Z_0 < -Z_\alpha$
H_0: $\mu = \mu_0$ H_1: $\mu > \mu_0$		$Z_0 > Z_\alpha$
H_0: $\mu_1 - \mu_2 = \gamma$ H_1: $\mu_1 - \mu_2 \neq \gamma$		$\lvert Z_0 \rvert > Z_{\alpha/2}$
H_0: $\mu_1 - \mu_2 = \gamma$ H_1: $\mu_1 - \mu_2 < \gamma$	$Z_0 \equiv \dfrac{\overline{X_1} - \overline{X_2} - \gamma}{\sqrt{\dfrac{\sigma_1^2}{n_1} + \dfrac{\sigma_2^2}{n_2}}}$	$Z_0 < -Z_\alpha$
H_0: $\mu_1 - \mu_2 = \gamma$ H_1: $\mu_1 - \mu_2 > \gamma$		$Z_0 > Z_\alpha$

Table B. Tests on means of normal distribution—variance unknown.

Hypothesis	Test Statistic	Criteria for Rejection
H_0: $\mu = \mu_0$ H_1: $\mu \neq \mu_0$		$\lvert t_0 \rvert > t_{\alpha/2,\, n-1}$
H_0: $\mu = \mu_0$ H_1: $\mu < \mu_0$	$t_0 = \dfrac{\overline{X} - \mu_0}{S/\sqrt{n}}$	$t_0 < -t_{\alpha,\, n-1}$
H_0: $\mu = \mu_0$ H_1: $\mu > \mu_0$		$t_0 > t_{\alpha,\, n-1}$
H_0: $\mu_1 - \mu_2 = \gamma$ H_1: $\mu_1 - \mu_2 \neq \gamma$	$t_0 = \dfrac{\overline{X_1} - \overline{X_2} - \gamma}{S_p \sqrt{\dfrac{1}{n_1} + \dfrac{1}{n_2}}}$ $v = n_1 + n_2 - 2$	$\lvert t_0 \rvert > t_{\alpha/2,\, v}$
H_0: $\mu_1 - \mu_2 = \gamma$ H_1: $\mu_1 - \mu_2 < \gamma$	$t_0 = \dfrac{\overline{X_1} - \overline{X_2} - \gamma}{\sqrt{\dfrac{S_1^2}{n_1} + \dfrac{S_2^2}{n_2}}}$	$t_0 < -t_{\alpha,\, v}$
H_0: $\mu_1 - \mu_2 = \gamma$ H_1: $\mu_1 - \mu_2 > \gamma$	$v = \dfrac{\left(\dfrac{S_1^2}{n_1} + \dfrac{S_2^2}{n_2}\right)^2}{\dfrac{\left(S_1^2/n_1\right)^2}{n_1 - 1} + \dfrac{\left(S_2^2/n_2\right)^2}{n_2 - 1}}$	$t_0 > t_{\alpha,\, v}$

In Table B, $S_p{}^2 = [(n_1 - 1)S_1{}^2 + (n_2 - 1)S_2{}^2]/v$

Table C. Tests on variances of normal distribution with unknown mean.

Hypothesis	Test Statistic	Criteria for Rejection
$H_0: \sigma^2 = \sigma_0^2$ $H_1: \sigma^2 \neq \sigma_0^2$		$\chi_0^2 > \chi_{\alpha/2,\,n-1}^2$ or $\chi_0^2 < \chi_{1-\alpha/2,\,n-1}^2$
$H_0: \sigma^2 = \sigma_0^2$ $H_1: \sigma^2 < \sigma_0^2$	$\chi_0^2 = \dfrac{(n-1)S^2}{\sigma_0^2}$	$\chi_0^2 < \chi_{1-\alpha/2,\,n-1}^2$
$H_0: \sigma^2 = \sigma_0^2$ $H_1: \sigma^2 > \sigma_0^2$		$\chi_0^2 > \chi_{\alpha,\,n-1}^2$
$H_0: \sigma_1^2 = \sigma_2^2$ $H_1: \sigma_1^2 \neq \sigma_2^2$	$F_0 = \dfrac{S_1^2}{S_2^2}$	$F_0 > F_{\alpha/2,\,n_1-1,\,n_2-1}$ $F_0 < F_{1-\alpha/2,\,n_1-1,\,n_2-1}$
$H_0: \sigma_1^2 = \sigma_2^2$ $H_1: \sigma_1^2 < \sigma_2^2$	$F_0 = \dfrac{S_2^2}{S_1^2}$	$F_0 > F_{\alpha,\,n_2-1,\,n_1-1}$
$H_0: \sigma_1^2 = \sigma_2^2$ $H_1: \sigma_1^2 > \sigma_2^2$	$F_0 = \dfrac{S_1^2}{S_2^2}$	$F_0 > F_{\alpha,\,n_1-1,\,n_2-1}$

ANTHROPOMETRIC MEASUREMENTS

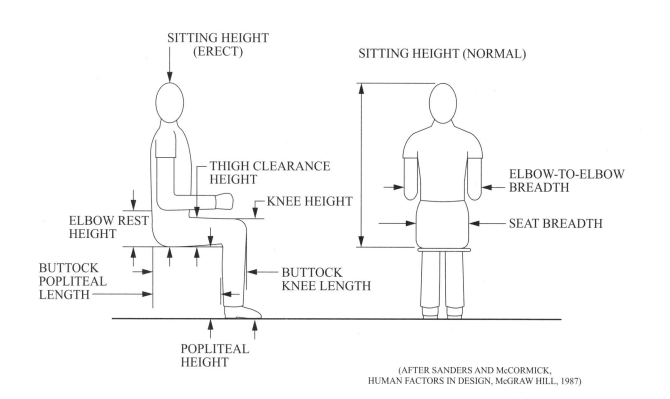

(AFTER SANDERS AND McCORMICK,
HUMAN FACTORS IN DESIGN, McGRAW HILL, 1987)

U.S. Civilian Body Dimensions, Female/Male, for Ages 20 to 60 Years (Centimeters)				
(See Anthropometric Measurements Figure)	Percentiles			
	5th	50th	95th	Std. Dev.
HEIGHTS				
Stature (height)	149.5 / 161.8	160.5 / 173.6	171.3 / 184.4	6.6 / 6.9
Eye height	138.3 / 151.1	148.9 / 162.4	159.3 / 172.7	6.4 / 6.6
Shoulder (acromion) height	121.1 / 132.3	131.1 / 142.8	141.9 / 152.4	6.1 / 6.1
Elbow height	93.6 / 100.0	101.2 / 109.9	108.8 / 119.0	4.6 / 5.8
Knuckle height	**64.3 / 69.8**	**70.2 / 75.4**	**75.9 / 80.4**	**3.5 / 3.2**
Height, sitting	78.6 / 84.2	85.0 / 90.6	90.7 / 96.7	3.5 / 3.7
Eye height, sitting	67.5 / 72.6	73.3 / 78.6	78.5 / 84.4	3.3 / 3.6
Shoulder height, sitting	49.2 / 52.7	55.7 / 59.4	61.7 / 65.8	3.8 / 4.0
Elbow rest height, sitting	18.1 / 19.0	23.3 / 24.3	28.1 / 29.4	2.9 / 3.0
Knee height, sitting	**45.2 / 49.3**	**49.8 / 54.3**	**54.5 / 59.3**	**2.7 / 2.9**
Popliteal height, sitting	35.5 / 39.2	39.8 / 44.2	44.3 / 48.8	2.6 / 2.8
Thigh clearance height	10.6 / 11.4	13.7 / 14.4	17.5 / 17.7	1.8 / 1.7
DEPTHS				
Chest depth	21.4 / 21.4	24.2 / 24.2	29.7 / 27.6	2.5 / 1.9
Elbow-fingertip distance	38.5 / 44.1	42.1 / 47.9	46.0 / 51.4	2.2 / 2.2
Buttock-knee length, sitting	51.8 / 54.0	56.9 / 59.4	62.5 / 64.2	3.1 / 3.0
Buttock-popliteal length, sitting	43.0 / 44.2	48.1 / 49.5	53.5 / 54.8	3.1 / 3.0
Forward reach, functional	64.0 / 76.3	71.0 / 82.5	79.0 / 88.3	4.5 / 5.0
BREADTHS				
Elbow-to-elbow breadth	31.5 / 35.0	38.4 / 41.7	49.1 / 50.6	5.4 / 4.6
Hip breadth, sitting	31.2 / 30.8	36.4 / 35.4	43.7 / 40.6	3.7 / 2.8
HEAD DIMENSIONS				
Head breadth	13.6 / 14.4	14.54 / 15.42	15.5 / 16.4	0.57 / 0.59
Head circumference	52.3 / 53.8	54.9 / 56.8	57.7 / 59.3	1.63 / 1.68
Interpupillary distance	5.1 / 5.5	5.83 / 6.20	6.5 / 6.8	0.4 / 0.39
HAND DIMENSIONS				
Hand length	16.4 / 17.6	17.95 / 19.05	19.8 / 20.6	1.04 / 0.93
Breadth, metacarpal	7.0 / 8.2	7.66 / 8.88	8.4 / 9.8	0.41 / 0.47
Circumference, metacarpal	16.9 / 19.9	18.36 / 21.55	19.9 / 23.5	0.89 / 1.09
Thickness, metacarpal III	2.5 / 2.4	2.77 / 2.76	3.1 / 3.1	0.18 / 0.21
Digit 1				
Breadth, interphalangeal	1.7 / 2.1	1.98 / 2.29	2.1 / 2.5	0.12 / 0.13
Crotch-tip length	4.7 / 5.1	5.36 / 5.88	6.1 / 6.6	0.44 / 0.45
Digit 2				
Breadth, distal joint	1.4 / 1.7	1.55 / 1.85	1.7 / 2.0	0.10 / 0.12
Crotch-tip length	6.1 / 6.8	6.88 / 7.52	7.8 / 8.2	0.52 / 0.46
Digit 3				
Breadth, distal joint	1.4 / 1.7	1.53 / 1.85	1.7 / 2.0	0.09 / 0.12
Crotch-tip length	7.0 / 7.8	7.77 / 8.53	8.7 / 9.5	0.51 / 0.51
Digit 4				
Breadth, distal joint	1.3 / 1.6	1.42 / 1.70	1.6 / 1.9	0.09 / 0.11
Crotch-tip length	6.5 / 7.4	7.29 / 7.99	8.2 / 8.9	0.53 / 0.47
Digit 5				
Breadth, distal joint	1.2 / 1.4	1.32 / 1.57	1.5 / 1.8	0.09/0.12
Crotch-tip length	4.8 / 5.4	5.44 / 6.08	6.2 / 6.99	0.44/0.47
FOOT DIMENSIONS				
Foot length	22.3 / 24.8	24.1 / 26.9	26.2 / 29.0	1.19 / 1.28
Foot breadth	8.1 / 9.0	8.84 / 9.79	9.7 / 10.7	0.50 / 0.53
Lateral malleolus height	5.8 / 6.2	6.78 / 7.03	7.8 / 8.0	0.59 / 0.54
Weight (kg)	46.2 / 56.2	61.1 / 74.0	89.9 / 97.1	13.8 / 12.6

ERGONOMICS—HEARING

The average shifts with age of the threshold of hearing for pure tones of persons with "normal" hearing, using a 25-year-old group as a reference group.

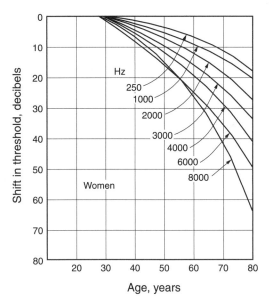

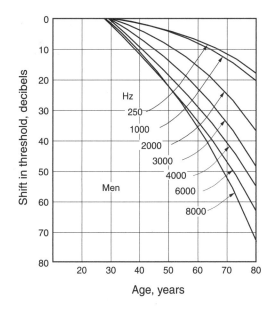

Equivalent sound-level contours used in determining the A-weighted sound level on the basis of an octave-band analysis. The curve at the point of the highest penetration of the noise spectrum reflects the A-weighted sound level.

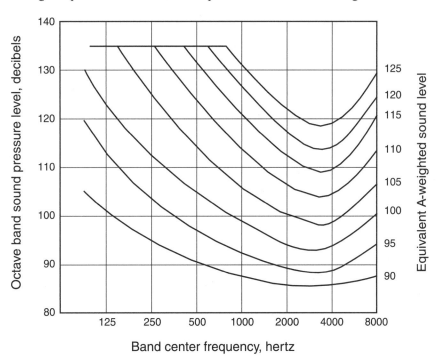

Estimated average trend curves for net hearing loss at 1,000, 2,000, and 4,000 Hz after continuous exposure to steady noise. Data are corrected for age, but not for temporary threshold shift. Dotted portions of curves represent extrapolation from available data.

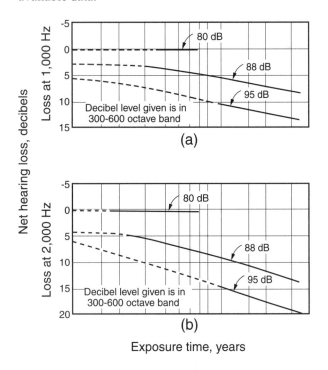

(a)

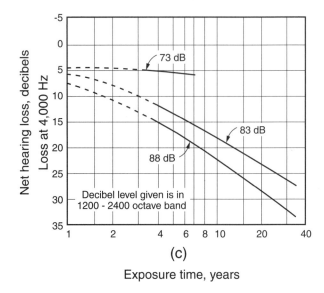

(c)

Exposure time, years

(b)

Exposure time, years

Tentative upper limit of effective temperature (ET) for unimpaired mental performance as related to exposure time; data are based on an analysis of 15 studies. Comparative curves of tolerable and marginal physiological limits are also given.

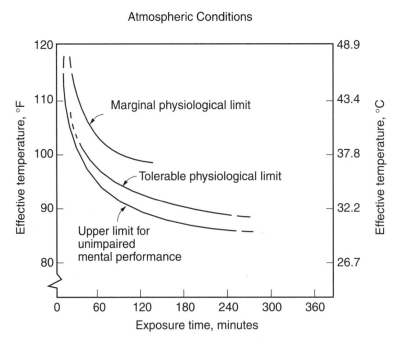

Effective temperature (ET) is the dry bulb temperature at 50% relative humidity, which results in the same physiological effect as the present conditions.

MECHANICAL ENGINEERING

MECHANICAL DESIGN AND ANALYSIS

Stress Analysis
See **MECHANICS OF MATERIALS** section.

Failure Theories
See **MECHANICS OF MATERIALS** section and the **MATERIALS SCIENCE** section.

Deformation and Stiffness
See **MECHANICS OF MATERIALS** section.

Components
Square Thread Power Screws: The torque required to raise, T_R, or to lower, T_L, a load is given by

$$T_R = \frac{Fd_m}{2}\left(\frac{l + \pi\mu d_m}{\pi d_m - \mu l}\right) + \frac{F\mu_c d_c}{2},$$

$$T_L = \frac{Fd_m}{2}\left(\frac{\pi\mu d_m - l}{\pi d_m + \mu l}\right) + \frac{F\mu_c d_c}{2}, \text{ where}$$

d_c = mean collar diameter,

d_m = mean thread diameter,

l = lead,

F = load,

μ = coefficient of friction for thread, and

μ_c = coefficient of friction for collar.

The efficiency of a power screw may be expressed as

$$\eta = Fl/(2\pi T)$$

Mechanical Springs
Helical Linear Springs: The shear stress in a helical linear spring is

$$\tau = K_s \frac{8FD}{\pi d^3}, \text{ where}$$

d = wire diameter,

F = applied force,

D = mean spring diameter

$K_s = (2C + 1)/(2C)$, and

$C = D/d$.

The deflection and force are related by $F = kx$ where the spring rate (spring constant) k is given by

$$k = \frac{d^4 G}{8D^3 N}$$

where G is the shear modulus of elasticity and N is the number of active coils. See Table of Material Properties at the end of the **MECHANICS OF MATERIALS** section for values of G.

Spring Material: The minimum tensile strength of common spring steels may be determined from

$$S_{ut} = A/d^m$$

where S_{ut} is the tensile strength in MPa, d is the wire diameter in millimeters, and A and m are listed in the following table:

Material	ASTM	m	A
Music wire	A228	0.163	2060
Oil-tempered wire	A229	0.193	1610
Hard-drawn wire	A227	0.201	1510
Chrome vanadium	A232	0.155	1790
Chrome silicon	A401	0.091	1960

Maximum allowable torsional stress for static applications may be approximated as

$$S_{sy} = \tau = 0.45 S_{ut} \text{ cold-drawn carbon steel}$$
$$\text{(A227, A228, A229)}$$

$$S_{sy} = \tau = 0.50 S_{ut} \text{ hardened and tempered carbon and}$$
$$\text{low-alloy steels (A232, A401)}$$

Compression Spring Dimensions

Type of Spring Ends		
Term	**Plain**	**Plain and Ground**
End coils, N_e	0	1
Total coils, N_t	N	$N + 1$
Free length, L_0	$pN + d$	$p(N + 1)$
Solid length, L_s	$d(N_t + 1)$	dN_t
Pitch, p	$(L_0 - d)/N$	$L_0/(N + 1)$

Term	**Squared or Closed**	**Squared and Ground**
End coils, N_e	2	2
Total coils, N_t	$N + 2$	$N + 2$
Free length, L_0	$pN + 3d$	$pN + 2d$
Solid length, L_s	$d(N_t + 1)$	dN_t
Pitch, p	$(L_0 - 3d)/N$	$(L_0 - 2d)/N$

Helical Torsion Springs: The bending stress is given as

$$\sigma = K_i[32Fr/(\pi d^3)]$$

where F is the applied load and r is the radius from the center of the coil to the load.

K_i = correction factor

$$= (4C^2 - C - 1)/[4C(C - 1)]$$

$C = D/d$

The deflection θ and moment Fr are related by

$$Fr = k\theta$$

where the spring rate k is given by

$$k = \frac{d^4 E}{64 DN}$$

where k has units of N•m/rad and θ is in radians.

Spring Material: The strength of the spring wire may be found as shown in the section on linear springs. The allowable stress σ is then given by

$$S_y = \sigma = 0.78 S_{ut} \text{ cold-drawn carbon steel}$$
$$\text{(A227, A228, A229)}$$

$$S_y = \sigma = 0.87 S_{ut} \text{ hardened and tempered carbon and}$$
$$\text{low-alloy steel (A232, A401)}$$

Ball/Roller Bearing Selection

The minimum required *basic load rating* (load for which 90% of the bearings from a given population will survive 1 million revolutions) is given by

$$C = PL^{1/a}, \text{ where}$$

C = minimum required basic load rating,
P = design radial load,
L = design life (in millions of revolutions), and
a = 3 for ball bearings, 10/3 for roller bearings.

When a ball bearing is subjected to both radial and axial loads, an equivalent radial load must be used in the equation above. The equivalent radial load is

$$P_{eq} = XVF_r + YF_a, \text{ where}$$

P_{eq} = equivalent radial load,
F_r = applied constant radial load, and
F_a = applied constant axial (thrust) load.

For radial contact, deep-groove ball bearings:
V = 1 if inner ring rotating, 1.2 if outer ring rotating,

If $F_a/(VF_r) > e$,

$$X = 0.56, \text{ and } Y = 0.840 \left(\frac{F_a}{C_0}\right)^{-0.247}$$

$$\text{where } e = 0.513 \left(\frac{F_a}{C_0}\right)^{0.236}, \text{ and}$$

C_0 = basic static load rating from bearing catalog.
If $F_a/(VF_r) \le e$, X = 1 and Y = 0.

Intermediate- and Long-Length Columns

The slenderness ratio of a column is $S_r = l/k$, where l is the length of the column and k is the radius of gyration. The radius of gyration of a column cross-section is, $k = \sqrt{I/A}$ where I is the area moment of inertia and A is the cross-sectional area of the column. A column is considered to be intermediate if its slenderness ratio is less than or equal to $(S_r)_D$, where

$$(S_r)_D = \pi \sqrt{\frac{2E}{S_y}}, \text{ and}$$

E = Young's modulus of respective member, and
S_y = yield strength of the column material.

For intermediate columns, the critical load is

$$P_{cr} = A \left[S_y - \frac{1}{E} \left(\frac{S_y S_r}{2\pi} \right)^2 \right], \text{ where}$$

P_{cr} = critical buckling load,
A = cross-sectional area of the column,
S_y = yield strength of the column material,
E = Young's modulus of respective member, and
S_r = slenderness ratio.

For long columns, the critical load is

$$P_{cr} = \frac{\pi^2 EA}{S_r^2}$$

where the variables are as defined above.

For both intermediate and long columns, the effective column length depends on the end conditions. The AISC recommended values for the effective lengths of columns are, for: rounded-rounded or pinned-pinned ends, $l_{eff} = l$; fixed-free, $l_{eff} = 2.1l$; fixed-pinned, $l_{eff} = 0.80l$; fixed-fixed, $l_{eff} = 0.65l$. The effective column length should be used when calculating the slenderness ratio.

Power Transmission

Shafts and Axles

Static Loading: The maximum shear stress and the von Mises stress may be calculated in terms of the loads from

$$\tau_{max} = \frac{2}{\pi d^3} \left[(8M + Fd)^2 + (8T)^2 \right]^{1/2},$$

$$\sigma' = \frac{4}{\pi d^3} \left[(8M + Fd)^2 + (48T)^2 \right]^{1/2}, \text{ where}$$

M = the bending moment,
F = the axial load,
T = the torque, and
d = the diameter.

Fatigue Loading: Using the maximum-shear-stress theory combined with the Soderberg line for fatigue, the diameter and safety factor are related by

$$\frac{\pi d^3}{32} = n\left[\left(\frac{M_m}{S_y} + \frac{K_f M_a}{S_e}\right)^2 + \left(\frac{T_m}{S_y} + \frac{K_{fs}T_a}{S_e}\right)^2\right]^{1/2}$$

where

d = diameter,

n = safety factor,

M_a = alternating moment,

M_m = mean moment,

T_a = alternating torque,

T_m = mean torque,

S_e = fatigue limit,

S_y = yield strength,

K_f = fatigue strength reduction factor, and

K_{fs} = fatigue strength reduction factor for shear.

Joining

<u>Threaded Fasteners:</u> The load carried by a bolt in a threaded connection is given by

$$F_b = CP + F_i \qquad\qquad F_m < 0$$

while the load carried by the members is

$$F_m = (1 - C)P - F_i \qquad F_m < 0, \text{ where}$$

C = joint coefficient,

$\quad = k_b/(k_b + k_m)$

F_b = total bolt load,

F_i = bolt preload,

F_m = total material load,

P = externally applied load,

k_b = the effective stiffness of the bolt or fastener in the grip, and

k_m = the effective stiffness of the members in the grip.

Bolt stiffness may be calculated from

$$k_b = \frac{A_d A_t E}{A_d l_t + A_t l_d}, \text{ where}$$

A_d = major-diameter area,

A_t = tensile-stress area,

E = modulus of elasticity,

l_d = length of unthreaded shank, and

l_t = length of threaded shank contained within the grip.

If all members within the grip are of the same material, *member stiffness* may be obtained from

$$k_m = dEAe^{b(d/l)}, \text{ where}$$

d = bolt diameter,

E = modulus of elasticity of members, and

l = grip length.

Coefficients A and b are given in the table below for various joint member materials.

Material	A	b
Steel	0.78715	0.62873
Aluminum	0.79670	0.63816
Copper	0.79568	0.63553
Gray cast iron	0.77871	0.61616

The approximate tightening torque required for a given preload F_i and for a steel bolt in a steel member is given by $T = 0.2\,F_i d$.

<u>Threaded Fasteners – Design Factors:</u> The bolt load factor is

$$n_b = (S_p A_t - F_i)/CP$$

The factor of safety guarding against joint separation is

$$n_s = F_i / [P(1 - C)]$$

<u>Threaded Fasteners – Fatigue Loading:</u> If the externally applied load varies between zero and P, the alternating stress is

$$\sigma_a = CP/(2A_t)$$

and the mean stress is

$$\sigma_m = \sigma_a + F_i/A_t$$

<u>Bolted and Riveted Joints Loaded in Shear:</u>

(a) FASTENER IN SHEAR

Failure by pure shear, (a)

$$\tau = F/A, \text{ where}$$

F = shear load, and

A = cross-sectional area of bolt or rivet.

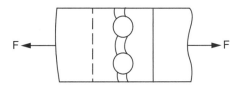

(b) MEMBER RUPTURE

Failure by rupture, (b)

$$\sigma = F/A, \text{ where}$$

F = load and

A = net cross-sectional area of thinnest member.

(c) MEMBER OR FASTENER CRUSHING

Failure by crushing of rivet or member, (c)

$$\sigma = F/A, \text{ where}$$

F = load and
A = projected area of a single rivet.

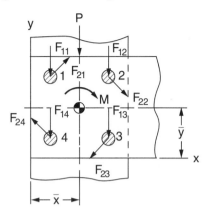

(d) FASTENER GROUPS

Fastener groups in shear, (d)

The location of the centroid of a fastener group with respect to any convenient coordinate frame is:

$$\bar{x} = \frac{\sum\limits_{i=1}^{n} A_i x_i}{\sum\limits_{i=1}^{n} A_i}, \quad \bar{y} = \frac{\sum\limits_{i=1}^{n} A_i y_i}{\sum\limits_{i=1}^{n} A_i}, \text{ where}$$

n = total number of fasteners,

i = the index number of a particular fastener,

A_i = cross-sectional area of the ith fastener,

x_i = x-coordinate of the center of the ith fastener, and

y_i = y-coordinate of the center of the ith fastener.

The total shear force on a fastener is the **vector** sum of the force due to direct shear P and the force due to the moment M acting on the group at its centroid.

The magnitude of the direct shear force due to P is

$$|F_{1i}| = \frac{P}{n}.$$

This force acts in the same direction as P.

The magnitude of the shear force due to M is

$$|F_{2i}| = \frac{Mr_i}{\sum\limits_{i=1}^{n} r_i^2}.$$

This force acts perpendicular to a line drawn from the group centroid to the center of a particular fastener. Its sense is such that its moment is in the same direction (CW or CCW) as M.

Press/Shrink Fits

The interface pressure induced by a press/shrink fit is

$$p = \frac{0.5\delta}{\dfrac{r}{E_o}\left(\dfrac{r_o^2 + r^2}{r_o^2 - r^2} + v_o\right) + \dfrac{r}{E_i}\left(\dfrac{r^2 + r_i^2}{r^2 - r_i^2} + v_i\right)}$$

where the subscripts i and o stand for the inner and outer member, respectively, and

p = inside pressure on the outer member and outside pressure on the inner member,

δ = the diametral interference,

r = nominal interference radius,

r_i = inside radius of inner member,

r_o = outside radius of outer member,

E = Young's modulus of respective member, and

v = Poisson's ratio of respective member.

See the **MECHANICS OF MATERIALS** section on thick-wall cylinders for the stresses at the interface.

The *maximum torque* that can be transmitted by a press fit joint is approximately

$$T = 2\pi r^2 \mu p l,$$

where r and p are defined above,

T = torque capacity of the joint,

μ = coefficient of friction at the interface, and

l = length of hub engagement.

MANUFACTURABILITY

Limits and Fits

The designer is free to adopt any geometry of fit for shafts and holes that will ensure intended function. Over time, sufficient experience with common situations has resulted in the development of a standard. The metric version of the standard is newer and will be presented. The standard specifies that uppercase letters always refer to the hole, while lowercase letters always refer to the shaft.

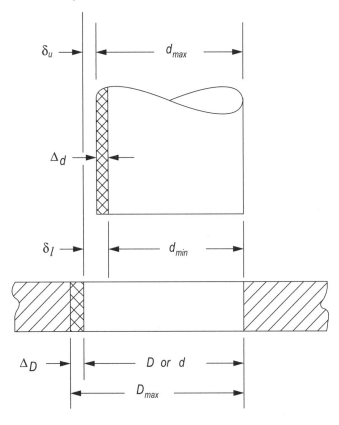

Definitions

Basic Size or *nominal size*, D or d, is the size to which the limits or deviations are applied. It is the same for both components.

Deviation is the algebraic difference between the actual size and the corresponding basic size.

Upper Deviation, δ_u, is the algebraic difference between the maximum limit and the corresponding basic size.

Lower Deviation, δ_l, is the algebraic difference between the minimum limit and the corresponding basic size.

Fundamental Deviation, δ_F, is the upper or lower deviation, depending on which is smaller.

Tolerance, Δ_D or Δ_d, is the difference between the maximum and minimum size limits of a part.

International tolerance (IT) grade numbers designate groups of tolerances such that the tolerance for a particular IT number will have the same relative accuracy for a basic size.

Hole basis represents a system of fits corresponding to a basic hole size. The fundamental deviation is H.

Some Preferred Fits

Clearance	*Free running fit*: not used where accuracy is essential but good for large temperature variations, high running speeds, or heavy journal loads.	H9/d9
	Sliding fit: where parts are not intended to run freely but must move and turn freely and locate accurately.	H7/g6
	Locational fit: provides snug fit for location of stationary parts but can be freely assembled and disassembled.	H7/h6
Transition	*Locational transition fit*: for accurate location, a compromise between clearance and interference.	H7/k6
Interference	*Location interference fit*: for parts requiring rigidity and alignment with prime accuracy of location but without special bore pressure requirements.	H7/p6
	Medium drive fit: for ordinary steel parts or shrink fits on light sections. The tightest fit usable on cast iron.	H7/s6
	Force fit: suitable for parts that can be highly stressed or for shrink fits where heavy pressing forces are impractical.	H7/u6

For the hole

$$D_{\max} = D + \Delta_D$$
$$D_{\min} = D$$

For a shaft with clearance fits d, g, or h

$$d_{\max} = d + \delta_F$$
$$d_{\min} = d_{\max} - \Delta_d$$

For a shaft with transition or interference fits $k, p, s,$ or u

$$d_{min} = d + \delta_F$$
$$d_{max} = d_{min} + \Delta_d$$

where

D = basic size of hole
d = basic size of shaft
δ_u = upper deviation
δ_l = lower deviation
δ_F = fundamental deviation
Δ_D = tolerance grade for hole
Δ_d = tolerance grade for shaft

The shaft tolerance is defined as $\Delta_d = |\delta_u - \delta_l|$

International Tolerance (IT) Grades

Lower limit < Basic Size ≤ Upper Limit
All values in mm

Basic Size	Tolerance Grade, (Δ_D or Δ_d)		
	IT6	IT7	IT9
0–3	0.006	0.010	0.025
3–6	0.008	0.012	0.030
6–10	0.009	0.015	0.036
10–18	0.011	0.018	0.043
18–30	0.013	0.021	0.052
30–50	0.016	0.025	0.062
Source: Preferred Metric Limits and Fits, ANSI B4.2-1978			

Deviations for shafts

Lower limit < Basic Size ≤ Upper Limit
All values in mm

Basic Size	Upper Deviation Letter, (δ_u)			Lower Deviation Letter, (δ_l)			
	d	g	h	k	p	s	u
0–3	–0.020	–0.002	0	0	+0.006	+0.014	+0.018
3–6	–0.030	–0.004	0	+0.001	+0.012	+0.019	+0.023
6–10	–0.040	–0.005	0	+0.001	+0.015	+0.023	+0.028
10–14	–0.095	–0.006	0	+0.001	+0.018	+0.028	+0.033
14–18	–0.050	–0.006	0	+0.001	+0.018	+0.028	+0.033
18–24	–0.065	–0.007	0	+0.002	+0.022	+0.035	+0.041
24–30	–0.065	–0.007	0	+0.002	+0.022	+0.035	+0.048
30–40	–0.080	–0.009	0	+0.002	+0.026	+0.043	+0.060
40–50	–0.080	–0.009	0	+0.002	+0.026	+0.043	+0.070
Source: Preferred Metric Limits and Fits, ANSI B4.2-1978							

As an example, 34H7/s6 denotes a basic size of $D = d = 34$ mm , an IT class of 7 for the hole, and an IT class of 6 and an "s" fit class for the shaft.

Maximum Material Condition (MMC)
The maximum material condition defines the dimension of a part such that the part weighs the most. The MMC of a shaft is at the maximum size of the tolerance while the MMC of a hole is at the minimum size of the tolerance.

Least Material Condition (LMC)
The least material condition or minimum material condition defines the dimensions of a part such that the part weighs the least. The LMC of a shaft is the minimum size of the tolerance while the LMC of a hole is at the maximum size of the tolerance.

KINEMATICS, DYNAMICS, AND VIBRATIONS
Kinematics of Mechanisms
Four-bar Linkage

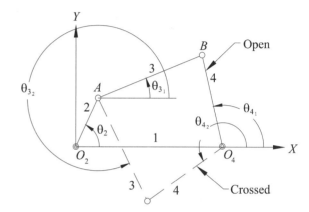

The four-bar linkage shown above consists of a reference (usually grounded) link (1), a crank (input) link (2), a coupler link (3), and an output link (4). Links 2 and 4 rotate about the fixed pivots O_2 and O_4, respectively. Link 3 is joined to link 2 at the moving pivot A and to link 4 at the moving pivot B. The lengths of links 2, 3, 4, and 1 are a, b, c, and d, respectively. Taking link 1 (ground) as the reference (X-axis), the angles that links 2, 3, and 4 make with the axis are θ_2, θ_3, and θ_4, respectively. It is possible to assemble a four-bar in two different configurations for a given position of the input link (2). These are known as the "open" and "crossed" positions or circuits.

Position Analysis. Given a, b, c, and d, and θ_2

$$\theta_{4_{1,2}} = 2 \arctan\left(\frac{-B \pm \sqrt{B^2 - 4AC}}{2A}\right)$$

where $A = \cos\theta_2 - K_1 - K_2\cos\theta_2 + K_3$
$B = -2\sin\theta_2$
$C = K_1 - (K_2 + 1)\cos\theta_2 + K_3$, and
$K_1 = \dfrac{d}{a}, K_2 = \dfrac{d}{c}, K_3 = \dfrac{a^2 - b^2 + c^2 + d^2}{2ac}$

In the equation for θ_4, using the minus sign in front of the radical yields the open solution. Using the plus sign yields the crossed solution.

$$\theta_{3_{1,2}} = 2 \arctan\left(\frac{-E \pm \sqrt{E^2 - 4DF}}{2D}\right)$$

where $D = \cos \theta_2 - K_1 + K_4 \cos \theta_2 + K_5$

$\quad E = -2\sin \theta_2$

$\quad F = K_1 + (K_4 - 1) \cos \theta_2 + K_5$, and

$\quad K_4 = \dfrac{d}{b}, K_5 = \dfrac{c^2 - d^2 - a^2 - b^2}{2ab}$

In the equation for θ_3, using the minus sign in front of the radical yields the open solution. Using the plus sign yields the crossed solution.

Velocity Analysis. Given a, b, c, and d, θ_2, θ_3, θ_4, and ω_2

$$\omega_3 = \frac{a\omega_2}{b}\frac{\sin(\theta_4 - \theta_2)}{\sin(\theta_3 - \theta_4)}$$

$$\omega_4 = \frac{a\omega_2}{c}\frac{\sin(\theta_2 - \theta_3)}{\sin(\theta_4 - \theta_3)}$$

$$V_{Ax} = -a\omega_2\sin \theta_2, \qquad V_{Ay} = a\omega_2\cos \theta_2$$

$$V_{BAx} = -b\omega_3\sin \theta_3, \qquad V_{BAy} = b\omega_3\cos \theta_3$$

$$V_{Bx} = -c\omega_4\sin \theta_4, \qquad V_{By} = c\omega_4\cos \theta_4$$

See also <u>Instantaneous Centers of Rotation</u> in the **DYNAMICS** section.

Acceleration analysis. Given a, b, c, and d, θ_2, θ_3, θ_4, and ω_2, ω_3, ω_4, and α_2

$$\alpha_3 = \frac{CD - AF}{AE - BD}, \quad \alpha_4 = \frac{CE - BF}{AE - BD}, \text{ where}$$

$$A = c\sin \theta_4, B = b\sin \theta_3$$

$$C = a\alpha_2\sin \theta_2 + a\omega_2^2\cos \theta_2 + b\omega_3^2\cos \theta_3 - c\omega_4^2\cos \theta_4$$

$$D = c\cos \theta_4, E = b\cos \theta_3$$

$$F = a\alpha_2\cos \theta_2 - a\omega_2^2\sin \theta_2 - b\omega_3^2\sin \theta_3 + c\omega_4^2\sin \theta_4$$

Gearing
Involute Gear Tooth Nomenclature

Circular pitch	$p_c = \pi d/N$
Base pitch	$p_b = p_c \cos \phi$
Module	$m = d/N$
Center distance	$C = (d_1 + d_2)/2$

where

$\quad N$ = number of teeth on pinion or gear

$\quad d$ = pitch circle diameter

$\quad \phi$ = pressure angle

Gear Trains: *Velocity ratio*, m_v, is the ratio of the output velocity to the input velocity. Thus, $m_v = \omega_{out}/\omega_{in}$. For a two-gear train, $m_v = -N_{in}/N_{out}$ where N_{in} is the number of teeth on the input gear and N_{out} is the number of teeth on the output gear. The negative sign indicates that the output gear rotates in the opposite sense with respect to the input gear. In a *compound gear train*, at least one shaft carries more than one gear (rotating at the same speed). The velocity ratio for a compound train is:

$$m_v = \pm \frac{\text{product of number of teeth on driver gears}}{\text{product of number of teeth on driven gears}}$$

A *simple planetary gearset* has a sun gear, an arm that rotates about the sun gear axis, one or more gears (planets) that rotate about a point on the arm, and a ring (internal) gear that is concentric with the sun gear. The planet gear(s) mesh with the sun gear on one side and with the ring gear on the other. A planetary gearset has two independent inputs and one output (or two outputs and one input, as in a differential gearset).

Often one of the inputs is zero, which is achieved by grounding either the sun or the ring gear. The velocities in a planetary set are related by

$$\frac{\omega_f - \omega_{\text{arm}}}{\omega_L - \omega_{\text{arm}}} = \pm m_v, \text{ where}$$

ω_f = speed of the first gear in the train,

ω_L = speed of the last gear in the train, and

ω_{arm} = speed of the arm.

Neither the first nor the last gear can be one that has planetary motion. In determining m_v, it is helpful to invert the mechanism by grounding the arm and releasing any gears that are grounded.

Dynamics of Mechanisms
Gearing
Loading on Straight Spur Gears: The load, W, on straight spur gears is transmitted along a plane that, in edge view, is called the *line of action*. This line makes an angle with a tangent line to the pitch circle that is called the *pressure angle* ϕ. Thus, the contact force has two components: one in the tangential direction, W_t, and one in the radial direction, W_r. These components are related to the pressure angle by

$$W_r = W_t \tan(\phi).$$

Only the tangential component W_t transmits torque from one gear to another. Neglecting friction, the transmitted force may be found if either the transmitted torque or power is known:

$$W_t = \frac{2T}{d} = \frac{2T}{mN},$$

$$W_t = \frac{2H}{d\omega} = \frac{2H}{mN\omega}, \text{ where}$$

W_t = transmitted force (newton),

T = torque on the gear (newton-mm),

d = pitch diameter of the gear (mm),

N = number of teeth on the gear,

m = gear module (mm) (same for both gears in mesh),

H = power (kW), and

ω = speed of gear (rad/sec).

Stresses in Spur Gears: Spur gears can fail in either bending (as a cantilever beam, near the root) or by surface fatigue due to contact stresses near the pitch circle. AGMA Standard 2001 gives equations for bending stress and surface stress. They are:

$$\sigma_b = \frac{W_t}{FmJ}\frac{K_a K_m}{K_v}K_S K_B K_I, \text{ bending and}$$

$$\sigma_c = C_p \sqrt{\frac{W_t}{FId}\frac{C_a C_m}{C_v}C_s C_f}, \text{ surface stress, where}$$

σ_b = bending stress,
σ_c = surface stress,
W_t = transmitted load,
F = face width,
m = module,
J = bending strength geometry factor,
K_a = application factor,
K_B = rim thickness factor,
K_I = idler factor,
K_m = load distribution factor,
K_s = size factor,
K_v = dynamic factor,
C_p = elastic coefficient,
I = surface geometry factor,
d = pitch diameter of gear being analyzed, and
C_f = surface finish factor.
C_a, C_m, C_s, and C_v are the same as K_a, K_m, K_s, and K_v, respectively.

Rigid Body Dynamics
See **DYNAMICS** section.

Natural Frequency and Resonance
See **DYNAMICS** section.

Balancing of Rotating and Reciprocating Equipment
Static (Single-plane) Balance

$$m_b R_{bx} = -\sum_{i=1}^{n} m_i R_{ix}, \quad m_b R_{by} = -\sum_{i=1}^{n} m_i R_{iy}$$

$$\theta_b = \arctan\left(\frac{m_b R_{by}}{m_b R_{bx}}\right)$$

$$m_b R_b = \sqrt{\left(m_b R_{bx}\right)^2 + \left(m_b R_{by}\right)^2}$$

where m_b = balance mass
R_b = radial distance to CG of balance mass
m_i = ith point mass
R_i = radial distance to CG of the ith point mass
θ_b = angle of rotation of balance mass CG with respect to a reference axis
x, y = subscripts that designate orthogonal components

Dynamic (Two-plane) Balance

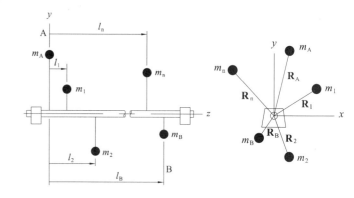

Two balance masses are added (or subtracted), one each on planes A and B.

$$m_B R_{Bx} = -\frac{1}{l_B}\sum_{i=1}^{n}m_i R_{ix}l_i, \quad m_B R_{By} = -\frac{1}{l_B}\sum_{i=1}^{n}m_i R_{iy}l_i$$

$$m_A R_{Ax} = -\sum_{i=1}^{n}m_i R_{ix} - m_B R_{Bx}$$

$$m_A R_{Ay} = -\sum_{i=1}^{n}m_i R_{iy} - m_B R_{By}$$

where
m_A = balance mass in the A plane
m_B = balance mass in the B plane
R_A = radial distance to CG of balance mass
R_B = radial distance to CG of balance mass
and θ_A, θ_B, R_A, and R_B are found using the relationships given in Static Balance above.

Balancing Equipment
The figure below shows a schematic representation of a tire/wheel balancing machine.

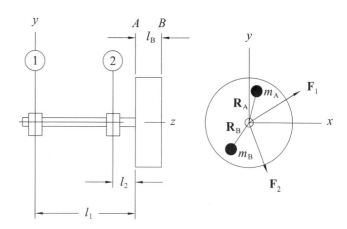

Ignoring the weight of the tire and its reactions at 1 and 2,
$$F_{1x} + F_{2x} + m_A R_{Ax}\omega^2 + m_B R_{Bx}\omega^2 = 0$$
$$F_{1y} + F_{2y} + m_A R_{Ay}\omega^2 + m_B R_{By}\omega^2 = 0$$
$$F_{1x}l_1 + F_{2x}l_2 + m_B R_{Bx}\omega^2 l_B = 0$$
$$F_{1y}l_1 + F_{2y}l_2 + m_B R_{By}\omega^2 l_B = 0$$

$$m_B R_{Bx} = \frac{F_{1x}l_1 + F_{2x}l_2}{l_B\,\omega^2}$$

$$m_B R_{By} = \frac{F_{1y}l_1 + F_{2y}l_2}{l_B\,\omega^2}$$

$$m_A R_{Ax} = -\frac{F_{1x} + F_{2x}}{\omega^2} - m_b R_{Bx}$$

$$m_A R_{Ay} = -\frac{F_{1y} + F_{2y}}{\omega^2} - m_b R_{By}$$

MATERIALS AND PROCESSING

Mechanical and Thermal Properties
See **MATERIALS SCIENCE** section.

Thermal Processing
See **MATERIALS SCIENCE** section.

Testing
See **MECHANICS OF MATERIALS** section.

MEASUREMENTS, INSTRUMENTATION, AND CONTROL

Mathematical Fundamentals
See DIFFERENTIAL EQUATIONS and LAPLACE TRANSFORMS in the **MATHEMATICS** section, and CONTROL SYSTEMS in the **MEASUREMENT and CONTROLS** section.

System Descriptions
See LAPLACE TRANSFORMS in the **MATHEMATICS** section, and CONTROL SYSTEMS in the **MEASUREMENT and CONTROLS** section.

Sensors and Signal Conditioning
See the Measurements segment of the **MEASUREMENT and CONTROLS** section and the Analog Filter Circuits segment of the **ELECTRICAL and COMPUTER ENGINEERING** section.

Data Collection and Processing
See the Sampling segment of the **MEASUREMENT and CONTROLS** section.

Dynamic Response
See CONTROL SYSTEMS in the **MEASUREMENT and CONTROLS** section.

THERMODYNAMICS AND ENERGY CONVERSION PROCESSES

Ideal and Real Gases
See **THERMODYNAMICS** section.

Reversibility/Irreversibility
See **THERMODYNAMICS** section.

Thermodynamic Equilibrium
See **THERMODYNAMICS** section.

Psychrometrics
See additional material in **THERMODYNAMICS** section.

HVAC—Pure Heating and Cooling

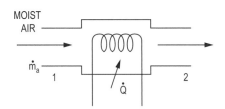

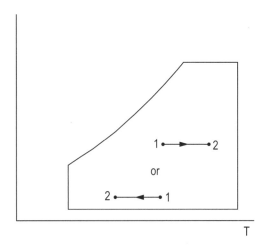

$$\dot{Q} = \dot{m}_a\left(h_2 - h_1\right) = \dot{m}_a c_{pm}\left(T_2 - T_1\right)$$

$$c_{pm} = 1.02\,\text{kJ}/(\text{kg} \cdot {}^\circ\text{C})$$

Cooling and Dehumidification

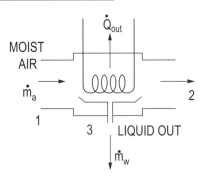

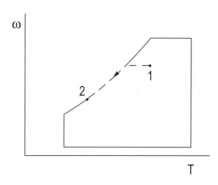

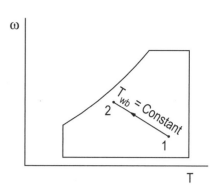

$$\dot{Q}_{out} = \dot{m}_a \left[(h_1 - h_2) - h_{f3}(\omega_1 - \omega_2) \right]$$

$$\dot{m}_w = \dot{m}_a (\omega_1 - \omega_2)$$

$$h_2 = h_1 + h_3(\omega_2 - \omega_1)$$

$$\dot{m}_w = \dot{m}_a (\omega_2 - \omega_1)$$

$$h_3 = h_f \text{ at } T_{wb}$$

Heating and Humidification

Adiabatic Mixing

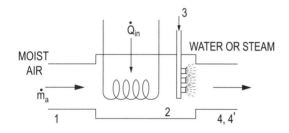

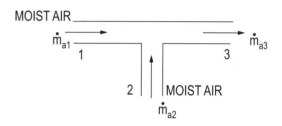

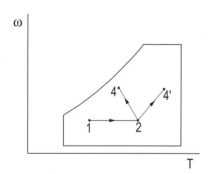

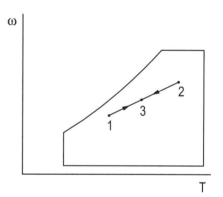

$$\dot{Q}_{in} = \dot{m}_a (h_2 - h_1)$$

$$\dot{m}_w = \dot{m}_a (\omega_{4'} - \omega_2) \text{ or } \dot{m}_w = (\omega_4 - \omega_2)$$

$$\dot{m}_{a3} = \dot{m}_{a1} + \dot{m}_{a2}$$

$$h_3 = \frac{\dot{m}_{a1} h_1 + \dot{m}_{a2} h_2}{\dot{m}_{a3}}$$

$$\omega_3 = \frac{\dot{m}_{a1} \omega_1 + \dot{m}_{a2} \omega_2}{\dot{m}_{a3}}$$

distance $\overline{13} = \dfrac{\dot{m}_{a2}}{\dot{m}_{a3}} \times$ distance $\overline{12}$ measured on

psychrometric chart

Adiabatic Humidification (evaporative cooling)

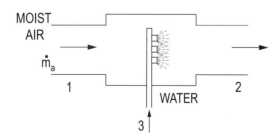

Performance of Components
Fans, Pumps, and Compressors
Scaling Laws

$$\left(\frac{Q}{ND^3}\right)_2 = \left(\frac{Q}{ND^3}\right)_1$$

$$\left(\frac{\dot{m}}{\rho ND^3}\right)_2 = \left(\frac{\dot{m}}{\rho ND^3}\right)_1$$

$$\left(\frac{H}{N^2D^2}\right)_2 = \left(\frac{H}{N^2D^2}\right)_1$$

$$\left(\frac{P}{\rho N^2D^2}\right)_2 = \left(\frac{P}{\rho N^2D^2}\right)_1$$

$$\left(\frac{\dot{W}}{\rho N^3D^5}\right)_2 = \left(\frac{\dot{W}}{\rho N^3D^5}\right)_1$$

where
Q = volumetric flow rate,
$\dot{m}$ = mass flow rate,
H = head,
P = pressure rise,
$\dot{W}$ = power,
ρ = fluid density,
N = rotational speed, and
D = impeller diameter.

Subscripts 1 and 2 refer to different but similar machines or to different operating conditions of the same machine.

Fan Characteristics

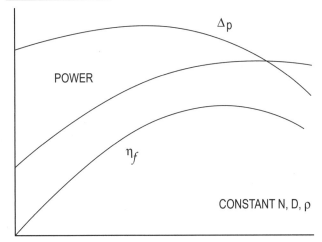

Typical *Backward Curved* Fans

$$\dot{W} = \frac{\Delta PQ}{\eta_f}, \text{ where}$$

$\dot{W}$ = fan power,
ΔP = pressure rise, and
η_f = fan efficiency.

Centrifugal Pump Characteristics

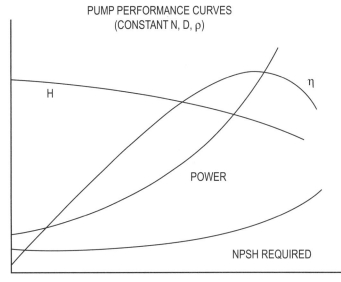

PUMP PERFORMANCE CURVES
(CONSTANT N, D, ρ)

FLOW RATE, Q

Net Positive Suction Head (*NPSH*)

$$NPSH = \frac{P_i}{\rho g} + \frac{V_i^2}{2g} - \frac{P_v}{\rho g}, \text{ where}$$

P_i = inlet pressure to pump,
V_i = velocity at inlet to pump, and
P_v = vapor pressure of fluid being pumped.

Fluid power $\dot{W}_{\text{fluid}} = \rho g H Q$

Pump (brake) power $\dot{W} = \dfrac{\rho g H Q}{\eta_{\text{pump}}}$

Purchased power $\dot{W}_{\text{purchased}} = \dfrac{\dot{W}}{\eta_{\text{motor}}}$

η_{pump} = pump efficiency (0 to 1)
η_{motor} = motor efficiency (0 to 1)
H = head increase provided by pump

Cycles and Processes
Internal Combustion Engines
Otto Cycle (see **THERMODYNAMICS** section)
Diesel Cycle

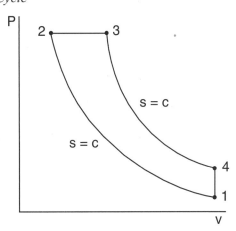

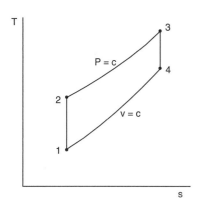

$$r = V_1/V_2$$

$$r_c = V_3/V_2$$

$$\eta = 1 - \frac{1}{r^{k-1}}\left[\frac{r_c^k - 1}{k(r_c - 1)}\right]$$

$$k = c_p/c_v$$

Brake Power

$$\dot{W}_b = 2\pi TN = 2\pi FRN, \text{ where}$$

$\dot{W}_b$ = brake power (W),

T = torque (N•m),

N = rotation speed (rev/s),

F = force at end of brake arm (N), and

R = length of brake arm (m).

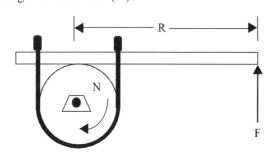

Indicated Power

$$\dot{W}_i = \dot{W}_b + \dot{W}_f, \text{ where}$$

$\dot{W}_i$ = *indicated power* (W), *and*

$\dot{W}_f$ = friction power (W).

Brake Thermal Efficiency

$$\eta_b = \frac{\dot{W}_b}{\dot{m}_f(HV)}, \text{ where}$$

η_b = brake thermal efficiency,

$\dot{m}_f$ = fuel consumption rate (kg/s), and

HV = heating value of fuel (J/kg).

Indicated Thermal Efficiency

$$\eta_i = \frac{\dot{W}_i}{\dot{m}_f(HV)}$$

Mechanical Efficiency

$$\eta_i = \frac{\dot{W}_b}{\dot{W}_i} = \frac{\eta_b}{\eta_i}$$

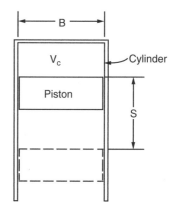

Displacement Volume

$$V_d = \frac{\pi B^2 S}{4}, \text{m}^3 \text{ for each cylinder}$$

$$\text{Total volume} = V_t = V_d + V_c, \text{m}^3$$

V_c = clearance volume (m³).

Compression Ratio

$$r_c = V_t/V_c$$

Mean Effective Pressure (*mep*)

$$mep = \frac{\dot{W}n_s}{V_d n_c N}, \text{ where}$$

n_s = number of crank revolutions per power stroke,

n_c = number of cylinders, and

V_d = displacement volume per cylinder.

mep can be based on brake power (*bmep*), indicated power (*imep*), or friction power (*fmep*).

Volumetric Efficiency

$$\eta_v = \frac{2\dot{m}_a}{\rho_a V_d n_c N} \qquad \text{(four-stroke cycles only)}$$

where

$\dot{m}_a$ = mass flow rate of air into engine (kg/s), and

ρ_a = density of air (kg/m³).

Specific Fuel Consumption (*SFC*)

$$sfc = \frac{\dot{m}_f}{\dot{W}} = \frac{1}{\eta HV}, \text{kg/J}$$

Use η_b and $\dot{W}_b$ for *bsfc* and η_i and $\dot{W}_i$ for *isfc*.

Gas Turbines
Brayton Cycle (Steady-Flow Cycle)

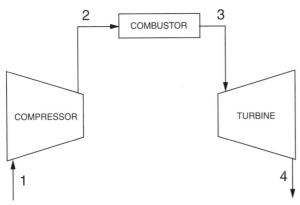

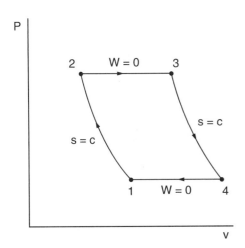

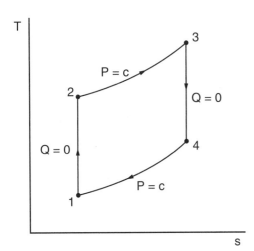

Steam Power Plants
Feedwater Heaters

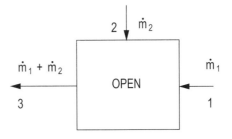

$$\dot{m}_1 h_1 + \dot{m}_2 h_2 = h_3 (\dot{m}_1 + \dot{m}_2)$$

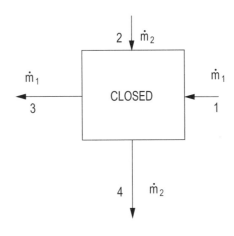

$$\dot{m}_1 h_1 + \dot{m}_2 h_2 = \dot{m}_1 h_3 + \dot{m}_2 h_4$$

Steam Trap

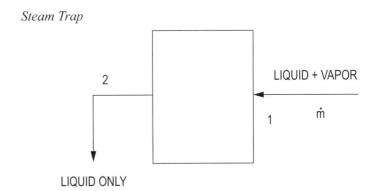

$$h_2 = h_1$$

$$w_{12} = h_1 - h_2 = c_p (T_1 - T_2)$$
$$w_{34} = h_3 - h_4 = c_p (T_3 - T_4)$$
$$w_{net} = w_{12} + w_{34}$$
$$q_{23} = h_3 - h_2 = c_p (T_3 - T_2)$$
$$q_{41} = h_1 - h_4 = c_p (T_1 - T_4)$$
$$q_{net} = q_{23} + q_{41}$$
$$\eta = w_{net}/q_{23}$$

Junction

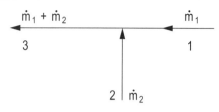

$$\dot{m}_1 h_1 + \dot{m}_2 h_2 = h_3 (\dot{m}_1 + \dot{m}_2)$$

Pump

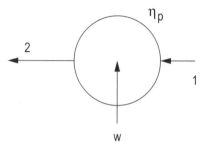

$$w = h_1 - h_2 = (h_1 - h_{2S})/\eta_P$$

$$h_{2S} - h_1 = v(P_2 - P_1)$$

$$w = -\frac{v(P_2 - P_1)}{\eta_p}$$

See also **THERMODYNAMICS** section.

Combustion and Combustion Products
See **THERMODYNAMICS** section.

Energy Storage
Energy storage comes in several forms, including chemical, electrical, mechanical, and thermal. Thermal storage can be either hot or cool storage. There are numerous applications in the HVAC industry where cool storage is utilized. The cool storage applications include both ice and chilled water storage. A typical chilled water storage system can be utilized to defer high electric demand rates, while taking advantage of cheaper off-peak power. A typical facility load profile is shown below.

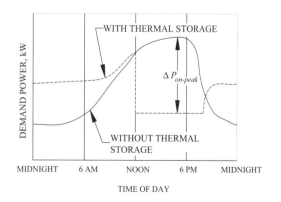

The thermal storage tank is sized to defer most or all of the chilled water requirements during the electric utility's peak demand period, thus reducing electrical demand charges. The figure above shows a utility demand window of 8 hours (noon to 8 pm), but the actual on-peak period will vary from utility to utility. The Monthly Demand Reduction (*MDR*), in dollars per month, is

$$MDR = \Delta P_{on\text{-}peak} R, \text{ where}$$

$\Delta P_{on\text{-}peak}$ = Reduced on-peak power, kW
R = On-peak demand rate, \$/kW/month

The *MDR* is also the difference between the demand charge without energy storage and that when energy storage is in operation.

A typical utility rate structure might be four months of peak demand rates (June – September) and eight months of off-peak demand rates (October – May). The customer's utility obligation will be the sum of the demand charge and the kWh energy charge.

FLUID MECHANICS AND FLUID MACHINERY

Fluid Statics
See **FLUID MECHANICS** section.

Incompressible Flow
See **FLUID MECHANICS** section.

Fluid Machines (Incompressible)
See **FLUID MECHANICS** section and **Performance of Components** above.

Compressible Flow
Mach Number
The local *speed of sound* in an ideal gas is given by:
$$c = \sqrt{kRT}, \text{ where}$$

$c \equiv$ local speed of sound

$k \equiv$ ratio of specific heats $= \dfrac{c_p}{c_v}$

$R \equiv$ gas constant

$T \equiv$ absolute temperature

This shows that the acoustic velocity in an ideal gas depends only on its temperature. The *Mach number* (Ma) is the ratio of the fluid velocity to the speed of sound.

$$\text{Ma} \equiv \frac{V}{c}$$

$V \equiv$ mean fluid velocity

Isentropic Flow Relationships
In an ideal gas for an isentropic process, the following relationships exist between static properties at any two points in the flow.

$$\frac{P_2}{P_1} = \left(\frac{T_2}{T_1}\right)^{\frac{k}{(k-1)}} = \left(\frac{\rho_2}{\rho_1}\right)^k$$

The stagnation temperature, T_0, at a point in the flow is related to the static temperature as follows:

$$T_0 = T + \frac{V^2}{2 \cdot c_p}$$

The relationship between the static and stagnation properties (T_0, P_0, and ρ_0) at any point in the flow can be expressed as a function of the Mach number as follows:

$$\frac{T_0}{T} = 1 + \frac{k-1}{2} \cdot \text{Ma}^2$$

$$\frac{P_0}{P} = \left(\frac{T_0}{T}\right)^{\frac{k}{(k-1)}} = \left(1 + \frac{k-1}{2} \cdot \text{Ma}^2\right)^{\frac{k}{(k-1)}}$$

$$\frac{\rho_0}{\rho} = \left(\frac{T_0}{T}\right)^{\frac{1}{(k-1)}} = \left(1 + \frac{k-1}{2} \cdot \text{Ma}^2\right)^{\frac{1}{(k-1)}}$$

Compressible flows are often accelerated or decelerated through a nozzle or diffuser. For subsonic flows, the velocity decreases as the flow cross-sectional area increases and vice versa. For supersonic flows, the velocity increases as the flow cross-sectional area increases and decreases as the flow cross-sectional area decreases. The point at which the Mach number is sonic is called the throat and its area is represented by the variable, A^*. The following area ratio holds for any Mach number.

$$\frac{A}{A^*} = \frac{1}{\text{Ma}}\left[\frac{1 + \frac{1}{2}(k-1)\text{Ma}^2}{\frac{1}{2}(k+1)}\right]^{\frac{(k+1)}{2(k-1)}}$$

where

$A \equiv$ area [length^2]
$A^* \equiv$ area at the sonic point (Ma = 1.0)

Normal Shock Relationships

A normal shock wave is a physical mechanism that slows a flow from supersonic to subsonic. It occurs over an infinitesimal distance. The flow upstream of a normal shock wave is always supersonic and the flow downstream is always subsonic as depicted in the figure.

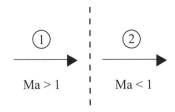

NORMAL SHOCK

The following equations relate downstream flow conditions to upstream flow conditions for a normal shock wave.

$$\text{Ma}_2 = \sqrt{\frac{(k-1)\text{Ma}_1^2 + 2}{2k\,\text{Ma}_1^2 - (k-1)}}$$

$$\frac{T_2}{T_1} = \left[2 + (k-1)\text{Ma}_1^2\right]\frac{2k\,\text{Ma}_1^2 - (k-1)}{(k+1)^2\text{Ma}_1^2}$$

$$\frac{P_2}{P_1} = \frac{1}{k+1}\left[2k\,\text{Ma}_1^2 - (k-1)\right]$$

$$\frac{\rho_2}{\rho_1} = \frac{V_1}{V_2} = \frac{(k+1)\text{Ma}_1^2}{(k-1)\text{Ma}_1^2 + 2}$$

$$T_{01} = T_{02}$$

Fluid Machines (Compressible)

Compressors

Compressors consume power to add energy to the working fluid. This energy addition results in an increase in fluid pressure (head).

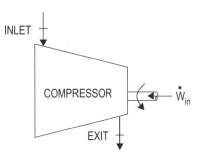

For an adiabatic compressor with $\Delta PE = 0$ and negligible ΔKE:

$$\dot{W}_{comp} = -\dot{m}\left(h_e - h_i\right)$$

For an ideal gas with constant specific heats:

$$\dot{W}_{comp} = -\dot{m}c_p\left(T_e - T_i\right)$$

Per unit mass:

$$w_{comp} = -c_p\left(T_e - T_i\right)$$

Compressor Isentropic Efficiency:

$$\eta_C = \frac{w_s}{w_a} = \frac{T_{es} - T_i}{T_e - T_i} \text{ where,}$$

$w_a \equiv$ actual compressor work per unit mass
$w_s \equiv$ isentropic compressor work per unit mass
$T_{es} \equiv$ isentropic exit temperature
    (see **THERMODYNAMICS** section)

For a compressor where ΔKE is included:

$$\dot{W}_{comp} = -\dot{m}\left(h_e - h_i + \frac{V_e^2 - V_i^2}{2}\right)$$

$$= -\dot{m}\left(c_p\left(T_e - T_i\right) + \frac{V_e^2 - V_i^2}{2}\right)$$

Adiabatic Compression:

$$\dot{W}_{comp} = \frac{P_i\, k}{(k-1)\rho_i\, \eta_c}\left[\left(\frac{P_e}{P_i}\right)^{1-1/k} - 1\right]$$

$\dot{W}_{comp}$ = fluid or gas power (W)

P_i = inlet or suction pressure (N/m^2)

P_e = exit or discharge pressure (N/m^2)

k = ratio of specific heats = c_p/c_v

ρ_i = inlet gas density (kg/m^3)

η_c = isentropic compressor efficiency

Isothermal Compression

$$\dot{W}_{comp} = \frac{\overline{R}\,T_i}{M\eta_c}\ln\frac{P_e}{P_i}$$

$\dot{W}_{comp}$, P_i, P_e, and η_c as defined for adiabatic compression

$\overline{R}$ = universal gas constant

T_i = inlet temperature of gas (K)

M = molecular weight of gas (kg/kmol)

<u>Turbines</u>

Turbines produce power by extracting energy from a working fluid. The energy loss shows up as a decrease in fluid pressure (head).

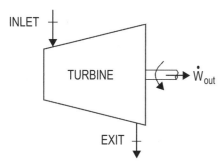

For an adiabatic turbine with $\Delta PE = 0$ and negligible ΔKE:

$$\dot{W}_{turb} = \dot{m}\left(h_i - h_e\right)$$

For an ideal gas with constant specific heats:

$$\dot{W}_{turb} = \dot{m}c_p\left(T_i - T_e\right)$$

Per unit mass:

$$w_{turb} = c_p\left(T_i - T_e\right)$$

Compressor Isentropic Efficiency:

$$\eta_T = \frac{w_a}{w_s} = \frac{T_i - T_e}{T_i - T_{es}}$$

For a compressor where ΔKE is included:

$$\dot{W}_{turb} = \dot{m}\left(h_e - h_i + \frac{V_e^2 - V_i^2}{2}\right)$$

$$= \dot{m}\left(c_p\left(T_e - T_i\right) + \frac{V_e^2 - V_i^2}{2}\right)$$

Operating Characteristics
See **Performance of Components** above.

Lift/Drag
See **FLUID MECHANICS** section.

Impulse/Momentum
See **FLUID MECHANICS** section.

HEAT TRANSFER

Conduction
See **HEAT TRANSFER** and **TRANSPORT PHENOMENA** sections.

Convection
See **HEAT TRANSFER** section.

Radiation
See **HEAT TRANSFER** section.

Composite Walls and Insulation
See **HEAT TRANSFER** section.

Transient and Periodic Processes
See **HEAT TRANSFER** section.

Heat Exchangers
See **HEAT TRANSFER** section.

Boiling and Condensation Heat Transfer
See **HEAT TRANSFER** section.

REFRIGERATION AND HVAC

Cycles
<u>Refrigeration and HVAC</u>
Two-Stage Cycle

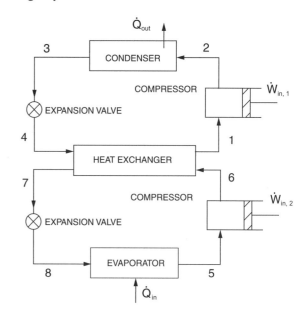

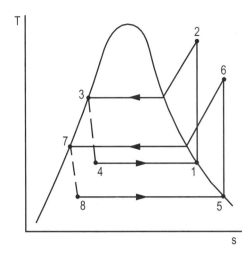

The following equations are valid if the mass flows are the same in each stage.

$$COP_{\text{ref}} = \frac{\dot{Q}_{\text{in}}}{\dot{W}_{\text{in},1} + \dot{W}_{\text{in},2}} = \frac{h_5 - h_8}{h_2 - h_1 + h_6 - h_5}$$

$$COP_{\text{HP}} = \frac{\dot{Q}_{\text{out}}}{\dot{W}_{\text{in},1} + \dot{W}_{\text{in},2}} = \frac{h_2 - h_3}{h_2 - h_1 + h_6 - h_5}$$

Air Refrigeration Cycle

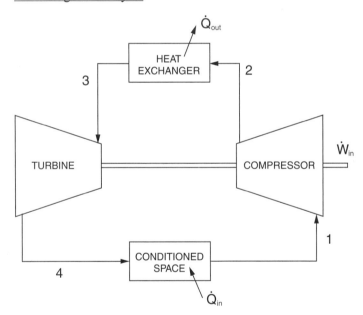

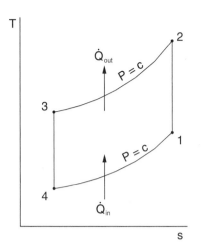

$$COP_{ref} = \frac{h_1 - h_4}{(h_2 - h_1) - (h_3 - h_4)}$$

$$COP_{HP} = \frac{h_2 - h_3}{(h_2 - h_1) - (h_3 - h_4)}$$

See also **THERMODYNAMICS** section.

Heating and Cooling Loads
Heating Load

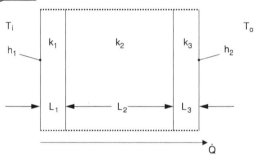

$$\dot{Q} = A(T_i - T_o)/R''$$

$$R'' = \frac{1}{h_1} + \frac{L_1}{k_1} + \frac{L_2}{k_2} + \frac{L_3}{k_3} + \frac{1}{h_2}, \text{ where}$$

$\dot{Q}$ = heat transfer rate,
A = wall surface area, and
R'' = thermal resistance.

Overall heat transfer coefficient = U
$\quad U = 1/R''$
$\quad \dot{Q} = UA\,(T_i - T_o)$

Cooling Load

$$\dot{Q} = UA\,(\text{CLTD}), \text{ where}$$

CLTD = effective temperature difference.

CLTD depends on solar heating rate, wall or roof orientation, color, and time of day.

<u>Infiltration</u>
Air change method

$$\dot{Q} = \frac{\rho_a c_p V n_{AC}}{3,600}(T_i - T_o), \text{ where}$$

ρ_a = air density,

c_p = air specific heat,

V = room volume,

n_{AC} = number of air changes per hour,

T_i = indoor temperature, and

T_o = outdoor temperature.

Crack method

$$\dot{Q} = 1.2CL(T_i - T_o)$$

where

C = coefficient, and

L = crack length.

See also **HEAT TRANSFER** section.

Psychrometric Charts
See **THERMODYNAMICS** section.

Coefficient of Performance (COP)
See section above and **THERMODYNAMICS** section.

Components
See **THERMODYNAMICS** section and above sections.

INDEX

Do not write in this book or remove any pages.

Do all scratch work in your exam booklet.

Do not write in this book or remove any pages.
Do all scratch work in your exam booklet.